Search and Foraging

Individual Motion and Swarm Dynamics

Search and Foraging

Individual Motion and Swarm Dynamics

Eugene Kagan

Ariel University and the Weizmann Institute of Science
Israel

Irad Ben-Gal

Tel Aviv University
Israel

CRC Press
Taylor & Francis Group
Boca Raton London New York

CRC Press is an imprint of the
Taylor & Francis Group, an **informa** business
A CHAPMAN & HALL BOOK

CRC Press
Taylor & Francis Group
6000 Broken Sound Parkway NW, Suite 300
Boca Raton, FL 33487-2742

First issued in hardback 2019

ISBN-13: 978-1-4822-4209-6 (hbk)

Visit the Taylor & Francis Web site at
http://www.taylorandfrancis.com

and the CRC Press Web site at
http://www.crcpress.com

To our teachers.

Contents

Preface

Starting from the pioneering ideas of von Neumann and Wiener, the progress in developing computers and intelligent machines is inspired by the studies of living organisms. Logical schemes, perceptron, neural networks, and storage modification machines—even in basic terminology, these fields of cybernetics follow biology. But what is a living organism? How can we recognize an intelligent behavior?

In 1962, at the rise of cybernetics, George Zopf wrote as follows*:

> To say that a system is intelligent is to say only that we cannot produce sufficient evidence for the determination of its behavior in certain "problem solving" situations [...]. Note how many computers drop in I.Q. as soon as their lid is lifted! [...] We must know in full deterministic detail what we are doing to build a complex machine; to call such a machine intelligent, requires that we forget or ignore our knowledge of just how it does what it does.

Probably, an intuitive awareness of this problem stimulated Turing to make an attempt in the mathematical description of morphogenesis† in such a form that being completely deterministic it allows complex unpredictable behavior. In those times, the Turing paper was considered as a kind of nonsense, but later, the Turing system became one of the basic models in the studies of self-organization and nonlinear dynamics, while the paper itself determined the essence of mathematical biology research.

We can observe a working brain and can obtain various characteristics, which indicate its functionality. But we certainly cannot directly observe the working mind or human intelligence. All we can do is build a mathematical model and hopefully implement it in a device demonstrating an activity looking like an activity of intelligence. If such a device passes the Turing test, then our goal is reached. However, when we consider the activity of living organisms different from humans, we even do not know what we are going to formalize and what the test is, which can check our models.

In 1969, introducing the Tsetlin works in mathematical biology, Israel Gelfand wrote as follows‡:

> What should be the degree of formalization in biology in the study of living systems? Considering quantum mechanics, one can distinguish two stages in its formation. The first stage took place when Bohr created the philosophy of quantum mechanics. At that time, the formulas did not yet exist, and even if they did, they were not quite as they should be or were completely wrong. The second stage was a period of a rapid growth, and quantum mechanics became an exact branch of physics with a large number of precise formulas. But this stage was possible only after the first stage had taken place. By comparison, in biology the first stage has not yet occurred.

Unfortunately, this more-than-40-year-old opinion is still correct. We have various mathematical models of different activities of humans and animals, but we still cannot present a testable device that implements the Kurzweil optimistic predictions,§ or, at least, can be compared with the living organism in its most basic activity. A fortunate exception is provided by the theory of foraging.

---

* Zopf, G. W. (1962). Attitude and context. In H. von Foerster, & G. W. Zopf (Ed.), *Trans. University of Illinois Symp. Self-Organization 1961* (pp. 325–346). Oxford, U.K.: Pergamon Press, p. 328.

† Turing, A. M. (1952). The chemical basis of morphogenesis. *Phil. Trans. R. Soc. B.*, *237*, 37–72.

‡ Tsetlin, M. L. (1973). *Automaton Theory and Modeling of Biological Systems.* New York: Academic Press, p. 130. First Russian edtion was published in Moscow in 1969.

§ Kurzweil, R. (1999). *The Age of Spiritual Machines. When Computers Exceed Human Intelligence.* New York: Penguin Books.

The theory of foraging addresses the behavior of individual animals and their swarms while seeking for food. We do not know whether their behavior is optimal from an abstract mathematical point of view, but as any natural behavior, it is certainly optimal from an evolutionary point of view. In the last years, Viswanathan and his colleagues initiated the mathematical modeling of the forager trajectories by the use of Lévy flights. In 2008, commenting the findings of this group, Mark Buchanan noted as follows*:

> They seemed to be on the track of a new principle of ecology when they showed that this way of moving is, under some conditions, theoretically the best way for animals to find scarce prey.

Probably it is the first model of external directly observable behavior of living organisms that is built without any specific knowledge about the internal activity of its brain and intelligence. On the other hand, there is a highly developed theory of search and screening, which was initiated in 1942 in response to the German submarine threat in the Atlantic. The pioneer of this theory, Bernard Koopman, wrote later† that the problem in consideration was

> … how best to search for an object when the amount of searching effort is limited and only probabilities of the object's possible positions are given.

Here the search agents are with artificial necessity so their abilities and structure are known and internal activity can be planned and programmed. In the case of search in stable unvarying environment, optimal search plans can be obtained by standard optimization techniques, but if the environment is changing during the search, then a global optimal solution cannot be found and the search plans have to be corrected online optimally—the task that is above the abilities of modern computers but that is naturally solved by living foragers.

Hence, if we will be able to build and program artificial search agents in such a way that they will demonstrate the same behavior as it is predicted by the foraging theory for living organisms, we will reach two goals. For cybernetics, we will obtain the techniques possible for the best online search planning in varying environments, and for biology, we will get reasonable insights regarding the internal activity of living organisms performing foraging tasks. The book deals with this task.

The book addresses foraging and search and screening theories and considers them in the same mathematical and algorithmic framework. It overviews the main ideas and methods of foraging and search theories; hopefully such overview will make the ideas of one theory available to the specialists in the other. In particular, the book considers Lévy flight models of individual foraging and corresponding diffusion models and algorithms of search and foraging in random environment both by single and by multiple agents, including results of laboratory verifications, and presents the active Brownian motion model for swarm dynamics with corresponding Fokker–Planck equations. The discourse is supported with numerical examples, which can be useful for practitioners in both theories.

We are in great debt to many people who influenced my understanding of search and foraging and underlying models. Prof. Vladimir Ryjov, Prof. Arkady Kolesnikov, Prof. Eshel Ben-Asher, and Prof. Shlomo Havlin were the people from whom we learned the ideas and philosophy of self-organization and applications of nonlinear dynamics. Algorithms of search were developed and programmed in collaboration with Prof. Evgeny Khmelnitsky, Gal Goren, and Moshe Israel. The ideas regarding ants' behavior were clarified during numerous discussions with Jennie Steshenko, who also collaborated in the laboratory experiments. The method of comparing

* Buchanan, M. (2008). The mathematical mirror to animal nature. *Nature, 453*, 714–716.
† Koopman, B. O. (1979). Search and its optimization. *Amer. Math. Monthly, 86*(7), 527–540, 527.

trajectories was formulated in collaboration with Dr. Alexander Novoselsky. During the work, we also had fruitful discussions with Prof. Hava Siegelmann, Prof. Boaz Golany, Dr. Alexander Rybalov, Sergey Hodorov, Alon Sela, and Rottem Botton. To all these people, we express our deep gratitude.

Eugene Kagan and Irad Ben-Gal
Ariel
Tel-Aviv
Rehovot, Israel

List of Figures

List of symbols

Notations

$a \subset \mathfrak{X}$ and $a(t)$	An observed area that is a subset of points, which is screened by the agent and over which an observation result is obtained. The observed area chosen at time t is denoted by $a(t)$. In the case of k sensors, $k > 1$, the agent screens k areas $a_1,\ldots,a_k$ simultaneously and k observation results are considered; it is assumed that $k \le 2$.
$\mathcal{A} = \{a_1, a_2, \ldots, a_m\}$	A set of observed areas $a_i \subset \mathfrak{X}$, $i = 1, 2, \ldots, m$.
$A^j \subset \mathfrak{X}$ and $\mathcal{A}^{0\ldots k}_{1\ldots m}$ $A \subset \mathfrak{X}$ and $\mathcal{A}^{0}_{1\ldots m}$	A patch of type $j = 1,2,\ldots,k$. In the foraging theory, it is assumed that the forager can move within a patch A^j. A set $\mathcal{A}^{0\ldots k}_{1\ldots m}$ of patches such that $\mathcal{A}^{0\ldots k}_{1\ldots m} = \{A^0, A_1^{j_1}, A_2^{j_2}, \ldots, A_m^{j_m}\}$, $j_i \in \{1,2,\ldots,k\}$, $i = 1,2,\ldots,m$, $A^0 = \mathfrak{X} \setminus \left(\bigcup_{i=1}^{m} A_i^{j_i}\right)$ is a partition of $\mathfrak{X}$ is called a patchy environment over $\mathfrak{X}$. In the stochastic setup with the unknown patch types, the patches are denoted by A (with bottom indices if needed) and the patchy environment is denoted by $\mathcal{A}^{0}_{1\ldots m} = \{A^0, A_1, A_2, \ldots, A_m\}$.
$\vec{\mathfrak{a}}$	A trajectory of the searcher that is a sequence $\vec{\mathfrak{a}} = a(0), a(1), a(2), \ldots, a(T)$ of observed areas chosen at times $t = 0, 1, 2, \ldots, T$. If $a(t) = \{\check{\mathfrak{x}}(t)\}$, $\check{\mathfrak{x}}(t) \in \mathfrak{X}$, for each t, then the notation $\vec{\mathfrak{a}} = \check{\mathfrak{x}}(0), \check{\mathfrak{x}}(1), \check{\mathfrak{x}}(2), \ldots, \check{\mathfrak{x}}(T)$ is used instead.
$b_j(A)$	A probability that the patch $A \subset \mathfrak{X}$ contains j units of prey.
$\mathfrak{P} \subset \mathfrak{X}$	A finite set of potential points defined with respect to the terrain topography $\mathcal{T}$.
$c(\mathfrak{x}, \kappa)$, $C(w,t)$, and C	$c(\mathfrak{x}, \kappa)$ is a cost of applying the search effort $\kappa(\mathfrak{x}, t)$ to the point $\mathfrak{x}$, $C(w,t)$ is the cost of the search strategy, and C is a constraint of total cost.
γ_0 and $\gamma(\mathfrak{x}, \mathrm{v})$	Linear and varying nonlinear friction, respectively.
$\mathcal{D}^a = d^a / d\lvert x\rvert^a$	The Riesz fractional derivative of order a.
δ	Step's length of the random walks and Lévy flights.
$\mathfrak{d}(t)$ and $\vec{\mathfrak{d}}$	$\mathfrak{d}(t)$ is a decision rule that specifies the next observed area $a(t+1)$ given that the current observed area $a(t)$ and the observation result are $a(a(t))$; $\vec{\mathfrak{d}} = \mathfrak{d}(0), \mathfrak{d}(1), \mathfrak{d}(2), \ldots, \mathfrak{d}(T)$ is a sequence of the decision rules also called search policy.
e(t)	Overall energy in the depot at time t.
$\epsilon(\mathfrak{x})$ and $\epsilon_i = \epsilon(\mathfrak{x}_i)$	$\epsilon(\mathfrak{x})$ is an overlook probability that is the probability of nondetecting the target in the point $\mathfrak{x} \in \mathfrak{X}$ given that it is in $\mathfrak{x}$ and the searcher checks $\mathfrak{x}$. For a discrete domain, this probability is denoted by ϵ_i, $i = 1,2,\ldots,n$.
F^{diss}, F^{stoch}, F^{poten}	Dissipative, stochastic, and potential forces, respectively.
$\mathfrak{f}_\alpha$ and $\mathfrak{g}_\alpha$	The functions that specify the dependence of prey concentration on the concentration of the predators and the predators' concentration with respect to the concentration of the prey, respectively. The bifurcation parameter is denoted by α.
$\mathcal{G}_t$	The Gaussian white noise; $\mathcal{E}(\mathcal{G}_t) = 0$ and $\mathcal{E}(\mathcal{G}_{t_2}\mathcal{G}_{t_1}) = \delta(t_2 - t_1)$.

H	The Hamiltonian function, which specifies the total energy $H(\mathfrak{x}_1,\mathfrak{x}_2,\ldots,\mathfrak{x}_m,\mathrm{p}_1,\mathrm{p}_2,\ldots,\mathrm{p}_m,t)$ of the system in time with respect to the locations $\mathfrak{x}_j$ of the particles or agents, $j = 1, 2,\ldots,m$, and their impulses $\mathrm{p}_j = \mathrm{m}_j\mathrm{v}_j$, where m_j is the mass of the jth particle and v_j its velocity.
$\eta_j = \eta(A^j)$	An encounter rate that is a proportion of the patches of type j, $j = 1, 2, \ldots, k$.
$i(t)$ and $\breve{i}(t)$	An expected and actual informational state of the searcher or a forager at time t, which represents the cumulated knowledge about its activity.
$\mathcal{I}_{\pm}^{a}$	The Riemann–Liouville fractional integral of order a.
$\kappa(\mathfrak{x},t)$, $K(t)$, and K	$\kappa(\mathfrak{x},t)$ is an amount of search effort applied to the point $\mathfrak{x}$ up to the time t; $K(t)$ is an amount of search effort distributed over a domain $\mathfrak{X}$ at time t; and $\mathrm{K} \in \mathbb{R}^+$ is a total search effort available to the searcher.
$l(\vec{\mathfrak{a}})$	A total length of the searcher's path $\vec{\mathfrak{a}}$.
l	An average distance between two successive target's locations.
$\mathcal{L}_t^{(\gamma)}$ and $\mathcal{L}_t^{(\gamma)2}$	1D and 2D Lévy flight or walk with parameter $\gamma \in (1,3]$. The value $a = \gamma - 1$ is called the Lévy index, which defines the distribution of the step lengths of the flight or walk.
$\mathcal{L}_t^{(c,\gamma)}$ and $\mathcal{L}_t^{(c,\gamma)2}$	Correlated or drifted 1D and 2D Lévy flight or walk with parameter $\gamma \in (1,3]$.
$\hat{m}$ and $\hat{\mu}$ $\hat{m}$ and $\hat{\mu}$	Drift coefficients of the searcher ($\hat{m}$ and $\hat{\mu}$) and of the target ($\hat{m}$ and $\hat{\mu}$). The varying drifts for both agents are denoted by m and constant drifts by μ with corresponding indices, if needed. The matrices of these coefficients are denoted by $\lVert\hat{m}\rVert$ and $\lVert\hat{\mu}\rVert$ for the searcher and by $\lVert\hat{m}\rVert$ and $\lVert\hat{\mu}\rVert$ for the target.
$\mathcal{M}_0$ and $\mathcal{M}$	Distance measures such that $\mathcal{M}_0(\mathcal{P},t)$ is the distance between the states of disturbed and not-disturbed processes, and $\mathcal{M}(\mathcal{P},t)$ is the distance between the states of the process at times $(t - dt)$ and t, where $\mathcal{P}$ is an abstract process.
$n(\vec{\mathfrak{a}})$	A number of visited areas, while the searcher follows the trajectory $\vec{\mathfrak{a}}$.
$n(A)$	An expected number of prey in the patch A.
$n(l)$	A number of Lévy flights between two successive target's locations.
$\xi_j(t_j)$ and $\xi : \mathbb{R}^+ \to \mathbb{R}^+$ $\xi(\mathfrak{x},t)$ $\xi : \mathfrak{X}\times[0,T] \to \mathbb{R}^+$	An expected net energy gained from the patch A^j of type j given that the forager spent t_j time units hunting in A^j, $j = 1, 2, \ldots, k$. The function $\xi : \mathbb{R}^+ \to \mathbb{R}^+$, which defines the values $\xi_j(t_j)$, is called gain function. In the case of prey gain, a quantity of prey eaten by the agents in the point $\mathfrak{x} \in \mathfrak{X}$ up to the time t is denoted by $\xi(\mathfrak{x},t)$ and is governed by the gain function $\xi : \mathfrak{X}\times[0,T] \to \mathbb{R}^+$.
$\vec{\mathfrak{o}}$	A trajectory of the searcher that is a sequence $\vec{\mathfrak{o}} = \hat{\mathfrak{x}}(0),\hat{\mathfrak{x}}(1),\hat{\mathfrak{x}}(2),\ldots,\hat{\mathfrak{x}}(T)$ of observed points occupied by the target at times $t = 0, 1, 2, \ldots, T$.
$p(\mathfrak{x},t)$ and $p(\mathfrak{x},\mathrm{v},t)$	A probability that the searcher or forager is located in the point $\mathfrak{x}$ at time t and, consequently, a probability that at time t it is located in the point $\mathfrak{x}$ and has a velocity v.
$p(\vec{\mathfrak{a}})$ and $p(\vec{\mathfrak{a}},t)$	A priory probability that the searcher will follow the trajectory $\vec{\mathfrak{a}}$. If it is needed, this probability is also denoted by $p(\vec{\mathfrak{a}},t)$ that stresses the time t of search or, in discrete time and space, the length of the trajectory.
$p(\vec{\mathfrak{a}},t)$ and $q(\vec{\mathfrak{a}},t)$ $p(\check{\mathfrak{x}},t)$ and $q(\check{\mathfrak{x}},t)$ $p_{\mathrm{i}}(\check{\mathfrak{x}},p\,\vert\,\check{\mathfrak{x}}_0,t_0)$ and $q_{\mathrm{i}}(\check{\mathfrak{x}},t\,\vert\,\check{\mathfrak{x}}_0,t_0)$	$p(\vec{\mathfrak{a}},t)\theta$ is a probability of detecting the target up to the time t while the search follows the trajectory $\vec{\mathfrak{a}}$, and $q(\vec{\mathfrak{a}},t) = 1 - p(\vec{\mathfrak{a}},t)$ is a probability of nondetecting the target up to the time t following this trajectory. If the searcher's trajectory is defined by the function $\check{\mathfrak{x}}$, then the probabilities of defecting and not detecting the target are denoted by $p(\check{\mathfrak{x}},t)$ and $q(\check{\mathfrak{x}},t)$, correspondingly. In the case of intermittent strategies, the probabilities $p_{\mathrm{i}}(\check{\mathfrak{x}},t\,\vert\,\check{\mathfrak{x}}_0, t_0)$ and $q_{\mathrm{i}}(\check{\mathfrak{x}},t\,\vert\,\check{\mathfrak{x}}_0, t_0)$ are defined for each motion phase $\mathrm{i} = 1,2$.

$p(\vec{\mathfrak{d}},t)$ and $q(\vec{\mathfrak{d}},t)$	$p(\vec{\mathfrak{d}},t)$ is a probability of detecting the target up to the time t using the search policy $\vec{\mathfrak{d}}$, and $q(\vec{\mathfrak{d}},t)=1-p(\vec{\mathfrak{d}},t)$ is a probability of nondetecting the target up to the time t using this policy.
$p(w,t)$ and $q(w,t)$	$p(w,t)$ is a probability of detecting the target up to the time t using the search strategy w, and $q(w,t) = 1-p(w,t)$ is a probability of nondetecting the target up to the time t using this strategy.
$r(\vec{\mathfrak{a}})$ and $r(\vec{\mathfrak{a}})$	A reward rate while the searcher follows the trajectory $\vec{\mathfrak{a}}$ and an expected reward rate of the search with respect to this trajectory.
$r_e(\vec{\mathfrak{d}})$ and $r_e(\vec{\mathfrak{d}} \mid A^j)$ $r_e(\mathfrak{x})$	A reward rate also called net energy intake rate. Given that the forager is in the patch A^j, the net energy intake rate is denoted by $r_e(\vec{\mathfrak{d}} \mid A^j)$. If the energy intake depends on the location of the forager, it is denoted by $r_e(\mathfrak{x})$.
$r_e^*(\mathcal{A}_{1\ldots m}^{0\ldots k})$	A maximal average reward rate over the patchy environment $\mathcal{A}_{1\ldots m}^{0\ldots k}$.
$r_e(\vec{\mathfrak{d}})$ and $r_e(\vec{\mathfrak{d}} \mid A^j)$	An expected reward rate also called an expected net energy intake rate. Given that the forager is in the patch, the expected reward rate is denoted by $r_e(\vec{\mathfrak{d}} \mid A^j)$.
$r_c(\vec{\mathfrak{d}})$ and $r_c(\vec{\mathfrak{d}} \mid A)$	An expected capture rate that is a ratio of expected number of prey captured in the randomly chosen patch to the expected time of foraging. Given that the forager is in the patch A, the expected capture rate net energy intake rate is denoted by $r_c(\vec{\mathfrak{d}} \mid A)$.
r_{vis}	A vision distance or visibility radius that is a maximal distance between the searcher's and the target's locations at which the searcher can detect the target.
r_{sense}	A sensing distance or sensing radius that is a maximal distance at which the agent can sense the presence of the target; it is assumed that $\mathrm{r}_{sense} \geq \mathrm{r}_{vis}$.
$\mathrm{r}_{signal}^{min}$ and $\mathrm{r}_{signal}^{max}$	Minimal and maximal signaling distances or signaling radiuses that define a ring around the agent called signaler. It is assumed that only the agents that are located at the distance larger than $\mathrm{r}_{signal}^{min}$ and smaller than $\mathrm{r}_{signal}^{max}$ receive the signals from the signaler.
r_{dom}	Radius of the domain (in the case of circle domain).
$R^{0\ldots t}(\mathring{\mathfrak{i}}(t))$	A reward obtained by the searcher or forager while at time t it is in the state $\mathring{\mathfrak{i}}(t)$.
$\mathcal{R}_{-t}^{t\ldots T}(\mathring{\mathfrak{i}}(t))$	An expected reward, which will be obtained by the forager up to time t given that at starting time t the state of the forager is $\mathring{\mathfrak{i}}(t)$.
$\mathcal{R}(\mathring{\mathfrak{i}}(t))$	A potential reward that can be obtained from the patch A given that at time t the forager's state is $\mathring{\mathfrak{i}}(t)$ that is the expected reward in the patch excluding the potential reward, which can be obtained in the other patches during the same foraging time.
$\hat{\rho} = \hat{\rho}_{ij\,n\times n}$ $\check{\rho} = \check{\rho}_{ij\,n\times n}$	Transition probability matrices of a discrete time Markov process, which governs the target's and the searcher's movements in discrete domain correspondingly; $\hat{\rho}_{ij} = Pr\{\hat{\mathfrak{x}}(t+1) = \mathfrak{x}_j \mid \hat{\mathfrak{x}}(t) = \mathfrak{x}_i\}$, $\hat{\mathfrak{x}}(t), \hat{\mathfrak{x}}(t+1) \in \mathfrak{X}$, $\check{\rho}_{ij} = Pr\{\check{\mathfrak{x}}(t+1) = \mathfrak{x}_j \mid \check{\mathfrak{x}}(t) = \mathfrak{x}_i\}, \check{\mathfrak{x}}(t), \check{\mathfrak{x}}(t+1) \in \mathfrak{X}, i,j = 1,2,\ldots,n, t = 0,1,2,\ldots$
$\mathbb{S}(p(E))$ and $\tilde{\mathbb{S}}(p(E))$	Entropy and normalized entropy of the distribution $p(E)$ of the energy E.
$\check{s}$ and $\check{\sigma}$ $\hat{s}$ and $\hat{\sigma}$	Diffusion rate coefficients of the searcher ($\check{s}$ and $\check{\sigma}$) and of the target ($\hat{s}$ and $\hat{\sigma}$). The varying rates for both agents are denoted by s and constant drifts by σ with corresponding indices, if needed. The matrices of these rates are denoted by $\hat{s}$ and $\check{\sigma}$ for the searcher and by $\hat{s}$ and $\hat{\sigma}$ for the target.

$\sigma\left(\mathcal{W}_{t_1},\mathcal{W}_{t_2}\right)$ $\sigma_c\left(\mathcal{W}_{t_1}^{(c)},\mathcal{W}_{t_2}^{(c)}\right)$ $\sigma_H\left(\mathcal{W}_{t_1}^{(H)},\mathcal{W}_{t_2}^{(H)}\right)$	Covariance functions of the Wiener process $\mathcal{W}_t$, correlated Wiener process $\mathcal{W}_t^{(c)}$, and fractional Wiener process $\mathcal{W}_t^{(H)}$, respectively. The process $\mathcal{W}_t$ corresponds to the simple Brownian motion, the process $\mathcal{W}_t^{(c)}$ to the correlated Brownian motion, and the process $\mathcal{W}_t^{(H)}$ to the fractional Brownian motion.
$\mathfrak{s}(t)$	Internal state of the controlled mobile agent at time t; $\mathfrak{s}(t)\in[0,1]$.
$\mathcal{S}_t$ and $\mathcal{S}_t^2$	Random walks in one and two dimensions, respectively. The 1D random walk is also denoted by $\mathcal{S}_t^1$, and the 2D random walk is defined by a pair $\mathcal{S}_t^2=\left(\mathcal{S}_t,\mathcal{S}_t\right)$ of 1D random walks.
ς_{ij}	Transition rate, which specifies the probability of transition from point $\mathfrak{x}_i$ to point $\mathfrak{x}_j$, $i,j = 1, 2,\ldots,n$.
t and T	Time. For infinite horizon $t\in[0,\infty)$, and for finite horizon $t\in[0,T]$, $0 < T < \infty$. In the case of discrete time, it is assumed that $t = 0, 1, 2,\ldots$ or $t = 0, 1, 2,\ldots,T$.
$t(A^j)$	A residence time during which the forager hunts in the patch A^j, $j = 1, 2, \ldots, k$.
$t_1,t_2(t),t_3(t,t),\ldots$	The "giving up" times, which represent the strategy of forager and define when the forager leaves the patch in the case of unsuccessful search.
$t(A)$	An expected time of foraging in the patch A.
$t(w)$ and $t\left(u(\mathfrak{x})\right)$	Expected time of search using the strategy w. If the strategy and, consequently, the expected time are defined by a certain probability $u(\mathfrak{x})$, then instead of $t(w)$, the notation $t\left(u(\mathfrak{x})\right)$ is used.
$t = t(t_1,t_2)$	Expected search time with intermittent strategies; t_1 is an expected time of slow motion with screening and t_2 is an expected time of fast motion without screening.
$t_{-t}^{t\ldots\infty}(\mathbb{i}(t))$	An expected time, which will be spent in the patch after time t given that at this time the forager's state is $\mathbb{i}(t)$.
$\mathcal{T}:\mathfrak{X}\rightarrow\mathbb{R}^+$	A smooth function over the domain $\mathfrak{X}$, $0\le\mathcal{T}(\mathfrak{x})<\infty$ for any point $\mathfrak{x}\in\mathfrak{X}$, which represents a topography of the terrain.
$u:\mathfrak{X}\times[0,T]\rightarrow\mathbb{R}^+$	A probability density function, which defines the conditional probabilities of the target's or prey's location in the points $\mathfrak{x}$ of the domain $\mathfrak{X}$ at time t given that it was not detected up to this time.
$u(\mathfrak{x},t)$	A conditional probability that the target or prey is located in the point $\mathfrak{x}$ at time t given that it was not detected up to this time. At the beginning of search at time $t = 0$, it is assumed that $u(\mathfrak{x},0)=u(\mathfrak{x})$, $\mathfrak{x}\in\mathfrak{X}$.
$\hat{u}:\mathfrak{X}\times[0,T]\rightarrow\mathbb{R}^+$	A probability density function, which defines the joint probabilities that the target or prey is located certain points $\mathfrak{x}$ of the domain $\mathfrak{X}$ at times t and was not detected up to the time t. Without the search, function $\hat{u}$ governs a free motion of the target or prey over a domain. In the terms of population dynamics, the function $\hat{u}$ defines the probabilities of finding at least one individual in the point $\mathfrak{x}$ at time t.
$\hat{u}(\mathfrak{x},t)$	A joint probability that the target or prey is located in the point $\mathfrak{x}$ at time t and was not detected up to this time. At the beginning of search at time $t = 0$, it is assumed that $\hat{u}(\mathfrak{x},0)=u(\mathfrak{x},0)=u(\mathfrak{x})$, $\mathfrak{x}\in\mathfrak{X}$, is an initial probability of the target's location in point $\mathfrak{x}_i$, and $\hat{u}(\mathfrak{x},T)$ is the probability that the target is located in point $\mathfrak{x}$ at the end of the search and survived during all search period. In terms of population dynamics, $\hat{u}(\mathfrak{x},t)$ is a probability of finding at least one animal in point $\mathfrak{x}$ at time t.

$\hat{u}(\vec{o})$ and $\hat{u}(\vec{o},t)$	A priory probability that the target will follow the trajectory $\vec{o}$. If it is needed, this probability is also denoted by $\hat{u}(\vec{o},t)$ that stresses the time t of search or, in discrete time and space, the length of the trajectory.
$U(u;\hat{s},\hat{m})$	A differential operator, which specifies spatial evolution of the target's location probabilities. Functions $\hat{s}_{ij}$, $i,j = 1, 2$, specify the diffusion rate and functions $\hat{m}$, $i = 1, 2$, define a drift. If $\hat{s}_{ij} = \hat{\sigma}_{ij} = const$, then they are interpreted as deviation values, and if $\hat{m}_i = \hat{\mu}_i = const$, then they stand for mean values.
$\mathcal{U}$	Potential field over a domain with respect to the target's location probabilities.
$v : \mathfrak{X} \times [0,T] \to \mathbb{R}^+$	A probability density function, which defines the probabilities of searcher's or forager's location in the points $\mathfrak{x}$ of the domain $\mathfrak{X}$ at time t.
v(t), $\check{\text{v}}(t)$, and $\hat{\text{v}}(t)$	Velocity of the particle, searcher, and target at time t, respectively.
$V(v;\check{s},\check{m})$	A differential operator, which specifies spatial evolution of the searcher's location probabilities. Functions $\check{s}_{ij}$, $i,j = 1, 2$, specify the diffusion rate and functions $\check{m}_i, i = 1,2$, define a drift. If $\check{s}_{ij} = \check{\sigma}_{ij} = const$, then they are interpreted as deviation values, and if $\check{m}_i = \check{\mu}_i = const$, then they stand for mean values.
$\mathcal{V}$	Potential field over a domain with respect to the searcher's location probabilities. Usually, this field is a combination of the potential field defined by the searcher's location probabilities and the interaction field, which, in its turn, includes attraction and repulsion fields.
$w : \mathfrak{X} \times [0,T] \to \mathbb{R}^+$	A search density function, which defines a distribution of the search efforts over a domain $\mathfrak{X}$ at time t; this distribution is also called search plan or allocation, and the function w is called search strategy.
W	A set of all possible search strategies w over the domain $\mathfrak{X}$ given the target's probability distribution $u(\mathfrak{x})$, $\mathfrak{x} \in \mathfrak{X}$.
$\mathcal{W}_t$ and $\mathcal{W}_t^2$	Wiener processes with defined Brownian motion in one and two dimensions, respectively. The 1D Wiener process is also denoted by $\mathcal{W}_t^1$, and the 2D Wiener process is defined by a pair $\mathcal{W}_t^2 = (\mathcal{W}_t, \mathcal{W}_t)$ of 1D Wiener processes.
$\mathcal{W}_t^{(c)}$ and $\mathcal{W}_t^{(c)2}$	Correlated Wiener processes with defined correlated Brownian motion in one and two dimensions, respectively. The 2D correlated Wiener process is defined by a pair $\mathcal{W}_t^{(c)2} = \left(\mathcal{W}_t^{(c_1)}, \mathcal{W}_t^{(c_2)}\right)$ of 1D processes such that at least one of the correlation coefficients c_1 or c_2 is greater than zero.
$\mathcal{W}_t^{(H)}$ and $\mathcal{W}_t^{(H)2}$	Fractional Wiener processes with defined fractional Brownian motion in one and two dimensions, respectively. The 2D fractional Wiener process is defined by a pair $\mathcal{W}_t^{(H)2} = \left(\mathcal{W}_t^{(H_1)}, \mathcal{W}_t^{(H_2)}\right)$ of 1D processes such that at least one of the Hurst parameters H_1 or H_2 differs from 1/2.
$\Phi(\check{\text{v}})$	Velocity potential over a domain.
$\mathfrak{X} \subset \mathbb{R}^2$	A 2D open domain over which the agents are moving. In the case of discrete domain, $\mathfrak{X} = \{\mathfrak{x}_1, \mathfrak{x}_2, \ldots, \mathfrak{x}_n\}$, where n is a number of points.
$\mathfrak{x} \in \mathfrak{X}$	A point of the domain $\mathfrak{X}$, which is specified by the pairs (x,y) of coordinates. For a discrete domain $\mathfrak{X} = \{\mathfrak{x}_1, \mathfrak{x}_2, \ldots, \mathfrak{x}_n\}$, each $\mathfrak{x}_i$, $i = 1, 2,\ldots,n$, is defined by a pair (i_x,i_y) of indices $i_x = 1, 2,\ldots,n_x$ and $i_y = 1, 2,\ldots,n_y$ such that $n = n_x \times n_y$ and $i = ((i_x - 1)n_x + i_y)$.
$\hat{\mathfrak{x}} : \mathfrak{X} \times [0,T] \to \mathfrak{X}$ and $\check{\mathfrak{x}} : \mathfrak{X} \times [0,T] \to \mathfrak{X}$	Location functions, which define the target's trajectory $\vec{o}$ and the searcher's trajectory $\vec{a}$, correspondingly. In the search in continuous time and space, these functions are also called trajectories.
$\hat{\mathfrak{x}}(t)$ and $\check{\mathfrak{x}}(t)$	$\hat{\mathfrak{x}}(t) \in \mathfrak{X}$ is a location of the target and $\check{\mathfrak{x}}(t) \in \mathfrak{X}$ is a location of the searcher at time t.

$\upsilon(\mathfrak{x},t)$	In the search and screening theory, $\upsilon(\mathfrak{x},t)$ stands for the detection rate that specifies the velocity, with which the detection probability in point $\mathfrak{x}$ changes in time. In the foraging theory, this value is considered as a constant υ that specifies the proportion of an area searcher per time unit.
$\varphi:\mathfrak{X}\times\mathbb{R}^{+}\rightarrow[0,1]$	A detection function such that $\varphi(\mathfrak{x},\kappa)$ is the probability of detecting the target in the point $\mathfrak{x}$ given that the target is located in this point and the search effort applied to this point is $\kappa(\mathfrak{x},t)$.
$\psi:\mathfrak{X}\times\mathfrak{X}\rightarrow\mathbb{R}^{+}$	A density function, which defines the probability $\psi(\hat{\mathfrak{x}}(t),\check{\mathfrak{x}}(t))dt$ of detecting the target during the period $[t,t+dt)$ while the target's location is $\hat{\mathfrak{x}}(t)$ and the searcher's location is $\check{\mathfrak{x}}(t)$.
$\psi:\mathbb{R}^{2}\rightarrow\mathbb{R}^{+}$	A function that defines the possibility $\psi(u,v)$ of detecting the target while the probability of the target's location in the point $\mathfrak{x}$ at time t is $u(\mathfrak{x},t)$ and the probability of the searcher's location in the point $\mathfrak{x}$ at time t is $v(\mathfrak{x},t)$. Usually, $\psi(u,v)$ is proportional to the probability of finding the target in these conditions.
$y(t)$	Output of the controller of the mobile agent at time t; $y(t)\in[0,1]$.
$z(a)$, $z(a,t)$, and $z(t)$	A result of observation of the area a. If it is necessary, the time t of observation is indicated and the observation result is denoted by $z(a,t)$. In the case of k sensors, $k>1$, an observation result $z(a_i,\mathfrak{x})$ is considered for each area a_i, $i=1,2\ldots,k$, separately; it is assumed that $k\leq 2$. If the observed area is not known and depends on the sensors and task, the observation result is denoted by $z(t)$, $z(t)\in[0,1]$.
$\oplus_{\theta}$	A uninorm aggregator $\oplus_{\theta}:[0,1]\times[0,1]\rightarrow[0,1]$ with neutral element $\theta\in[0,1]$.
$\otimes_{\vartheta}$	An absorbing norm aggregator $\otimes_{\vartheta}:[0,1]\times[0,1]\rightarrow[0,1]$ with absorbing element $\vartheta\in[0,1]$. This operator defines the $\otimes_{\vartheta}$-automaton, in which inputs are given by observation results $z(t)$, internal states are denoted by $\mathfrak{s}(t)$, and outputs are denoted by $y(t)$. It is assumed that $z(t)$, $\mathfrak{s}(t)$, $y(t)$ for any $t=0,1,2,\ldots$

1 Introduction

Historically, the first mathematical formulation of the problem, which addresses the activity of seeking a hidden object, chasing the prey, and catching the target, was considered a process of pursuit. It was required to find a following or tailing curve, while the persecutor sees the target and the speeds of both the persecutor and the target are known. According to Nahin (2007, p. 23), this problem was mentioned in 1693 in the paper by Gottfried W. Leibniz, who remarked that sometime in 1672–1676 he worked with Christian Huygens on this problem earlier formulated by Claude Perrault. Sixty years later, in 1732, Pierre Bouguer presented similar problem for the French Academy, and after its publication in 1735, the problem of pursuit became a problem of pure calculus.

Nowadays, the problems of search and foraging are studied in several fields of applied mathematics, in computer science, and in mathematical biology. In applied mathematics and mathematical biology, this problem is mainly considered as a problem of search for hidden static or dynamic object and is formulated in the probabilistic terms; in computer science, in contrast, it addresses combinatorial search in arrays of data (Knuth, 1998) and finding the shortest paths in the networks (Russell & Norvig, 2010).

1.1 GROUP TESTING

Two classical versions of the problem in the form of search for a hidden object appeared during World War II. The first one was formulated in 1942 as a problem of search for all fault units in a given pool. Initially, it required to find an optimal procedure for testing blood samples for the presence of an antigen. A year later, Dorfman (1943) solved this problem using the following group-testing strategy. A set of units is tested simultaneously, and if the test indicates a presence of antigen, then the set is partitioned into subsets, and each subset is tested separately. The procedure of partitioning and testing continues up to finding a unit or units with the antigen. In 1957, Sterrett (1957) extended the Dorfman procedure to the search with multiple targets, while the number of targets is unknown. Later, the online procedure of multiple-target search for known number of targets was suggested by Hwang; a description of this procedure and of other combinatorial search algorithms is given in the book by Du and Hwang (1993).

In general, a group-testing approach addresses mainly the problem of statistical decisions, which include a selection of the best action under the uncertainty with respect to certain payoffs and a determination of the size of the test samples with respect to the results of the previous tests. An implementation of this approach to the search problem results in the following procedure (Ahlswede & Wegener, 1987). The searcher acts on the set of possible locations of the target. At each step, the searcher chooses a subset of the locations set and checks whether the target is somewhere in this subset or not. The procedure continues recursively on the subsets where the target is detected. The search terminates when the searcher detects the target in a single-point set. In this procedure, the main problem is in the determination of the size and the location of the subsets with respect to a given constant or varying detection function.

An optimal solution of this problem with perfect detection was found in 1959 by Zimmerman (1959). Later, it was found that the Zimmerman procedure is equivalent to the Huffman procedure of optimal coding (Huffman, 1952), and the length of the testing procedure up to identification of the fault unit is analogous to the length of the binary code. In 1994, Abrahams (1994) generalized this procedure to the search by multiple searchers. In 2005, this procedure was distributed on the group-testing search with coalitional and noncoalitional decision making, and an online algorithm of search was suggested; detailed description of this model and an overview of the other group-testing search algorithms are presented in the book by Kagan and Ben-Gal (2013b).

1.2 SEARCH AND SCREENING

The second classical version of the search problem was originated by Koopman also in 1942 in response to the German submarine threat in the Atlantic. In a 1946 report (Koopman, 1946) and then in a series of papers (Koopman, 1956–1957) and in a book (Koopman, 1980), Koopman named this problem as the search and screening problem and presented its widely accepted formulation. Nowadays, the considerations of this problem are unified into the theory of search and screening, which according to Frost and Stone (2001, p. 13)

> ... is the study of how to most effectively employ limited resources when trying to find an object whose location is not precisely known. The goal is to deploy search assets to maximize the probability of locating the search object with the resources available. Sometimes this goal is stated in terms of minimizing the time to find the search object.

In this theory, it is assumed that the searcher acts under uncertainty and during the search accumulates information about the target location. The amount of available information is specified by a detection function, which defines the probability of detecting the target with respect to the applied search efforts. The most popular defection function is a Koopman function that has an exponential form with respect to search efforts and is concave in time.

Originally, the theory of search dealt with offline planning of search, and the solution of the problem was specified in the form of optimal distribution of search efforts. Such solution represents an assumption that the search is conducted by a group of search agents and a total search effort is large enough as it follows from the initial task. The basic case of the problem of search planning for static target using Koopman detection function was solved for different distributions of the target. Detailed consideration of analytical results and algorithms was published in 1975 by Stone (1975) and then in 1992—by Iida (1992). Recent results obtained in the theory and military applications were presented in the reports by Frost and Stone (2001) and by Cooper, Frost, and Robe (2003), and in the book by Washburn and Kress (2009).

In his book (Stone, 1975), Stone in particular presented the algorithm of building an optimal search plan for the search in discrete space and time. Based on the Stone algorithm with Koopman detection function, in the beginning of the 1980s, Brown (1980), Washburn (1980, 1983), and Eagle (1984) developed algorithms of optimal search planning for a Markovian target moving in discrete domain. Recently, Singh and Krishnamurthy (2003) generalized this approach and reported about the algorithm, which is applicable both for non-Koopman detection function and for a search planning in the case of infinite horizon.

In parallel to the search for a moving target in discrete space and time, the problem of search planning was considered in continuous time and space. In particular, in 1972 Hellman (1972) formulated a general equation of target's movement and found necessary conditions for the search optimality with respect to a given finite period of search. In 1977 Lukka (1977) restricted this problem by certain assumptions regarding motion abilities of the searcher and the target; using these assumptions, he derived necessary conditions for the optimality of search paths. Later, in 1981 more general models were studied by Mangel (1981). A detailed consideration of the models and results in search planning in continuous space and time was published in 1985 in the book by Hellman (1985); introductory presentation with examples of applications was presented by Washburn (1989). Following this direction, in 1991 Ohsumi (1991) considered a search for a target moving according to a diffusion process and, using smoothness and concavity of the Koopman detection function, found optimal search paths, which provide maximum probability of detecting the target in a finite fixed period for some special cases of search.

Last years, on the basis of the ideas of informational group-testing search (Kagan & Ben-Gal, 2006), the search in nonfixed period with varying termination time was studied and heuristic near-optimal algorithms were suggested. In particular, in 2010 Kagan, Goren, and Ben-Gal (2010) suggested an online algorithm of search for static and moving targets in discrete time and space. The algorithm required perfect detection function, which is necessary non-Koopman. Two years later, this algorithm was modified for the search with imperfect detection, including application of the Koopman detection function (Chernikhovsky, Kagan, Goren, & Ben-Gal, 2012; Kagan, Goren, & Ben-Gal, 2012). In the same year, Israel, Khmelnitsky, and Kagan (2012) applied a discrete variant of the Ohsumi model for the search over a terrain with shadowing and suggested an online algorithm of search for static target and the target governed by not necessary unbiased diffusion process. All these algorithms act online and create near-optimal path of the searcher.

1.3 GAMES OF SEARCH

The theory of search games was originated in 1954 mainly by Isaacs and Karlin (1954) and Isaacs (1954) as a generalization of the search theory for the cases, in which the target can change its behavior and try to escape from been caught by the searcher. Initially, such games were called "aiming and evasion games," and nowadays, they are known as "pursuit–evasion games" or "search games." If the target is static, then the game is reduced to the search process as it is considered in the search and screening theory and is often called "the game of hiding." In such a case, the strategy of the hider includes a single step—hiding a target—while the searcher's strategy can include either a single step of choosing a trajectory from a certain predefined set or several steps of choosing positions during its walk over a domain.

The initial works in search games concentrated on the pursuit–evasion games in discrete domain (Isaacs, 1954; Isaacs & Karlin, 1954). In 1965, Isaacs in his classical monograph (Isaacs, 1965) considered the game of hiding and proved general theorem regarding the cost of such game and optimal strategies of the hider and the searcher. In addition, in this monograph Isaacs formulated generic "the princess and the monster" game. In this game, the monster searches for the princess in a totally dark room, and the time of search is considered as a monster's payoff. Both the princess and the monster are cognizant of the boundaries of the room. The game is terminated when the distance between the monster and the princess becomes smaller than a certain value, which is much smaller than the size of the room. It is assumed that the speed of the monster is constant and the princess has full freedom in her motion.

In the end of the 1970s, Gal obtained general solution of the princess and the monster game in convex domain; detailed description of this solution and over of the field of search games was presented in his book (Gal, 1980). Later, this game was generalized in several directions, in particular to the search in unbounded domain. The results of its treatment and the state of the art in the search game theory up to the beginning of the 2000s are presented in the book by Alpern and Gal (2003).

The consideration of the game of hiding in discrete domain with Koopman detection function of the searcher was conducted in 2000 by Garnaev (2000), who studied the game with and without payoffs. As it was demonstrated later (Kagan & Ben-Gal, 2013), the obtained solutions are the limiting ones for the Stone algorithm, which was obtained in the framework of the search theory (Stone, 1975).

1.4 FORAGING

In spite of different techniques and methods, the aforementioned approaches to the search problem are led by the common idea, namely, to define a behavior of the agents with the aim of their further development or specifying their activity. Foraging theory, in contrast, addresses the process of search from the opposite point of view; it starts with the observed search activity of the living organisms and concentrates on formal modeling of their behavior. Moreover, as it was indicated by Pyke in his critical review of the theory (Pyke, 1984, p. 523),

> Proponents of optimal foraging theory attempt to predict the behavior of animals while they are foraging.

In general, the foraging theory deals with two different problems: the first is a problem of search for a prey or food, and the second is a problem of making a decision to hunt or not to hunt the found prey. This discourse is restricted by considering the search activity.

The first attempts in formal consideration of foraging behavior were done at the end of the 1950s. The results of these studies were published in 1966 in the papers by MacArthur and Pianka (1966) and by Emlen (1966) that determined the main directions of further researches in foraging theory. In particular, for specifying a forager behavior, MacArthur and Pianka (1966) suggested to consider its movement in a patchy environment using certain optimization techniques.

Ten years later, Charnov in his famous paper (Charnov, 1976) formulated a model of optimal foraging by patches and under certain assumptions regarding predator's behavior and energy depot derived a condition that governs whether it has to stay in the current patch or leave it for search in the other patch. This result is widely known as marginal value theorem and forms a basis for classical optimal foraging theory. In 1977, Oaten (1977) distributed the Charnov approach to the foraging in stochastic environment, and in 1980, Green (1980) suggested a simple model of such foraging and considered its application. During the subsequent years, similar optimization techniques were applied for the analysis of foraging processes in different conditions; the resulting models appear, for example, in the papers by McNamara (1982), by Stephens and Charnov (1982), and by Mangel and Clark (1986). The methods and results obtained during this 20-year period of the development of optimal foraging theory were summarized in 1986 by Stephens and Krebs in their book (Stephens & Krebs, 1986). A detailed contemporary review of the mathematical methods and optimization techniques used in classical optimal foraging theory is given in the book by Pirolli (2007), and the state of the art in this theory is presented in recently published book (Stephens, Brown, & Ydenberd, 2007).

The optimal foraging theory mostly follows biological point of view and in the considered models applies several assumptions both regarding the foraging process itself and regarding a link between evolution and foraging. In a concentrated form, these assumptions are listed by Pyke in his already mentioned critical review (Pyke, 1984). Analyses of these assumptions and historical overview of the theory up to the last time appear in the first section of the recent book (Stephens et al., 2007).

In parallel to the studies concentrated specifically on the foraging behavior, there were suggested several models of animals' movement based on mathematical random walk processes. Probably, the first results in this direction formulated in the terms of trajectories were reported in 1952 by Wilkinson (1952) who investigated a possibility of random search in the birds' wandering. Following the Wilkinson results and based on the work published in 1951 by Skellam (1951), Patlak (1953) in 1953 suggested detailed mathematical techniques for modeling of animals' migration in the terms of Brownian random walks.

These models formed a basis for recent considerations of foraging following the methodology different from the optimization techniques, which are used in classical foraging theory. In particular, in 1983 Hoffman (1983) studied the optimality of Brownian search or foraging following Brownian random walk in terms of search theory and stochastic processes. Five years later, Bovet and Benhamou (1988) applied a correlated Brownian motion for modeling of foraging in stochastic environment and indicated a good correspondence between the modeled trajectories and the observed trajectories of foraging ants. Detailed consideration of such models and underlying theories was published in 1998 in the book by Turchin (1998).

Another approach to the studies of forager motion was suggested by the group led by Viswanathan. In the research published in 1996 (Viswanathan et al., 1996), it was found that the trajectories of albatrosses are better described by Lévy flights rather than by Brownian walks. This finding initiated intensive research of Lévy flights and Lévy walks applied to animal motion and to modeling of individual trajectories of the foragers. In 1999, the same research group considered an optimality of search following Lévy flights (Viswanathan et al., 1999), and then this behavior was studied in a wide context of foraging activity (Viswanathan et al., 2000; da Luz et al., 2001; Bartumeus, Da Luz, Viswanathan, & Catalan, 2005; Viswanathan, Raposo, & da Luz, 2008), in particular, in comparison with the Brownian walks' search (Bartumeus & Catalan, 2009; Reynolds, 2009; Nurzaman et al., 2011). A review of the results obtained in this direction up to the last time, including a detailed literature review, was published in 2011 by Viswanathan et al. in his book (Viswanathan, da Luz, Raposo, & Stanley, 2011).

In spite of a great success of Lévy flights and Lévy walks in the models of animals foraging, during the last years, there were distinguished several situations, which do not meet the results provided by these models. The main critics of Lévy flights models are summarized in the series of papers published by Plank and colleagues (Codling, Plank, & Benhamou, 2008; Plank & James, 2008; James, Plank, & Edwards, 2011). Such inconsistencies gave a rise to considerations of biased Lévy walks (Marthaler, Bertozzi, & Schwartz, 2004) and Lévy flights in random environment (Humphries et al., 2010; Ferreira, Raposo, Viswanathan, & da Luz, 2012).

Another alternative to the Lévy models was suggested in 2005–2007 by Bénichou and colleagues (Benichou, Coppey, Moreau, Suet, & Voituriez, 2005; Bénichou, Loverdo, Moreau, & Voituriez, 2006, 2007). This approach deals with the intermittent search strategies, which combine the strategies of optimal search and screening with the strategies specified by optimal foraging by patches. The resulting walks consist of the movements with low and high velocities and can model the motion in different environments (Condamin, Benichou, Tejedor, Voituriez, & Klafter, 2007; Benichou, Chevalier, Klafter, Meyer, & Voituriez, 2010). A recent overview of this approach appears in arXiv (Benichou, Loverdo, Moreau, & Voituriez, 2011).

1.5 GOAL AND STRUCTURE OF THIS BOOK

A brief outline of the approaches to the problem of optimal search and foraging does not pretend to present a complete overview of the methods and results that were obtained during more than 60 years of studies in operations research and mathematical biology. However, it, on one hand, indicates a similarity between the problems, which are addressed by these approaches, but, on the other hand, demonstrates a difference between the ideas and insights used in the search theory and the theory of foraging. This book makes an attempt to bridge the gap between search and foraging theories and to show how the ideas and methods developed in one theory can be used in the other.

The main goal of this book is to provide a formal framework that allows considerations of the search and foraging behavior using the same models and algorithms. The book starts with the models and optimization techniques used in the search theory and in optimal foraging theory. Then it considers the models of individual foraging and search algorithms, which result in the similar foraging behavior. Finally, it addresses a dynamics of swarms and a search by multiple agents.

In this book, general models are considered in the terms of search and foraging in continuous space and time, while for numerical simulations and algorithms, these models are formulated for the actions in discrete time over gridded domain rather than in the terms of numerical solutions of the problems. Such approach allows a consideration of both exact models of the agents' behavior and computer emulations of their activity.

The rest of this book is organized as follows:

Chapter 2 introduces general definitions and notation used in the search and foraging theories and considers classical approaches to the search and screening. The discourse follows optimization and algorithmic techniques of search for static target and moving targets in random environment. These techniques form a basis for further considerations of the models of individual search and foraging and of swarm dynamics.

Charter 3 deals with classical theory of optimal foraging. It starts with two general models of spatial dynamics of animal's population and then considers the deterministic and stochastic models foraging in patchy environment. It demonstrates the relation between the methods used in the foraging theory and the theory of search and screening that allows a consideration of these theories in the same theoretical and algorithmic framework.

Chapter 4 presents the models of individual foraging and the searcher's trajectories. It considers general equations of random walks and their modifications, as well as the models of Lévy flights and underlying diffusion equations and optimization techniques. The presentation indicates a relation between individual and classical search and considers the algorithms of probabilistic search and foraging, which implement this relation.

Chapter 5 addresses the models of coalitional search and swarm dynamics. As an introduction, it presents algorithms of search by multiple searchers and gives some general remarks on application of shared memory and information flow between the searchers. Then, based on the previously developed techniques, it presents an algorithm of foraging by swarms and provides general equations of probabilistic search and foraging.

Chapter 6 includes some remarks on the swarm robotic systems and usage for implementation of the presented models of search and foraging.

Chapter 7 concludes the discourse and outlines its main topics.

2 Methods of Optimal Search and Screening

Starting from initial works by Dorfman (1943) and Koopman (1946), the theory of search addresses a problem how to organize an optimal process of search or testing. The main interest of this theory is a distribution of search efforts given a probability distribution of the target's location in a certain domain (Hellman, 1985; Stone, 1975; Washburn, 1989). In the case of moving target, the theory also deals with the trajectories of the searcher over a domain (Hellman, 1985; Kagan & Ben-Gal, 2013). This chapter presents formal methods and algorithms of the search and screening theory both in continuous and in discrete time and space.

2.1 LOCATION PROBABILITIES AND SEARCH DENSITY

Let us consider a situation, in which a hider secretes an object in a point of a domain and then a seeker has to find the hidden object. According to the goal of the seeker, this situation is traditionally defined as either a search game (Isaacs, 1965), where the seeker strives to find the object in the least time (while the hider strives to maximize this time), or as a problem related to the search theory (Stone, 1975), where it is required to maximize the probability of detecting the hidden object during a given period. In the search theory, the hidden object is usually called target.

The process of search for the target starts with the known probability distribution, which defines the probabilities of target's location in the points of the domain. In the worst case, target's location probabilities are specified by the uniform distribution that corresponds to absence of initial information about location of the target.

In the classical search theory (Koopman, 1946; Stone, 1975; Washburn, 1989), it is assumed that the search is conducted by a large search system so that the "seeker" is represented by a group of searchers rather than by a sole search agent. In such system, the searchers share information and are coordinated and controlled in such a way that their joint actions can be represented by a single function of coordinates and time called search density function. In other words, the system acts as a search unit distributed over a domain. A part of this search unit applied to the point of the domain at given time is called search effort, and the search density function defines a distribution of search efforts over a domain in time (Hellman, 1985). When the goal is to find a distribution of search efforts such that during given period of search, the probability of target's detection reaches its maximum.

It is clear that if the search system consists of infinitely large numbers of searchers, then the most probability of target's detection is obtained by distributing search efforts proportionally to the target's location probabilities. In the case of search in gridded domain, the same holds true if the number of searchers is significantly larger than the number of cells in the grid. However, in practice, the number of searchers in the group is finite and relatively small, and an initial insight regarding distribution of search efforts fails. Additional difficulties are related to the fact that the detection abilities of the searchers are usually restricted and probability of detecting the target in the point of its location increases with the time of observation of this point.

Usually, general abilities of the search system including the number of searchers and available search period are characterized by a positive value called total search effort. Then, while the total search effort increases, the distribution of search efforts converges to the distribution of target

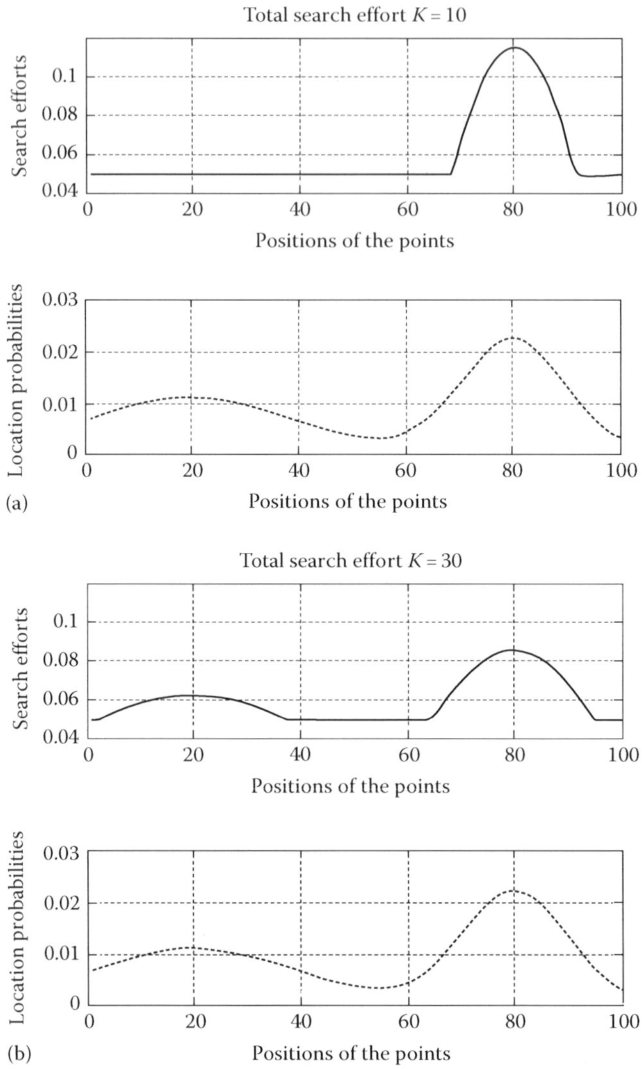

FIGURE 2.1 Location probabilities and search efforts: (a) total search effort $K = 10$ and (b) total search effort $K = 30$.

location probabilities. An example of target's location probabilities and search efforts for two different values of total search efforts are shown in Figure 2.1.

In the example shown in the figure, it is assumed that the target is located in one of 100 points along a line, and its location probabilities are given by two normal distributions with the means in the points 20 and 80 with standard deviations 20 and 10, respectively. The search efforts were calculated using the Stone algorithm (Stone, 1975); later, this algorithm will be considered in details. It is seen that if the total search effort is relatively small (case of $K = 10$ in Figure 2.1a), the most search efforts are applied to the most probable location of the target. While total search effort becomes greater (case of $K = 30$ in Figure 2.1b), the distribution of search efforts becomes closer to the distribution of the target's location probabilities.

An additional meaning of the search density function is the following: Assume that there is an abstract search agent seeking for a target over a domain. Then, starting from the distribution of the search efforts, at each time, there can be defined a distribution of the location probabilities of this searcher in the points of the domain. In the simplest case, such probabilities are obtained by a normalization of the search efforts over a domain. As a result, in parallel to the probability density of target's location, at each time, it is obtained a probability density of the searcher's location. The same holds in the simplest one-step search game, where location probabilities of the searcher specify the searcher's mixed strategy (Garnaev, 2000). In the search for moving targets, the situation is more complicated and searcher's location probabilities are defined according to the dynamics of the target; a brief consideration of the algorithms applied for such search is presented as follows.

2.2 SEARCH FOR A STATIC TARGET

There are several approaches to the search for a static target. The section starts with the constrained search and then considers group-testing search and sequential search. Finally, it describes the method of search by areas.

2.2.1 Constrained Search

Assume that a target is hidden in some point $\mathfrak{x} = (x, y)$ of a 2D open domain $\mathfrak{X} \subset \mathbb{R}^2$, and it is required to plan the search such that it provides maximum probability of detecting the target during given period [0, *T*]. The searcher is aware of the distribution of the target's location probabilities and deals with its probability density function $u : \mathfrak{X} \times [0, T] \to \mathbb{R}^+$. The goal of the searcher is to find a search density function $w : \mathfrak{X} \times [0, T] \to \mathbb{R}^+$, which defines a distribution of the total amount $K \in \mathbb{R}^+$ of available search efforts in time over the domain.

For each point $\mathfrak{x} \in \mathfrak{X}$, the density function u specifies a probability $u(\mathfrak{x}, t)d\mathfrak{x}$ that the target is in the neighborhood $[\mathfrak{x}, \mathfrak{x} + d\mathfrak{x})$ of the point $\mathfrak{x}$ at time $t \in [0, T]$ given that it was not found up to this time. Similarly, the value $u(\mathfrak{x}, t)dt$ is a probability that the target is located in the point $\mathfrak{x}$ during the period $[t, t + dt)$ given that it was not found up to the time t. It is assumed that at the beginning of search at time $t = 0$, the values $u(\mathfrak{x}, 0) = u(\mathfrak{x})$, $\mathfrak{x} \in \mathfrak{X}$, are specified a priori by a known function $u : \mathfrak{X} \to \mathbb{R}^+$; as earlier, for each point $\mathfrak{x} \in \mathfrak{X}$, the value $u(\mathfrak{x})d\mathfrak{x}$ is a probability of target's location in the neighborhood $[\mathfrak{x}, \mathfrak{x} + d\mathfrak{x})$ of the point $\mathfrak{x}$. Since the target is certainly in the domain $\mathfrak{X}$, for the function u, it holds true that $\int_{\mathfrak{X}} u(\mathfrak{x}, t)d\mathfrak{x} = 1$ for each $t \in [0, T]$. In the case of a gridded or discrete domain $\mathfrak{X} = \{\mathfrak{x}_1, \mathfrak{x}_2, \ldots, \mathfrak{x}_n\}$ consisting of $n \in \mathbb{N}$ cells, each point $\mathfrak{x}_i$, $i = 1, 2, \ldots, n$, is defined by a pair (i_x, i_y) of indices $i_x = 1, 2, \ldots, n_x$ and $i_y = 1, 2, \ldots, n_y$ such that $n = n_x \times n_y$ and $i = ((i_x - 1)n_x + i_y)$. Then u is a probability mass function, and $u(\mathfrak{x}_i, t)$ is a probability that the target is located in the point $\mathfrak{x}_i$ at time t given that it was not detected up to t; by the same as the aforementioned reason, for the probabilities $u(\mathfrak{x}_i, t)$, it holds true that $\sum_{i=1}^{n} u(\mathfrak{x}_i, t) = 1$ for each $t \in [0, T]$.

Given certain total search effort K, the search density function w specifies a distribution of the search efforts over a domain $\mathfrak{X}$ at time $t \in [0, T]$. This distribution is often called search plan or allocation, and the function w is called search strategy; in the following, these terms are used interchangeably. The search effort distributed over a domain at time t is $K(t) = \int_{\mathfrak{X}} w(\mathfrak{x}, t)d\mathfrak{x}$, and the search effort applied to the point $\mathfrak{x} \in \mathfrak{X}$ up to the time t is defined by $\kappa(\mathfrak{x}, t) = \int_0^t w(\mathfrak{x}, \tau)d\tau$. It is clear that $\int_{\mathfrak{X}} \int_0^T w(\mathfrak{x}, t)dtd\mathfrak{x} \le K$. For discrete domain $\mathfrak{X} = \{\mathfrak{x}_1, \mathfrak{x}_2, \ldots, \mathfrak{x}_n\}$ and discrete time $t = 0, 1, 2, \ldots, T$, the meanings of these functions are obvious: $w(\mathfrak{x}_i, t)$ is a proportion of the search effort applied to the point $\mathfrak{x}_i$ at time t; $K(t) = \sum_{i=1}^{n} w(\mathfrak{x}_i, t)$ is an amount of search effort distributed over a domain at time t; and $\kappa(\mathfrak{x}_i, t) = \sum_{\tau=0}^{t} w(\mathfrak{x}_i, \tau)$ is a search effort applied to the point $\mathfrak{x}_i$ from the beginning of search up to the time t.

The searcher starts at $t = 0$ with the distribution $u(\mathfrak{x}, 0) = u(\mathfrak{x})$, $\mathfrak{x} \in \mathfrak{X}$, of target location probabilities and implements a certain search strategy w; the value $w(\mathfrak{x}, t)dt$ is a probability of detecting the target in the point $\mathfrak{x}$ during the period $\left[t, t+dt\right)$ given that the target is located in $\mathfrak{x}$. According to the strategy w, to each point $\mathfrak{x}$ at time $t \in \left(0, T\right]$ is applied the search effort $\kappa\left(\mathfrak{x}, t\right)$, and the target location probabilities change with respect to these efforts. The dynamics of these changes are defined by the following equation (Hellman, 1985):

$$\frac{\partial u\left(\mathfrak{x}, t\right)}{\partial t} = -\left(w\left(\mathfrak{x}, t\right) - \int_{\mathfrak{X}} w\left(\mathfrak{x}, t\right) u\left(\mathfrak{x}, t\right) d\mathfrak{x} \right) u\left(\mathfrak{x}, t\right), \quad \mathfrak{x} \in \mathfrak{X}, \quad t \in \left(0, T\right]. \tag{2.1}$$

Assume that the density function w is already defined. Then, given an initial distribution $u\left(\mathfrak{x}, 0\right) = u\left(\mathfrak{x}\right)$, $\mathfrak{x} \in \mathfrak{X}$, solution of Equation 2.1 is

$$u\left(\mathfrak{x}, t\right) = u\left(\mathfrak{x}\right) \exp\left(-\int_0^t w\left(\mathfrak{x}, \tau\right) d\tau \right) \times \left(\int_{\mathfrak{X}} u\left(\mathfrak{x}\right) \exp\left(-\int_0^t w\left(\mathfrak{x}, \tau\right) d\tau \right) d\mathfrak{x} \right)^{-1}. \tag{2.2}$$

The value $\hat{u}\left(\mathfrak{x}, t\right) = u\left(\mathfrak{x}\right) \exp\left(-\int_0^t w\left(\mathfrak{x}, \tau\right) d\tau \right)$ is a joint probability that the target is located in the point $\mathfrak{x}$ and was not detected up to the time t.

Denote by $q(w, t)$ the probability of nondetecting the target by the use of the search strategy w up to the time t. Thus, $\hat{u}\left(\mathfrak{x}, t\right) = q\left(w, t\right) u\left(\mathfrak{x}, t\right)$, and from Equation 2.2, it follows that

$$q\left(w, t\right) = \int_{\mathfrak{x}} u\left(\mathfrak{x}\right) \exp\left(-\int_0^t w\left(\mathfrak{x}, \tau\right) d\tau \right) d\mathfrak{x}. \tag{2.3}$$

The same result can be obtained by integrating both sides of the equation $\hat{u}\left(\mathfrak{x}, t\right) = q\left(w, t\right) u\left(\mathfrak{x}, t\right)$ by $\mathfrak{x}$ and using the condition that $\int_{\mathfrak{X}} u\left(\mathfrak{x}, t\right) d\mathfrak{x} = 1$ for each t.

Since $q(w, t)$ is the probability of nondetecting the target, the probability of detecting the target during the period $[0, t]$ using the strategy w is defined as

$$p\left(w, t\right) = 1 - q\left(w, t\right). \tag{2.4}$$

Finally, using the definitions of search effort $\kappa\left(\mathfrak{x}, t\right) = \int_0^t w\left(\mathfrak{x}, \tau\right) d\tau$ applied in the point $\mathfrak{x}$ up to the time t, for the full search period $[0, T]$, one obtains

$$q\left(w, T\right) = \int_{\mathfrak{x}} u\left(\mathfrak{x}\right) \exp\left(-\kappa\left(\mathfrak{x}, T\right)\right) d\mathfrak{x} \quad \text{and} \quad p\left(w, T\right) = 1 - q\left(w, T\right). \tag{2.5}$$

Additional condition, which represents a "conversation" of search efforts, is specified as follows:

$$\int_{\mathfrak{x}} \kappa\left(\mathfrak{x}, T\right) d\mathfrak{x} = \int_0^T K\left(t\right) dt. \tag{2.6}$$

Thus, according to the goal of the searcher, it is required to find such search strategy w that maximizes the probability $p(w, t)$ or minimizes the probability $q(w, t)$ of nondetecting the target. Additional details regarding the outlined reasoning earlier can be found in the book (Hellman, 1985) by Hellman.

Let us consider the probability $p(w, t)$ of detecting the target up to the time t. From Equations 2.3 and 2.5 and condition $\int_{\mathfrak{X}} u(\mathfrak{x},t)d\mathfrak{x} = 1$, $t \in [0, T]$, it follows that

$$p(w,t) = 1 - q(t) = 1 - \int_{\mathfrak{x}} u(\mathfrak{x})\exp(-\kappa(\mathfrak{x},t))d\mathfrak{x} = \int_{\mathfrak{x}} u(\mathfrak{x})(1 - \exp(-\kappa(\mathfrak{x},t)))d\mathfrak{x}. \tag{2.7}$$

In the last integral, the value $u(\mathfrak{x})$ is a probability of target's location in the point $\mathfrak{x}$ at the beginning of search. Thus, the value $1 - \exp(-\kappa(\mathfrak{x},t))$ is the probability of detecting the target in the point $\mathfrak{x}$.

In general, the probability of detecting the target, also called detection probability, is defined by the detection function $\varphi : \mathfrak{X} \times \mathbb{R}^{+} \to [0,1]$, such that $\varphi(\mathfrak{x},\kappa)$ is the probability of detecting the target in the point $\mathfrak{x}$ given that the target is located in this point and the search effort applied to this point is $\kappa(\mathfrak{x},t)$. Usually, it is assumed that this function is strictly increasing and has continuous derivative by the second argument and such that $\varphi(\mathfrak{x},0) = 0$ for each point $\mathfrak{x}$. The detection function, which meets these requirements, is called regular (Stone, 1975). In particular, the exponential detection function

$$\varphi(\mathfrak{x},\kappa) = 1 - \exp(-\kappa(\mathfrak{x},t)), \tag{2.8}$$

which is implemented in Equation 2.7, is regular. The formula (2.8) was suggested by Koopman (1946, 1956–1957) and is called the random search formula or the Koopman formula (Stone, 1975).

Using the detection function φ, the probability of detecting the target during the period $[0,t)$ using the search strategy w is defines as follows:

$$p(w,t) = \int_{\mathfrak{x}} u(\mathfrak{x})\varphi(\mathfrak{x},\kappa)d\mathfrak{x}, \quad \kappa(\mathfrak{x},t) = \int_0^t w(\mathfrak{x},\tau)d\tau. \tag{2.9}$$

Let W be a set of all possible search strategies w over the domain $\mathfrak{X}$ given the target's probability distribution $u(\mathfrak{x})$, $\mathfrak{x} \in \mathfrak{X}$. Then, according to the goal of the searcher, it is required to find a search strategy $w^* \in W$ such that the probability $p(w^*, T)$ is maximal over all strategies $w \in W$. Hellman (1985) demonstrated that the solutions of such maximization problem exist for a wide class of probability density functions and of the detection functions; however, an exact form of such solutions can be found only in some special cases.

Example 2.1

To illustrate the problem, let us consider the discrete case and formulate the problem in the terms of one-step search game (Garnaev, 2000). In this case, at time $t = 0$, the hider secretes the target in some point $\mathfrak{x}$ of discrete domain $\mathfrak{X} = \{\mathfrak{x}_1, \mathfrak{x}_2, \ldots, \mathfrak{x}_n\}$, and the searcher has one attempt to find the target at time $t = 1$. In the theory of search games, the target's location probabilities $u(\mathfrak{x}_i) = u(\mathfrak{x}_i, 0)$ are considered as the hider's strategy and the search efforts $\kappa(\mathfrak{x}_i) = \kappa(\mathfrak{x}_i, 1)$ as the searcher's strategy; by definition, it is assumed that $\kappa(\mathfrak{x}_i, 0) = 0$ for all $i = 1, 2, \ldots, n$. In addition, it is assumed that the detection function φ is defined by the Koopman formula (2.8) with two additional parameters

$\alpha_i \in (0,1)$ and $\beta_i \in \mathbb{R}^+$, $i = 1, 2, \ldots, n$, which represent the conditions of search, and such that instead of formula (2.8), the probability $\phi(\mathfrak{x}_i, \kappa)$ of detecting the target in the point $\mathfrak{x}_i$ is defined as

$$\phi(\mathfrak{x}_i, \kappa) = \alpha_i \left(1 - \exp\left(-\beta_i \kappa(\mathfrak{x}_i)\right)\right). \tag{2.10}$$

Let ν^* be a root of the equation

$$\sum_{i=1}^{n} \frac{1}{\beta_i} \ln\left(\frac{\alpha_i}{\alpha_i - \nu}\right) = \sum_{i=1}^{n} \kappa(\mathfrak{x}_i) = K, \tag{2.11}$$

such that $\nu^* \in (0, \min_{i=1,\ldots,n} \alpha_i)$. Then, because of linearity of ϕ with respect to α_i and its concavity with respect to β_i, the optimal strategies of the hider $u^*(\mathfrak{x}_i)$ and of the searcher $\kappa^*(\mathfrak{x}_i)$ are defined as follows (Garnaev, 2000):

$$u^*(\mathfrak{x}_i) = \frac{1}{\beta_i(\alpha_i - \nu^*)} \left(\sum_{j=1}^{n} \frac{1}{\beta_j(\alpha_j - \nu^*)} \right)^{-1}, \tag{2.12}$$

$$\kappa^*(\mathfrak{x}_i) = \frac{1}{\beta_i} \ln \frac{\alpha_i}{\alpha_i - \nu^*}, \quad i = 1, 2, \ldots, n. \tag{2.13}$$

If the search conditions in all points $\mathfrak{x} \in \mathfrak{X}$ are equal, that is, $\alpha_i = \alpha$ and $\beta_i = \beta$, $i = 1, 2, \ldots, n$, then the optimal strategies of the hider and the searcher are

$$u^*(\mathfrak{x}_i) = \frac{1}{n} \quad \text{and} \quad \kappa^*(\mathfrak{x}_i) = \frac{1}{n} K. \tag{2.14}$$

correspondingly. In other words, if the search conditions are the same for all points of the domain, the hider has to choose the target's location randomly according to the uniform distribution, and the searcher has to distribute the search efforts equivalently to all points of the domain.

It is seen that in the case of search game, the strategies of the hider and the searcher are defined by the search conditions. If, in contrast, the search parameters α_i and β_i are defined by the target's location probabilities $u(\mathfrak{x}_i)$, $i = 1, 2, \ldots, n$,

$$\alpha_i = f\left(u(\mathfrak{x}_i)\right) \quad \text{and} \quad \beta_i = g\left(u(\mathfrak{x}_i)\right), \tag{2.15}$$

then instead of the search game, one obtains the one-step search process. The similar as in (2.12) and (2.13) form of solution of the search problem for the continuous space was obtained by Hellman (1985) and for discrete space by Kagan and Ben-Gal (2013), on the basis of the general Stone theorems (Stone, 1975). ■

Let us consider the most known version of the search problem, also called the constrained search. In addition to the probability (2.9), let $c : \mathfrak{X} \times \mathbb{R}^+ \to \mathbb{R}^+$ be a cost function such that the value $c(\mathfrak{x}, \kappa)$ is the cost of the search effort $\kappa(\mathfrak{x}, t)$ applied to the point $\mathfrak{x}$. Then, the cost $C(w, t)$ of the search strategy w up to the time t is

$$C(w, t) = \int_{\mathfrak{X}} c(\mathfrak{x}, \kappa) d\mathfrak{x}, \quad \kappa(\mathfrak{x}, t) = \int_0^t w(\mathfrak{x}, \tau) d\tau. \tag{2.16}$$

In the case of discrete domain $\mathfrak{X} = \{\mathfrak{x}_1, \mathfrak{x}_2, \ldots, \mathfrak{x}_n\}$ and discrete time t = 0, 1, 2,…,T, instead of Equations 2.9 and 2.16, the probability $p(w, t)$ and the cost $C(w, t)$ are defined as follows:

$$p(w,t) = \sum_{i=1}^{n} u(\mathfrak{x}_i)\varphi(\mathfrak{x}_i, \kappa), \tag{2.17}$$

$$C(w,t) = \sum_{i=1}^{n} c(\mathfrak{x}_i, \kappa), \tag{2.18}$$

where, according to the definitions given in the beginning of the section, $\kappa(\mathfrak{x}_i, t) = \sum_{\tau=0}^{t} w(\mathfrak{x}_i, \tau)$.

Let $C \in \mathbb{R}^+$ be the maximal possible cost of the search. The search problem is formulated as follows.

Problem 2.1: (Constrained Static Target Search) (Stone, 1975)

Find the strategy w^* such that

$$p(w^*, T) = \max_{w \in W} \{p(w,T) | C(w,T) \le C\} \quad \text{and} \quad C(w^*, T) \le C. \tag{2.19}$$

The solution of this problem was found by Stone (1975) following the techniques for solving separable optimization problems with constraints. Let us briefly consider this solution. Formal reasoning and proofs are presented in the Stone book (1975); detailed consideration of the search in discrete space and time is given in the book (Kagan & Ben-Gal, 2013) by Kagan and Ben-Gal.

For Problem 2.1, denote by

$$\mathcal{L}(w) = p(w,T) - \lambda C(w,T) \tag{2.20}$$

the Lagrange function (or Lagrangian) with Lagrange multipliers $\lambda \ge 0$. Then, for $C < \infty$, it follows that if

$$p(w^*, T) - \lambda C(w^*, T) \ge p(w,T) - \lambda C(w,T), \tag{2.21}$$

then (2.19) holds. According to the definitions of the probability $p(w, T)$ and the cost $C(w, T)$, for each point $\mathfrak{x} \in \mathfrak{X}$, the Lagrangian (2.20) is defined as follows:

$$\ell(\mathfrak{x}, \lambda, \kappa) = u(\mathfrak{x})\varphi(\mathfrak{x}, \kappa) - \lambda c(\mathfrak{x}, \kappa), \tag{2.22}$$

where $\kappa(\mathfrak{x}, T)$ is the search effort applied to the point $\mathfrak{x}$ following the strategy w.

Denote by $\kappa^*(\mathfrak{x}, T)$ the search effort applied to the point $\mathfrak{x}$ according to the strategy w^*. Then, the necessary and sufficient conditions for the optimality of the strategy w^* are the following (Stone, 1975):

Continuous space:

$$\ell(\mathfrak{x}, \lambda, \kappa^*) = \max_{w \in W} \{\ell(\mathfrak{x}, \lambda, \kappa)\} \quad \text{for almost every } \mathfrak{x} \in \mathfrak{X}; \tag{2.23}$$

and there exists such finite Lagrange multiplier λ and such strategy w^* that result in (2.23).

Discrete space:

$$\ell(\mathfrak{x}_i,\lambda,\kappa^*) = \max_{w\in W}\{\ell(\mathfrak{x}_i,\lambda,\kappa)\} \quad i = 1,2,\ldots,n; \tag{2.24}$$

detection function φ is concave with respect to κ, cost function c is convex with respect to κ, and there exists such finite Lagrange multiplier λ and such strategy w^* that result in (2.24). It is clear that if the detection function φ is defined by the Koopman formula (2.8) and $c(\mathfrak{x},\kappa(\mathfrak{x},t)) = \kappa(\mathfrak{x},t)$ for all $\mathfrak{x}$ and t, then the necessary condition holds.

Example 2.2: (Kagan & Ben-Gal, 2013; Stone, 1975)

Let $\mathfrak{X} = \{\mathfrak{x}_1,\mathfrak{x}_2,\ldots,\mathfrak{x}_n\}$ be a discrete domain with the target's location probabilities $u(\mathfrak{x}_i,t)$, i = 1, 2,..., n, t = 0, 1, 2,..., T. Assume that the detection function is defined by the Koopman formula (2.8) and that the cost is equal to the search effort, that is, $c(\mathfrak{x}_i,\kappa(\mathfrak{x}_i,t)) = \kappa(\mathfrak{x}_i,t)$, i = 1, 2,..., n, t = 0, 1, 2,..., T. Then the pointwise Lagrangian (2.22) is

$$\ell(\mathfrak{x}_i,\lambda,\kappa) = u(\mathfrak{x}_i,t)\left(1-\exp(-\kappa(\mathfrak{x}_i,t))\right) - \lambda\kappa(\mathfrak{x}_i,t). \tag{2.25}$$

The Lagrangian ℓ reaches its maximum for the search effort $\kappa(\mathfrak{x}_i,t)$ such that $\dfrac{\partial\ell}{\partial\kappa} = u(\mathfrak{x}_i,t)\exp(-\kappa(\mathfrak{x}_i,t)) - \lambda = 0$ and $\dfrac{\partial^2\ell}{\partial\kappa^2} = -u(\mathfrak{x}_i,t)\exp(-\kappa(\mathfrak{x}_i,t)) < 0$; thus

$$\kappa^*(\mathfrak{x}_i,t) = \ln\left(\frac{u(\mathfrak{x}_i,t)}{\lambda}\right), \quad \lambda > 0, \quad i = 1,2,\ldots,n. \tag{2.26}$$

Let us find the value of the Lagrange multiplier λ. By the use of (2.26), one obtains

$$\varphi(\mathfrak{x}_i,\kappa) = 1-\exp(-\kappa^*(\mathfrak{x}_i,t)) = 1-\frac{\lambda}{u(\mathfrak{x}_i,t)}. \tag{2.27}$$

Then, from (2.17), it follows that

$$p(w,t) = \sum_{i=1}^{n} u(\mathfrak{x}_i)\varphi(\mathfrak{x}_i,\kappa) = \sum_{i=1}^{n} u(\mathfrak{x}_i)\left(1-\frac{\lambda}{u(\mathfrak{x}_i,t)}\right) = 1-n\lambda; \tag{2.28}$$

and since $p(w,t) \le 1$, the multiplier λ is bounded as $\lambda \le \dfrac{1}{n}$. Similarly, since $c(\mathfrak{x}_i,\kappa(\mathfrak{x}_i,t)) = \kappa(\mathfrak{x}_i,t)$, i = 1, 2,..., n, t = 0, 1, 2,..., T, from (2.18), one obtains

$$C(w,t) = \sum_{i=1}^{n} c(\mathfrak{x}_i,\kappa) = \sum_{i=1}^{n} \kappa^*(\mathfrak{x}_i,t) = \sum_{i=1}^{n} \ln\left(\frac{u(\mathfrak{x}_i,t)}{\lambda}\right) = \sum_{i=1}^{n} \ln(u(\mathfrak{x}_i,t)) - n\ln(\lambda). \tag{2.29}$$

Since $\sum_{i=1}^{n}\kappa(\mathfrak{x}_i,t) = K(t)$, where $K(t)$ is the search effort distributed over a domain at time t, Equation 2.29 leads to the following value of λ:

$$\lambda = \exp\left(\frac{1}{n}\left(\sum_{i=1}^{n}\ln(u(\mathfrak{x}_i,t)) - K(t)\right)\right). \tag{2.30}$$

Finally, substitution of the value (2.30) into Equation 2.26 gives

$$\kappa^*(\mathfrak{x}_i,t) = \ln\left(u(\mathfrak{x}_i,t)\right) - \frac{1}{n}\left(\sum_{i=1}^{n}\ln\left(u(\mathfrak{x}_i,t)\right) - K(t)\right). \tag{2.31}$$

As it was expected, Equation 2.31 has the same form as Equation 2.13. Also, it is clear that if $u(\mathfrak{x}_1,t) = u(\mathfrak{x}_2,t) = \ldots = u(\mathfrak{x}_n,t) = \frac{1}{n}$, then $\kappa^*(\mathfrak{x}_i,t) = \ln\left(\frac{1}{n}\right) - \frac{n\ln(1/n) - K(t)}{n} = \left(\frac{1}{n}\right)K(t)$, as it was expected. ■

The reasoning given by Examples 2.1 and 2.2 forms a basis for the Stone algorithm. This algorithm acts on the discrete domain $\mathfrak{X} = \{\mathfrak{x}_1, \mathfrak{x}_2, \ldots, \mathfrak{x}_n\}$ and, given the initial target's location probabilities $u(\mathfrak{x}_i) = u(\mathfrak{x}_i, 0)$, $i = 1, 2, \ldots, n$, and total search effort K, creates the distribution of search efforts $\kappa^*(\mathfrak{x}_i, T)$ over the domain $\mathfrak{X}$. The algorithm terminates at time T, which indicates that all the available effort K is exhaust. The algorithm implements the Koopman detection function in the form (2.10) with $\beta_i = 1$, $i = 1, 2, \ldots, n$.

The idea of the algorithm is the following: At first, it defines the part $K(1)$ of search effort proportionally to the logarithmic difference between two maximal target's location probabilities and allocates this search effort $K(1)$ to the point with maximal probability. Then, it considers the third point with the greatest target's location probability and defines the part $K(2)$ by considering the second and the third points. This search effort $K(2)$ is allocated to the second point. The process continues until the total search effort K is not exhaust, that is, while $K(1) + K(2) + \ldots \leq K$. The algorithm is outlined as follows.

Algorithm 2.1: (Stone, 1975)

Given discrete domain $\mathfrak{X} = \{\mathfrak{x}_1, \mathfrak{x}_2, \ldots, \mathfrak{x}_n\}$, total search effort K, target's location probabilities $u(\mathfrak{x}_i)$ and search parameters α_i, $i = 1, 2, \ldots, n$, do:

1. Enumerate the points $\mathfrak{x}_i$ by the decreasing order: $\alpha_1 u(\mathfrak{x}_1) \geq \alpha_2 u(\mathfrak{x}_2) \geq \ldots \geq \alpha_n u(\mathfrak{x}_n)$.
2. For all $i = 1, 2, \ldots, n$ set $\kappa(\mathfrak{x}_i) = 0$.
3. Set EffortExhaust = false.
4. Set $i = 1$;
5. While $i < n$ do:
6. Set $h_i = \ln\left(\alpha_i u(\mathfrak{x}_i)\right) - \ln\left(\alpha_{i+1} u(\mathfrak{x}_{i+1})\right)$.
7. If $i = 1$ set $c(\mathfrak{x}_i) = h_i \sum_{j=1}^{i} 1/\alpha_j$.
8. Else set $c(\mathfrak{x}_i) = c(\mathfrak{x}_{i-1}) + h_i \sum_{j=1}^{i} 1/\alpha_j$.
9. If $c(\mathfrak{x}_i) \geq K$ set EffortExhaust = true; break.
10. For all $j = 1, 2, \ldots, i$ set $\kappa(\mathfrak{x}_j) = \kappa(\mathfrak{x}_j) + h_i/\alpha_j$.
11. Set $i = i + 1$.
12. End while.
13. If EffortExhaust = true set $a = \left(K - c(\mathfrak{x}_{i-1})\right) / \left(c(\mathfrak{x}_i) - c(\mathfrak{x}_{i-1})\right)$.
14. Else set $a = \left(K - c(\mathfrak{x}_{n-1})\right) / \sum_{i=1}^{n} 1/\alpha_i$; set $h_n = 1$.
15. For all $j = 1, 2, \ldots, i$ set $\kappa(\mathfrak{x}_j) = \kappa(\mathfrak{x}_j) + a h_i/\alpha_j$.
16. For all $j = i + 1, \ldots, n$ set $\kappa(\mathfrak{x}_j) = 0$.
17. Return $\kappa(\mathfrak{x}_i)$, $i = 1, 2, \ldots, n$.

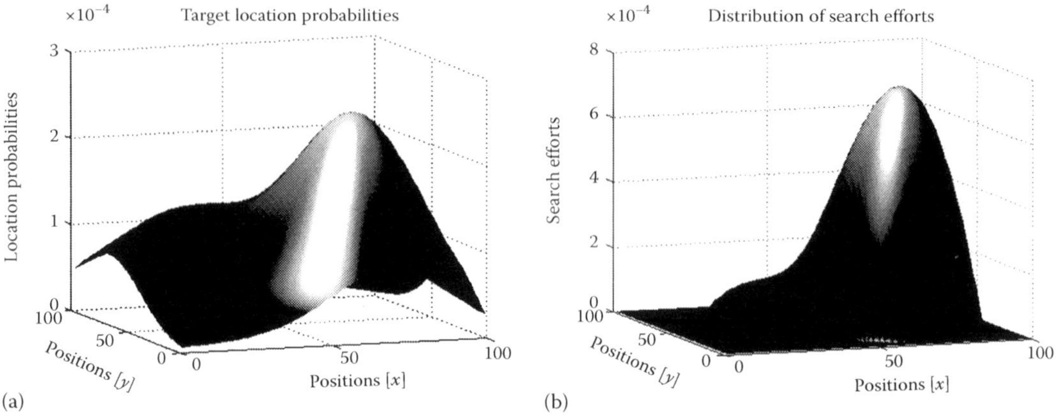

FIGURE 2.2 (a) Initial target's location probabilities and (b) distribution of search efforts created by Algorithm 2.1.

This algorithm was presented by Stone in his book (Stone, 1975); a detailed description of the algorithm and underlying reasoning is given in the book (Kagan & Ben-Gal, 2013) by Kagan and Ben-Gal. The activity of the search is illustrated by Figure 2.2.

In the figure, the domain $\mathfrak{X}$ includes $n = 100 \times 100$, total search effort is $K = 1000$, and detection parameters are $\alpha_i = 1$ for all $i = 1, 2, \ldots, n$. The target's location probabilities are defined by two binormal distributions with the centers in the points $\mathfrak{x}_1 = (30, 70)$ and $\mathfrak{x}_2 = (70, 30)$, and deviations $\sigma = (\sigma_x, \sigma_y)$ for the axes x and y are $\sigma_1 = (30, 30)$ and $\sigma_2 = (20, 20)$, respectively. According to the Stone Algorithm 2.1, the most of search efforts are distributed around the point $\mathfrak{x}_2$ over the points with greater location probabilities, and the remaining search efforts are distributed around the point $\mathfrak{x}_1$ over the points with less location probabilities. The values of the search efforts distributed over the points are governed by the logarithms of the location probabilities, and over the points with small probabilities zero search efforts are distributed.

In the continuous case, the solution of Problem 2.1 is more complicated and strongly depends on properties of the implemented functions. However, if w is piecewise continuous and the other functions are smooth, then one of the possible search strategies has the form similar to Equations 2.13 and 2.31. The reasoning in this case is the following (Hellman, 1985).

Assume that at the final time T, the search effort $\kappa(\mathfrak{x}, T)$ applied to each point $\mathfrak{x} \in \mathfrak{X}$ is nonzero. According to the considered search strategies, the search efforts are distributed over the domain $\mathfrak{X}$ in such a way that for each time t, there is a subset $\mathfrak{X}^+(t) \subset \mathfrak{X}$ such that the search efforts applied to the points $\mathfrak{x} \in \mathfrak{X}^+(t)$ up to the time t are positive, while the search efforts applied to the points $\mathfrak{x} \in \mathfrak{X} \backslash \mathfrak{X}^+(t)$ are zero, that is, $\kappa(\mathfrak{x}, t) > 0$ for $\mathfrak{x} \in \mathfrak{X}^+(t)$ and $\kappa(\mathfrak{x}, t) = 0$ for $\mathfrak{x} \in \mathfrak{X} \backslash \mathfrak{X}^+(t)$. In other words, for each point $\mathfrak{x} \in \mathfrak{X}$, there exists such time $t(\mathfrak{x}) \in [0, T]$ that before this time, the search effort applied to $\mathfrak{x}$ is zero, that is, $\kappa(\mathfrak{x}, t(\mathfrak{x})) = \int_0^{t(\mathfrak{x})} w(\mathfrak{x}, \tau)\,d\tau = 0$, while after $t(\mathfrak{x})$, it is $\int_{t(\mathfrak{x})}^{T} w(\mathfrak{x}, \tau)\,d\tau > 0$. Recall that the joint probability that the target is located in the point $\mathfrak{x}$ and was not detected up to the time t is

$$\hat{u}(\mathfrak{x}, t) = u(\mathfrak{x}) \exp\left(-\int_0^t w(\mathfrak{x}, \tau)\,d\tau \right); \tag{2.32}$$

this probability already appeared in Equation 2.3 defining the probability $q(w, t)$. Then,

$$u(\mathfrak{x}) \exp\left(-\int_{t(\mathfrak{x})}^{T} w(\mathfrak{x}, \tau)\,d\tau \right) = \hat{p} = \text{const}, \quad \mathfrak{x} \in \bigcup_{t \in [0, T]} \mathfrak{X}^+(t), \tag{2.33}$$

and

$$\int_{t(\mathfrak{x})}^{T} w(\mathfrak{x},\tau)\,d\tau = \ln\frac{u(\mathfrak{x})}{\hat{p}}. \tag{2.34}$$

Denote by $\mathfrak{X}_{\hat{p}} = \bigcup_{t\in[0,T]} \mathfrak{X}^{+}(t)$. Then, according to Equation 2.6, from (2.34), it follows that

$$\int_{\mathfrak{X}_{\hat{p}}} \kappa(\mathfrak{x},T)\,d\mathfrak{x} = \int_{\mathfrak{X}_{\hat{p}}}\int_{t(\mathfrak{x})}^{T} w(\mathfrak{x},\tau)\,d\tau\,d\mathfrak{x} = \int_{\mathfrak{X}_{\hat{p}}} \ln\frac{u(\mathfrak{x})}{\hat{p}}\,d\mathfrak{x} = \int_{0}^{T} K(t)\,dt. \tag{2.35}$$

The obtained equation forms a basis for the definition of optimal strategy. For exact specification of the strategy, there is a required value of the constant $\hat{p}$ or, equivalently, the subset $\mathfrak{X}_{\hat{p}}$, and time $t(\mathfrak{x})$, which cannot be specified explicitly.

To obtain the simplest search strategy, assume that $\mathfrak{X}_{\hat{p}} = \mathfrak{X}$, $t(\mathfrak{x}) = 0$ for all $\mathfrak{x} \in \mathfrak{X}$ and that all available total search effort K is exhaust at the time T, that is, $\int_{0}^{T} K(t)\,dt = K$. Then, according to Equation 2.35, $\hat{p} = \exp\left(\int_{\mathfrak{X}} \ln u(\mathfrak{x})\,d\mathfrak{x} - K\right)$, and following Equation 2.34, the search efforts are distributed (Hellman, 1985):

$$\kappa(\mathfrak{x},T) = \ln u(\mathfrak{x}) - \int_{\mathfrak{X}} \ln u(\mathfrak{x})\,d\mathfrak{x} + K, \quad \mathfrak{x} \in \mathfrak{X}, \tag{2.36}$$

The distribution of search efforts obtained for the same as in Figure 2.2a target's location probabilities is shown in Figure 2.3a. For comparison, in Figure 2.3b, it is shown the distribution of search efforts obtained by the Stone Algorithm 2.1 with total search effort is K = 10,000.

It is seen that the form of the distribution obtained by the use of Equation 2.36 follows the distribution of target's location probabilities and is close to the distribution created by the Stone Algorithm 2.1 with large total search effort.

The presented techniques are widely used in the search and screening theory for planning of search by the large search system without considering trajectories of the single agents. Certainly, after specification of the distribution of search efforts, the trajectories of the individual agents can

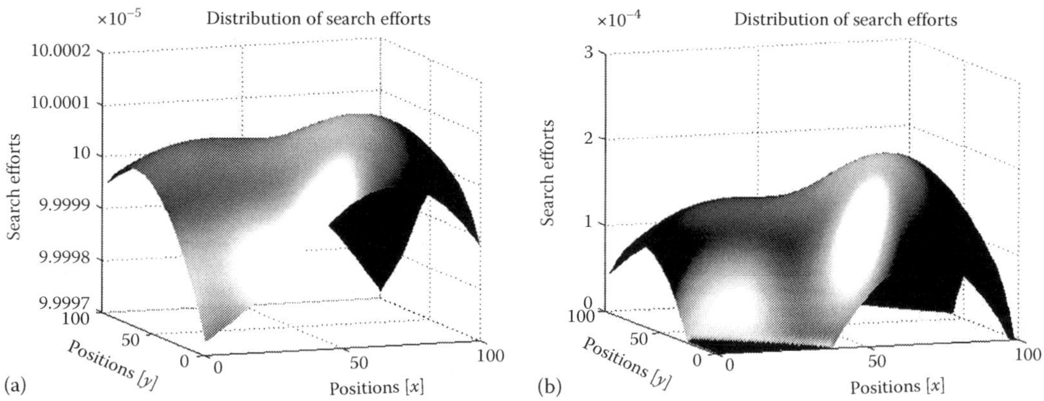

FIGURE 2.3 Distribution of search efforts: (a) according to Equation 2.36 and (b) according to the Algorithm 2.1 with total search effort K = 10,000.

be defined according to this distribution. However, such trajectories are far from optimal, and the consideration of search by a single agent requires special study. Some results in this direction were presented by Hellman (1985) with the credit to Lukka (1974) and by Washburn (1989). In this book, in contrast, the search trajectory of a single search agent will be considered in the framework of search for a moving target, which is concentrated on the trajectories of the agents, and then by the use of the methods of foraging theory.

2.2.2 Group-Testing Search

In the aforementioned methods, the search plan for further observations of the points $\mathfrak{x} \in \mathfrak{X}$ is obtained in the form of distribution of the search efforts over the domain $\mathfrak{X}$. Now let us consider a group-testing approach, which starts with target's location probabilities and specify the search plan in the form of the search tree.

In the group-testing search, it is assumed that the search system is able to observe an area $a \subset \mathfrak{X}$ of points and obtain observation result $z(a) \in \{0, 1\}$, that is, $z(a) = 1$ is the target located in a point $\mathfrak{x} \in a$ and $z(a) = 0$ if $\mathfrak{x} \notin a$. Notice that the observation result differs from the detection probability and represents the result of search as it is obtained by the searcher in the immediate observation of the area a.

Let $\mathfrak{X} = \{\mathfrak{x}_1, \mathfrak{x}_2, \dots, \mathfrak{x}_n\}$, $n \in \mathbb{N}$, be a set of points and let the target be located in some point $\mathfrak{x} \in \mathfrak{X}$. According to the group-testing approach, assume that the search is conducted by the search system, which observes the observed areas $a \subset \mathfrak{X}$ and obtains observation results $z(a) \in \{0, 1\}$. If $z(a) = 1$ continues search in a; otherwise, it continues with the points, which are not in a. Since the searcher seeks for a single target, the search terminates when the observed area a includes a single point, that is, $a = \{\mathfrak{x}\}$, and for this area, the searcher obtains observation result $z(a) = z(\{\mathfrak{x}\}) = 1$.

As earlier, the search planning starts with the target's location probabilities $u(\mathfrak{x}_i)$, $i = 1, 2, \dots, n$, $\sum_{i=1}^{n} u(\mathfrak{x}_i) = 1$, and results in a set $\mathcal{A} = \{a_1, a_2, \dots, a_m\}$ of search areas $a_j \subset \mathfrak{X}$, $j = 1, 2, \dots, m$, and a sequence $\vec{\mathfrak{d}} = \mathfrak{d}(0), \mathfrak{d}(1), \mathfrak{d}(2), \dots, \mathfrak{d}(T)$ of decision rules $\mathfrak{d}(t \mid z)$, which specify the observed areas $a(t)$ given the previous observed areas $a(t-1)$ and observation results $z(a(t-1))$; the rule $\mathfrak{d}(0)$ is interpreted as a general decision to start the search in the set $\mathfrak{X}$. The sequence $\vec{\mathfrak{d}}$ is called policy; if each decision rule unambiguously defines the next observed area, then $\vec{\mathfrak{d}}$ is also called strategy. The goal of the searcher is to minimize the number of observations required to find the target.

Usually, the search plan is represented by the binary decision tree, which specifies the observed areas as they obtained using the decision rules applied to the observation results. The search plan in the form of decision tree is shown in Figure 2.4.

In the figure, the search starts with the full set $\mathfrak{X}$, that is, at time $t = 0$, the observed area is $a(0) = \mathfrak{X}$. If the observation of $a(0)$ results in $z(a(0)) = 0$, then the search terminates since the target is not in the set $\mathfrak{X}$. Otherwise, the search continues and for the time $t = 1$, the decision rule $\mathfrak{d}(1 \mid z(a(0)) = 1)$ is applied. The result of application of this rule is the observed area $a_1(1)$. If observation of this area results in $z(a_1(1)) = 1$, then the search continues in this area since the target is in $a_1(1)$; otherwise, the target is in $\mathfrak{X} \backslash a_1(1)$, so if $z(a_1(1)) = 0$, then the search continues in $\mathfrak{X} \backslash a_1(1)$. Given the observation result $z(a_1(1)) = 1$, the application of decision rule $\mathfrak{d}(2 \mid z(a_1(1)) = 1)$ results in the observed area $a_{11}(2)$, while given the observation result $z(a_1(1)) = 0$, that is, equivalent to the result $z(\mathfrak{X} \backslash a_1(1)) = 1$, the decision rule $\mathfrak{d}(2 \mid z(a_1(1)) = 0)$ gives the observed area $a_{10}(2)$. In the same manner, the search continues in the areas $a_{11}(2)$ and $a_{10}(2)$ up to the termination with finding the target.

For binary observation results, at each time t, the decision rule applied to the observed area $a(t)$ unambiguously defines two areas $a_1(t + 1)$ and $a_2(t + 1)$ such that $a_1(t + 1) \bigcup a_2(t + 1) = a(t)$. Thus, the search plan is defined by the binary partitioning tree $\mathcal{T}_u[\mathfrak{X}]$, in which the vertices correspond to the observed areas and the edges specify the transitions from the areas at time t to the areas at

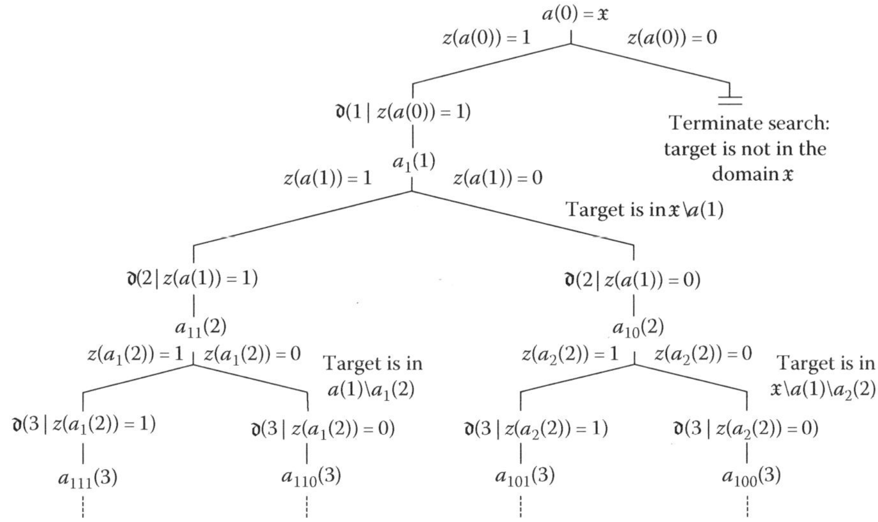

FIGURE 2.4 A fragment of the search plan in the form of binary decision tree.

time $t + 1$. The areas associated with the vertices at each level of the tree $\mathcal{T}_u\left[\mathfrak{X}\right]$ form a partition of the set $\mathfrak{X}$. Usually, it is assumed that the target is in the set $\mathfrak{X}$ so $z\left(\mathfrak{X}\right) = 1$, and the search plan starts with splitting the set in two areas $a_1(1)$ and $a_0(1)$ such that $a_1\left(1\right)\bigcup a_0\left(1\right) = \mathfrak{X}$.

For a given set $\mathfrak{X}$ and target's location probabilities $u\left(\mathfrak{x}_i\right)$, $i = 1, 2,\ldots, n$, there can be defined several binary partitioning trees. To define an optimal tree, denote by $l\left(\mathfrak{x}_i\right)$, $i = 1, 2,\ldots, n$, a number of edges in the tree $\mathcal{T}_u\left[\mathfrak{X}\right]$ from the root to the leaf $a = \left\{\mathfrak{x}_i\right\}$. For the values $l\left(\mathfrak{x}_i\right)$, the following inequality holds (Ahlswede & Wegener, 1987):

$$\sum_{i=1}^{n} 2^{-l(x_i)} \leq 1. \tag{2.37}$$

In the information theory (Cover & Thomas, 1991), the inequality (2.37) is known as Kraft inequality.

Given the probabilities $u\left(\mathfrak{x}_i\right)$, $i = 1, 2,\ldots, n$, the average length of the branches in the tree $\mathcal{T}_u\left[\mathfrak{X}\right]$ is defined as follows:

$$L\left(\mathcal{T}_u\right) = \sum_{i=1}^{n} u\left(\mathfrak{x}_i\right) l\left(x_i\right). \tag{2.38}$$

In other words, the number $L\left(\mathcal{T}_u\right)$ defines an average number of observations required to find a target starting from observing the full set $\mathfrak{X}$. Thus, according to the goal of the searcher to minimize the number of observations, the problem of group-testing search for a single target is as follows:

Problem 2.2: (Group-Testing Search for Single Target) (Ahlswede & Wegener, 1987)

Given the probabilities $u\left(\mathfrak{x}_i\right)$, $i = 1, 2,\ldots, n$, over all partitioning trees on the set $\mathfrak{X}$, find the partitioning tree with minimal average branch length with respect to Kraft inequality, that is,

$$\mathcal{T}_u^*\left[\mathfrak{X}\right] = \underset{\mathcal{T}_u\left[\mathfrak{X}\right]}{\operatorname{argmin}} \left\{ L\left(\mathcal{T}_u\right) \mid \sum_{i=1}^{n} 2^{-l(x_i)} \leq 1 \right\}. \tag{2.39}$$

The Lagrange function for Problem 2.2 is the following:

$$\mathcal{L}(\mathcal{T}_u) = L(\mathcal{T}_u) - \lambda \sum_{i=1}^{n} 2^{-l(x_i)}, \tag{2.40}$$

where $\lambda \geq 0$ is a Lagrange multiplier and the sum $\sum_{i=1}^{n} 2^{-l(x_i)}$ has the same meaning as the cost C in the Lagrangian (2.20).

Solution of Problem 2.2 is provided by the Zimmerman algorithm (Zimmerman, 1959). This algorithm starts with target's location probabilities $u(\mathfrak{x}_i)$, $i = 1, 2, \ldots, n$, and results in the binary partitioning tree $\mathcal{T}_u^*[\mathfrak{X}]$ with minimal average branch length.

Algorithm 2.2: (Zimmerman, 1959)

Given a set $\mathfrak{X} = \{\mathfrak{x}_1, \mathfrak{x}_2, \ldots, \mathfrak{x}_n\}$ and target's location probabilities $u(\mathfrak{x}_i)$, $i = 1, 2, \ldots, n$, do:

1. Start with the set $\mathcal{A} = \{a_1, a_2, \ldots, a_n\}$ of single-point observed areas $a_i = \{\mathfrak{x}_i\}$, $i = 1, 2, \ldots, n$.
2. While the set $\mathcal{A}$ includes more than one element do:
3. Choose two observed areas a' and a'' such that the probabilities $u(a') = \sum_{\mathfrak{x} \in a'} u(\mathfrak{x})$ and $u(a'') = \sum_{\mathfrak{x} \in a''} u(\mathfrak{x})$ are minimal over all areas from $\mathcal{A}$ (ties are broken randomly).
4. Set combined observed area $a''' = a' \bigcup a''$.
5. Include the leaves associated with the areas a', a'', and a''' and the edges (a', a''') and (a'', a''') into the tree $\mathcal{T}_u^*[\mathfrak{X}]$.
6. Exclude the areas a' and a'' from the set $\mathcal{A}$ and include the area a''' into the set $\mathcal{A}$.
7. End while.
8. Return the tree $\mathcal{T}_u^*[\mathfrak{X}]$.

This algorithm creates an optimal (in the sense of Problem 2.2) partitioning tree $\mathcal{T}_u^*[\mathfrak{X}]$, that is, the search plan with minimal number of observations. The algorithm was suggested by Zimmerman (1959). Later, it was clarified that this algorithm is equivalent to the algorithm of optimal coding, which was found seven years earlier by Huffman (1952). The actions of the Zimmerman algorithm and corresponding Huffman code are illustrated by the following example.

Example 2.3

Assume that the set $\mathfrak{X} = \{\mathfrak{x}_1, \mathfrak{x}_2, \ldots, \mathfrak{x}_6\}$ consists of $n = 6$ points and that the target's location probabilities are the following:

$\mathfrak{x}_i$	$\mathfrak{x}_1$	$\mathfrak{x}_2$	$\mathfrak{x}_3$	$\mathfrak{x}_4$	$\mathfrak{x}_5$	$\mathfrak{x}_6$
$u(\mathfrak{x}_i)$	0.05	0.1	0.15	0.15	0.2	0.35

Then, according to the Algorithm 2.2, at first are combined the points $\mathfrak{x}_1$ and $\mathfrak{x}_2$ with minimal probabilities and obtained the area $\{\mathfrak{x}_1, \mathfrak{x}_2\}$ with the probability $u(\{\mathfrak{x}_1, \mathfrak{x}_2\}) = u(\mathfrak{x}_1) + u(\mathfrak{x}_2) = 0.05 + 0.1 = 0.15$. At the second step, there are two points, $\mathfrak{x}_3$ and $\mathfrak{x}_4$, and the area $\{\mathfrak{x}_1, \mathfrak{x}_2\}$ with the same minimal probabilities $u(\mathfrak{x}_3) = u(\mathfrak{x}_4) = u(\{\mathfrak{x}_1, \mathfrak{x}_2\}) = 0.15$, and the algorithm breaks these ties randomly. Assume that it chooses to combine $\mathfrak{x}_3$ and $\mathfrak{x}_4$; then the obtained area is $\{\mathfrak{x}_3, \mathfrak{x}_4\}$ with probability $u(\{\mathfrak{x}_3, \mathfrak{x}_4\}) = u(\mathfrak{x}_3) + u(\mathfrak{x}_4) = 0.15 + 0.15 = 0.3$. Now the algorithm

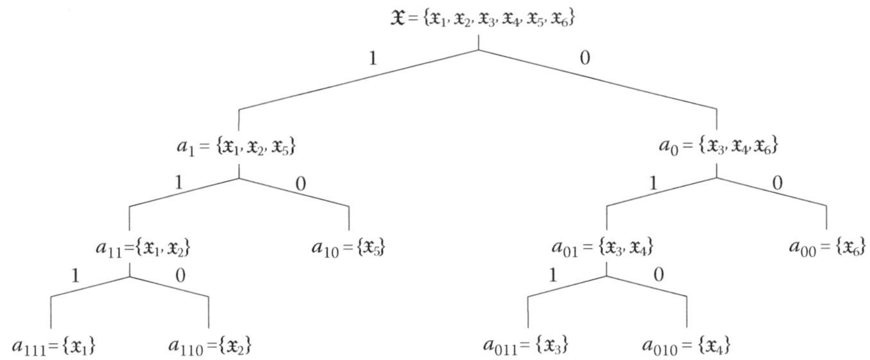

FIGURE 2.5 Example of Huffman coding tree.

considers the area $\{x_1, x_2\}$ with $u(\{x_1, x_2\}) = 0.15$, the point x_5 with $u(x_5) = 0.2$, the area $\{x_3, x_4\}$ with $u(\{x_3, x_4\}) = 0.3$, and the point x_6 with $u(x_6) = 0.35$ and combines $\{x_1, x_2\}$ and x_5. The resulting area is $\{x_1, x_2, x_5\}$ with the probability $u(\{x_1, x_2, x_5\}) = 0.35$. Then, by the same manner, it obtains the area $\{x_3, x_4, x_6\}$ with the probability $u(\{x_3, x_4, x_6\}) = 0.65$. The last combination of the areas $\{x_1, x_2, x_5\}$ and $\{x_3, x_4, x_6\}$ results in the full set $\mathfrak{X}$, and the algorithm terminates. The obtained binary tree is shown in Figure 2.5.

To associate the Huffman code words with the points x_i, $i = 1, 2, \ldots, 6$, assume that the branches of the obtained tree are labeled by "0" and "1." In the figure, the right branches are labeled by "0" and the left branches by "1." The code word associated with the point is combined from the labels of branches starting from the root and ending at the corresponding leaf of the tree. According to the tree shown in Figure 2.5, the code words and their lengths are the following (in the figure, they appear as indices of the corresponding areas):

x_i	x_1	x_2	x_3	x_4	x_5	x_6
Code word	111	110	011	010	10	00
$l(x_i)$	3	3	3	3	2	2

According to Equation 2.38, the average length of the branches in the tree is 2.45. ■

In 1995, Abrahams (1994) extended the Zimmerman algorithm to the search by several searchers, which conduct the search simultaneously and interchange information regarding obtained results. Detailed consideration of these algorithms is given in the book (Kagan & Ben-Gal, 2013) by Kagan and Ben-Gal.

2.2.3 Sequential Search

In the sequential search, it is assumed that the search is conducted by a single searcher, which is able to observed only one point at each time. The searcher starts with target's location probabilities and specifies the search plan in the form of a sequence of points, which are observed sequentially. In other words, the searcher dears with the single-point observed areas and screens the set $\mathfrak{X} = \{x_1, x_2, \ldots, x_n\}$, $n \in \mathbb{N}$, up to detecting the target in some point $x \in \mathfrak{X}$. As earlier, it is assumed that the observations result in the binary values such that the value $z(a) = z(\{x\}) = 1$ indicates that the target is in the point x and the search terminates, while the value $z(a) = z(\{x\}) = 0$ indicates that the target is not in x and the search continues. A search strategy $\vec{d}$ in this case is interpreted as a sequence, which defines the order of their observation, and the goal of the searcher is to specify such strategy that maximizes the probability of finding the target. Notice that this goal differs from the goal of the group-testing search and follows the methods search and screening considered in Section 2.2.1.

Similarly to Section 2.2.1, let $\mathfrak{D}$ be a set of all possible strategies $\vec{\mathfrak{d}}$ of sequential search over the domain $\mathfrak{X}$, and denote by $p(\vec{\mathfrak{d}},T)$ the probability of detecting the target up to the time T using the strategy $\vec{\mathfrak{d}}$. Since at each time $t = 1, 2,\ldots, T$ the searcher conducts a single observation, the value T is equivalent to the available number of observations. Then, given the target's location probabilities $u(\mathfrak{x}_i)$, $i = 1, 2,\ldots, n$, over a domain $\mathfrak{X}$, the problem is formulated as follows.

Problem 2.3: (Sequential Search for a Single Target) (Chew, 1967)

Find the strategy $\vec{\mathfrak{d}}^*$ such that

$$p(\vec{\mathfrak{d}}^*,T) = \max_{\vec{\mathfrak{d}}\in\mathfrak{D}} \left\{ p(\vec{\mathfrak{d}},T) \mid T < \infty \right\}. \tag{2.41}$$

Certainly, this problem is a variant of Problem 2.1 with unit costs $c(\mathfrak{x}_i) = 1$.

Recall that $\kappa(\mathfrak{x}_i,t)$, $t = 1, 2,\ldots, T$, is the search effort applied to the point $\mathfrak{x}_i$ up to the time t, $i = 1, 2,\ldots, n$. In the sequential search, the search effort is interpreted as a number of observations of the point $\mathfrak{x}_i$ up to t. Denote by $\epsilon_i \in (0, 1)$ an overlook probability of nondetecting the target in the point $\mathfrak{x}_i$ given that it is in $\mathfrak{x}_i$, $i = 1, 2,\ldots, n$. In general, these probabilities do not relate with the detection function φ and can be considered as parameters of the search (cf. Example 2.1). However, given a detection function φ, the overlook probabilities are defined as $\epsilon_i = 1 - \varphi(\mathfrak{x}_i,1)$, where $\varphi(\mathfrak{x}_i,1)$ is the probability of detecting the target by a single-time observation of the point $\mathfrak{x}_i$ given that the target is in $\mathfrak{x}_i$.

Then, the detection of the target is governed by a geometric distribution with the parameter $1-\alpha_i$, and optimal strategy $\vec{\mathfrak{d}}^*$, which solves Problem 2.3, is defined as follows (Chew, 1967):

$$u(\mathfrak{x}_j)\epsilon_j^{\kappa(\mathfrak{x}_j,T)-1}(1-\epsilon_j) = \max_{i=1,\ldots,n} \left\{ u(\mathfrak{x}_i)\epsilon_i^{\kappa(\mathfrak{x}_i,T)-1}(1-\epsilon_i) \right\}, \quad j = 1,2,\ldots,n. \tag{2.42}$$

In other words, given discrete domain $\mathfrak{X} = \{\mathfrak{x}_1,\mathfrak{x}_2,\ldots,\mathfrak{x}_n\}$, target's location probabilities $u(\mathfrak{x}_i)$, and overlook probabilities ϵ_i, it is required to choose the points $\mathfrak{x}_i$ such that the probability $u(\mathfrak{x}_i)\epsilon_i^{\kappa(\mathfrak{x}_i,T)-1}(1-\epsilon_i)$, $i = 1, 2,\ldots, n$, reaches its maximum.

Later, the problem of sequential search for a single target was extended to the sequential search with fictive targets (Kadane, 1968) and to the search, in which each point can include several targets (Assaf & Zamir, 1985).

Similarly to the aforementioned methods, the obtained solution does not assume that the searcher follows certain trajectory over a domain and the choice of the points is governed by Equation 2.42 independently on the order of the points in the domain. Since the resulting solution includes the points $\mathfrak{x}_i$ with the corresponding numbers $\kappa(\mathfrak{x}_i,T-1)$ of observations, it provides a distribution of search efforts as it was required in the search and screening methods considered in Section 2.2.1.

2.2.4 Search by Areas

Let us apply the methods of search and screening considered in Section 2.2.1 to the planning of search following the observed areas as it is used in the group-testing search. The consideration follows the book (Hellman, 1985) who gives a credit to Leipata (1976).

Assume that the continuous domain $\mathfrak{X} \subset \mathbb{R}^2$ is divided to m observed areas $a_j \subset \mathfrak{X}$, $j = 1, 2, \ldots, m$, such that the set $\mathcal{A} = \{a_1,a_2,\ldots,a_m\}$ is a partition of $\mathfrak{X}$, that is, $a_i \bigcap a_j = \varnothing$ if $i \neq j$ and $\bigcup_{j=1}^{m} a_j = \mathfrak{X}$.

Similar to the group-testing techniques, for each observed area a_j, it is assumed that to all points of a_j at time t, the searcher applies the same search effort, that is, $\kappa(\mathfrak{x}',t)=\kappa(\mathfrak{x}'',t)$, $\mathfrak{x}',\mathfrak{x}''\in a_j$, $j = 1, 2,\ldots, m$. Then, instead of Equation 2.5, the probability $q(w, T)$ of nondetecting the target during the period $[0, T]$ using the strategy w is defined as follows (Hellman, 1985):

$$q(w,T)=\sum_{j=1}^{m}\int_{a_j}u(\mathfrak{x})\exp(-\kappa(\mathfrak{x},T))d\mathfrak{x},\quad j=1,2,\ldots,m, \tag{2.43}$$

where a search strategy w specifies a distribution of the search efforts over the observed areas and, according to the assumption regarding the efforts, is constant inside each area a_j, $j = 1, 2, \ldots, m$.

Denote by $\kappa(a_j,T)$ the search effort applied to the points $\mathfrak{x}\in a_j$ and by $u(a_j)=\int_{a_j}u(\mathfrak{x})d\mathfrak{x}$ the probability that at the beginning of search at time $t = 0$, the target is located somewhere in the area a_j. Then, from Equation 2.43, it follows that

$$q(w,T)=\sum_{j=1}^{m}u(a_j)\exp(-\kappa(a_j,T)),\quad j=1,2,\ldots,m. \tag{2.44}$$

Thus, the searcher is required to find the strategy w, for which the probability $q(w, T)$ reaches its minimum. Since the probability of detecting the target during the period $[0, T]$ is $p(w, T) = 1-q(w, T)$, this problem is the same as Problem 2.1 of constrained search, where cost C is associated with the total search effort K.

Similarly to the sequential search, let us interpret the search effort $\kappa(a_j, t)$, $j = 1, 2,\ldots, m$, as a number of observations of the area a_j up to $t \in [0, T]$. Then, $\varphi(a_j, \kappa)$ stands for the probability of detecting the target during the κth observation of the area a_j given that the target is somewhere in a_j. Consequently, the probability of detecting the target by the search in all m observed areas is (Hellman, 1985)

$$p(w,T)=\sum_{j=1}^{m}u(a_j)\sum_{k=1}^{\kappa(a_j,T)}\varphi(a_j,k)\prod_{i=1}^{k-1}(1-\varphi(a_j,i)). \tag{2.45}$$

The constraints in this case are specified as follows. Denote by $t(a_j)$ the time required for observation of the area a_j, $j = 1, 2,\ldots, m$. Then the cost of the strategy w, which defines a distribution of the search efforts over the observed areas is

$$C(w,T)=\sum_{j=1}^{m}\kappa(a_j,T)t(a_j). \tag{2.46}$$

Such a specification of the total cost C represents the dependence of search on time and machinery resources. By the same token, define the cost of the search as $C = K \times T$. Then, according to Problem 2.1 of constrained search, it is required to find the strategy w^* specifying the observation numbers $\kappa^*(a_j,T)$ such that the probability $p(w^*,T)$ is maximal and the cost is constrained as

$$C(w^*,T)\le C. \tag{2.47}$$

This problem is solved by the following recursive definition of the observation numbers $\kappa^*(a_j,T)$ (Hellman, 1985):

$$\begin{aligned}
&\kappa^*(a_k,T) = n_k(C),\\
&\kappa^*(a_{k-1},T) = n_{k-1}\left(C - \kappa^*(a_k,T)t(a_k)\right),\\
&\ldots\\
&\kappa^*(a_1,T) = n_1\left(C - \sum_{j=2}^{m}\kappa^*(a_j,T)t(a_j)\right),
\end{aligned} \tag{2.48}$$

where n_j are integer valued functions, which define the numbers of observations of the areas a_j, $j = 1, 2,\ldots, m$, given the available cost. An example of such functions applied to the points $\mathfrak{x} \in \mathfrak{X}$ of the domain appears in the Algorithm 2.1.

Finally, assume that the probability of detecting the target in the area a_j, $j = 1, 2,\ldots, m$, does not depend on the observation number; in other words, similarly to the sequential search, the observations do not add information about the target's location and are defined as $1 - \epsilon_j = \varphi(a_j, 1)$, where ϵ_j is an overlook probability of nondetecting the target in the area a_j given that it is somewhere in a_j. Then,

$$\varphi(a_j,k)\prod_{i=1}^{k-1}\left(1-\varphi(a_j,i)\right) = (1-\epsilon_j)\epsilon_j^{k-1}, \tag{2.49}$$

$$\sum_{k=1}^{\kappa(a_j,T)}(1-\epsilon_j)\epsilon_j^{k-1} = 1-\epsilon_j^{\kappa(a_j,T)}. \tag{2.50}$$

Thus Equation 2.45, which defines the probability $p(w, T)$ is reduced as follows:

$$p(w,T) = \sum_{j=1}^{m}u(a_j)\left(1-\epsilon_j^{\kappa(a_j,T)}\right). \tag{2.51}$$

Here, detection probability for the area a_j is $\varphi(a_j,\kappa) = 1-\epsilon_j^{\kappa(a_j,T)}$ that in the terms of Koopman detection function (see Equation 2.8) results in

$$\varphi(a_j,\kappa) = 1-\exp\left(-\kappa(a_j,T)\ln\frac{1}{\epsilon_j}\right).$$

The comparison of this probability with Equation 2.10 shows that $\beta_j = \ln\frac{1}{\epsilon_j}$, which gives a sense for these search parameters formally introduced in Example 2.1.

The obtained results follow the line of sequential search where search efforts $\kappa(a_j,T)$ are interpreted as integer numbers of observations. Now assume that these numbers are real, and the numbers of observations are obtained by the use of certain search parameters β_j like in the last remark. In addition, assume that the searcher applies all available efforts and time, that is, instead of condition (2.47), the following equality is used:

$$C(w^*,T) = C. \tag{2.52}$$

Then, the search efforts $\kappa^*(a_j,T)$ are defined as follows (Hellman, 1985). Given the observation times $t(a_j)$ and the parameters $\beta_j = \ln\frac{1}{\epsilon_j}$, denote $\alpha_j = \frac{\beta_j}{t(a_j)}$, $j = 1, 2,\ldots, m$, and assume that the areas a_j are enumerated following the decreasing order

$$\alpha_1(1-\epsilon_1) \geq \alpha_2(1-\epsilon_2) \geq \ldots \geq \alpha_m(1-\epsilon_m). \tag{2.53}$$

Let k be a greatest value of the index i, $i = 1, 2,\ldots,m$, such that

$$\alpha_i(1-\epsilon_i) - \left(\exp\left[C \times \prod_{j=1}^{i}\left(\alpha_j(1-\epsilon_j)\right)^{-\frac{1}{\alpha_j}}\right]\right)^{\frac{1}{\sum_{j=1}^{i} 1/\alpha_j}} > 0. \tag{2.54}$$

Then, the search efforts $\kappa^*(a_i,T)$ are the following:

$$\kappa^*(a_i,T) = \begin{cases} \frac{1}{\beta_i}\left(\frac{1}{\sum_{j=1}^{k} 1/\alpha_j}\left(C - \sum_{j=1}^{k}\frac{1}{\alpha_j}\ln\alpha_j(1-\epsilon_j)\right) + \ln\frac{1}{\alpha_i(1-\epsilon_i)}\right), & i = 1,2,\ldots,k, \\ 0, & i = k+1,\ldots,m. \end{cases} \tag{2.55}$$

It is clear that for the single-point observed areas $a_i = \{x_i\}$, $i = 1, 2,\ldots, n$, Equation 2.55 defines the distribution of search efforts over a discrete domain $\mathcal{X} = \{x_1, x_2,\ldots, x_n\}$ similar to the distribution created by the Stone Algorithm 2.1. However, since Equation 2.55 implements constant detection probabilities independently on the applied search efforts, the obtained distribution is less sensitive to the target's location probabilities. An example of the search efforts distribution calculated according to Equation 2.55 for single-point observed areas is shown in Figure 2.6a. In the figure, it is assumed that the target's location probabilities are the same as in Figure 2.2a and that the overlook probabilities are $\epsilon_i = 1 - \varphi(\{x_i\}, 1)$ and the times required for observations of single-point areas are $t(\{x_i\}) = 1$, $i = 1, 2,\ldots, n$.

The other way of specifying the search efforts $\kappa(a_i,T)$ for the search with constant detection probabilities follows from the aforementioned relation between the Zimmerman algorithm

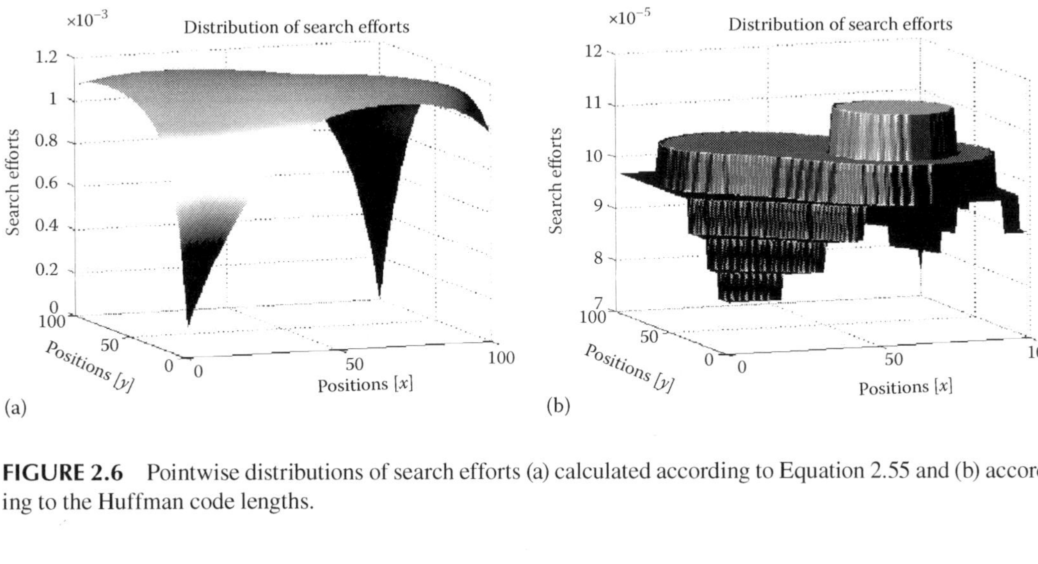

FIGURE 2.6 Pointwise distributions of search efforts (a) calculated according to Equation 2.55 and (b) according to the Huffman code lengths.

(Zimmerman, 1959) and Huffman code (Huffman, 1952) (see Example 2.3). Recall that the Zimmerman Algorithm 2.2 applied to the discrete set $\mathfrak{X} = \{\mathfrak{x}_1, \mathfrak{x}_2, \ldots, \mathfrak{x}_n\}$ with probabilities $u(\mathfrak{x}_i)$, $i = 1, 2, \ldots, n$, results in the binary partitioning tree $\mathcal{T}_u[\mathfrak{X}]$ such that the average number of observations defined by Equation 2.38 reaches its minimum. In the terms of Huffman coding (Huffman, 1952), the number $l(\mathfrak{x}_i)$ of edges in the tree $\mathcal{T}_u[\mathfrak{X}]$ from the root to the leaf $\{\mathfrak{x}_i\}$, i = 1, 2,..., n, is exactly the number of bits required to code a symbol associated with $\{\mathfrak{x}_i\}$ by optimal code. In other words, the number $l(\mathfrak{x}_i)$ specifies a proportion of efforts needed for minimal unambiguous representation of $\mathfrak{x}_i$ given a full set $\mathfrak{X}$ and probabilities $u(\mathfrak{x}_i)$, $i = 1, 2, \ldots, n$. In the terms of information theory (see Example 2.3), the sequence of bits is called code word of the symbol $\mathfrak{x}_i$ and the number $l(\mathfrak{x}_i)$ is a length of the code word. It is clear that since the coding does not depend on the nature of the leaf, it can be applied to the areas a_j with arbitrary number of points and built according to the probabilities $u(a_j)$, $j = 1, 2, \ldots, m$.

The last informal step is a specification of the search efforts applied to the points $\mathfrak{x}_i$ or to the areas a_j that are counted proportionally to the numbers $l(\mathfrak{x}_i)$ and $l(a_i)$, respectively. In the simplest case, the relation between the lengths $l(\mathfrak{x}_i)$ and the search efforts $\kappa(\mathfrak{x}_i, T)$, $i = 1, 2, \ldots, n$, can be defined as follows:

$$\kappa(\mathfrak{x}_i, T) = k\frac{1}{l(\mathfrak{x}_i)}, \quad k = \frac{K}{\sum_{i=1}^{n} 1/l(\mathfrak{x}_i)}, \tag{2.56}$$

where K is total available search effort. Since the Huffman coding associates the code words with the same length to different points, the obtained distribution is piecewise linear that allows to use these techniques for specification of the observed areas. An example of the search efforts distribution calculated according to Equation 2.56 for the same target's location probabilities (see Figure 2.2a) is shown in Figure 2.6b. The Huffman code was created using the routines written by Danaher (1998). Certainly, instead of Equation 2.56, another relation between the lengths and search efforts including the dependence on the observation times can be used.

The earlier methods result in the search strategy and starting from initial target's location probabilities provide the distributions of search efforts over a domain $\mathfrak{X}$. The order of the observations depends on a general goal of the searcher. In particular, if the searcher minimizes the time of search, in the case of discrete domain, the order of observations is defined by the strategy given by Equation 2.42. In general case of search by areas, if

$$\varphi(a_j, k) \geq \varphi(a_j, k+1)\left(1 - \varphi(a_j, k)\right), \quad k = 1, \ldots, \kappa^*(a_j, T) - 1, \quad j = 1, 2, \ldots, m, \tag{2.57}$$

then it follows the decreasing order of the values (Hellman, 1985)

$$\frac{1}{t(a_j)} u(a_j)\varphi(a_j, k)\prod_{i=1}^{k-1}\left(1 - \varphi(a_j, i)\right), \quad k = 1, \ldots, \kappa^*(a_j, T), \quad j = 1, 2, \ldots, m, \tag{2.58}$$

where the variables have the same meaning as in Equation 2.45 and the numbers $\kappa^*(a_j, T)$ are defined by Equation 2.48.

The order of observations defines the trajectory of the search over the domain up to detecting the target. In the search for a static target, such trajectory can be defined at the beginning of the search according to the initial target's location probabilities. In the search for a moving target, specification of the searcher's trajectory is more complicated; the next section includes a brief presentation of the methods such search.

2.3 SEARCH FOR A MOVING TARGET

The previous section dealt with the search for a static target, so it was assumed that the target's location probabilities changed in time only because of the application of the search efforts (see Equation 2.1), while without the search they are still equivalent to their initial values $u(\mathfrak{x},0)=u(\mathfrak{x}), \mathfrak{x}\in\mathfrak{X}$. In the case of moving target, the situation is different, and the changes of the target's location probabilities in time are governed both by the searcher's actions and by rules of target's dynamics. In the theory of search games (Alpern & Gal, 2003; Gal, 1980; Garnaev, 2000), it is assumed that the target is informed about the searcher's actions and the target's goal is to escape its finding. In contrast, in the search and screening theory (Hellman, 1985; Kagan & Ben-Gal, 2013), it is assumed that the target is not informed about the search process and moves independently on the searcher's activity. The following discourse follows the second assumption. Also, since the methods of search in discrete and in continuous domains are rather different, these two cases are considered separately.

2.3.1 Basic Two-Cell Models

Let us start with the two-cell model of search with discrete time; this pioneering model was suggested by Pollock (1970) in 1970 and has inspired further development of the algorithms of search for moving target in discrete domain.

Let $\mathfrak{X}=\{\mathfrak{x}_1,\mathfrak{x}_2\}$ and assume that the target moves between two points $\mathfrak{x}_1$ and $\mathfrak{x}_2$ according to a Markov process with transition probabilities matrix $\hat{\rho}=\|\hat{\rho}_{ij}\|_{2\times 2}$ such that

$$\hat{\rho}_{ij}=Pr\{\hat{\mathfrak{x}}(t+1)=\mathfrak{x}_j \mid \hat{\mathfrak{x}}(t)=\mathfrak{x}_i\},\quad i,j=1,2,\quad t=0,1,2,\ldots, \tag{2.59}$$

where $\hat{\mathfrak{x}}(t)$ and $\hat{\mathfrak{x}}(t+1)$ are locations of the target at the times t and $t+1$, correspondingly.

The searcher is allowed to observe one of the points $\mathfrak{x}_1$ and $\mathfrak{x}_2$ before the target's transition and to apply its detection function such that the overlook probability is $\epsilon_i=\epsilon(\mathfrak{x}_i)$, $i=1,2$. It is assumed that the observations are independent; so similar to the group-testing search (see Section 2.2.2),

$$\epsilon_i=1-\varphi(\mathfrak{x}_i,1). \tag{2.60}$$

In addition, since $\mathfrak{X}$ includes two points, from the assumption that the target is certainly somewhere in the domain $\mathfrak{X}$ it follows that for any $t=0,1,2,\ldots$, given the probability $u(\mathfrak{x}_1,t)$ that the target is in the point $\mathfrak{x}_1$, the probability that it is in the second point is $u(\mathfrak{x}_2,t)=1-u(\mathfrak{x}_1,t)$ and vice versa.

Then, given the probability $u(\mathfrak{x}_1)=u(\mathfrak{x}_1,0)$ that the target is in the point $\mathfrak{x}_1$ at the beginning of search, target's transition matrix $\hat{\rho}$, and overlook probabilities ϵ_1 and ϵ_2, there rise two problems (Pollock, 1970):

1. What strategy that is an order of observations results in certain detection of the target in minimal time, and what is this time?
2. Given a search period, what strategy provides maximal probability of target's detection, and what is the value of this probability?

It is clear that the first problem differs from the problems considered in the search for a static target, while the second problem is a variation of Problem 2.3 of sequential search; however, because of the target's movement, its solution is rather different. The indicated problems are solved as follows (Pollock, 1970).

Denote by $u(\mathfrak{x}_1 \mid i)$ a probability that after the first transition, the target is in the point $\mathfrak{x}_1$ given that the observation of the point $\mathfrak{x}_i$ before the transition failed, that is,

$$u(\mathfrak{x}_1 \mid i)=Pr\{\mathfrak{x}(1)=\mathfrak{x}_1 \mid z(\{\mathfrak{x}_i\},0)=0\},\quad i=1,2, \tag{2.61}$$

where $z(\{\mathfrak{x}_i\},0)$, $i = 1, 2$, is a result of observation of the point $\mathfrak{x}_i$ at time $t = 0$. The values of these probabilities are defined as follows:

$$u(\mathfrak{x}_1 \mid 1) = \frac{\left[u(\mathfrak{x}_1)\epsilon_1\hat{\rho}_{11} + u(\mathfrak{x}_2)\hat{\rho}_{21}\right]}{\left[1-(1-\epsilon_1)u(\mathfrak{x}_1)\right]}, \tag{2.62}$$

$$u(\mathfrak{x}_1 \mid 2) = \frac{\left[u(\mathfrak{x}_1)\hat{\rho}_{11} + u(\mathfrak{x}_2)\epsilon_2\hat{\rho}_{21}\right]}{\left[1-(1-\epsilon_2)u(\mathfrak{x}_2)\right]}, \tag{2.63}$$

where the relation $u(\mathfrak{x}_2) = 1 - u(\mathfrak{x}_1)$ is used.

Following the first problem, denote by $t(u(\mathfrak{x}_1))$ an expected time required for certain detection of the target given the probability $u(\mathfrak{x}_1)$, so the value $t(u(\mathfrak{x}_1 \mid i))$ is an expected time required for certain detection of the target after the first observation given the probability $u(\mathfrak{x}_1 \mid i)$, $i = 1, 2$. Then,

$$t(u(\mathfrak{x}_1)) = 1 + \begin{cases} (1-u(\mathfrak{x}_1)(1-\epsilon_1))t(u(\mathfrak{x}_1 \mid 1)), & \text{first observation is of the point } \mathfrak{x}_1, \\ (1-u(\mathfrak{x}_2)(1-\epsilon_2))t(u(\mathfrak{x}_1 \mid 2)), & \text{first observation is of the point } \mathfrak{x}_2. \end{cases} \tag{2.64}$$

where the first addendum specifies that it is required at least one observation before the first transition and $(1-u(\mathfrak{x}_i)(1-\epsilon_i))$ stands for the probability of unsuccessful observation of the point $\mathfrak{x}_i$, $i = 1, 2$, during the first observation. Consequently, an optimal strategy provides a minimum between the expected times required by these two observations, that is (Pollock, 1970),

$$t(u(\mathfrak{x}_1)) = 1 + \min \begin{cases} (1-u(\mathfrak{x}_1)(1-\epsilon_1))t(u(\mathfrak{x}_1 \mid 1)), \\ (1-u(\mathfrak{x}_2)(1-\epsilon_2))t(u(\mathfrak{x}_1 \mid 2)). \end{cases} \tag{2.65}$$

Alternatively, this strategy can be obtained as follows:

$$t(u(\mathfrak{x}_1)) = 1 + \begin{cases} (1-u(\mathfrak{x}_1)(1-\epsilon_1))t(u(\mathfrak{x}_1 \mid 1)), & \text{if } u(\mathfrak{x}_1) \geq u^*(\mathfrak{x}_1), \\ (1-u(\mathfrak{x}_2)(1-\epsilon_2))t(u(\mathfrak{x}_1 \mid 2)), & \text{if } u(\mathfrak{x}_1) \leq u^*(\mathfrak{x}_1), \end{cases} \tag{2.66}$$

where $u^*(\mathfrak{x}_1)$ is a threshold probability depending on the parameters of the problem. An exact value of the threshold probability $u^*(\mathfrak{x}_1)$ can be calculated only for very specific cases of the problem. However, in general, it can be approximated recursively using the following equation (Schweitzer, 1971):

$$u_{t+1}(\mathfrak{x}_1) = \frac{1}{(1-\epsilon_2)}\left(\frac{\left[f(u(\mathfrak{x}_1), u_t(\mathfrak{x}_1)) - 1\right]}{\left[f(u(\mathfrak{x}_1 \mid 2), u_t(\mathfrak{x}_1)) - \epsilon_2\right]}\right), \tag{2.67}$$

where the values $f(u(\mathfrak{x}_1), u_t(\mathfrak{x}_1))$ and $f(u(\mathfrak{x}_1 \mid 2), u_t(\mathfrak{x}_1))$ are defined by the following recursion:

$$f(u(\mathfrak{x}_1), b) = 1 + \begin{cases} (1-u(\mathfrak{x}_1)(1-\epsilon_1))f(u(\mathfrak{x}_1 \mid 1), b), & \textit{if } u(\mathfrak{x}_1) \geq b, \\ (1-u(\mathfrak{x}_2)(1-\epsilon_2))f(u(\mathfrak{x}_1 \mid 2), b), & \textit{if } u(\mathfrak{x}_1) \leq b, \end{cases} \tag{2.68}$$

As it was demonstrated by Schweitzer (1971), the values $u_t(\mathfrak{x}_1)$ converge with t to the threshold probabilities $u^*(\mathfrak{x}_1)$ for any probabilities $u(\mathfrak{x}_1)$ and any overlook probabilities $\epsilon_i < 1$, $i = 1, 2$.

Now, consider the second problem. Denote by $p(u(\mathfrak{x}_1),T)$ a probability of detecting the target by T observations given the probability $u(\mathfrak{x}_1)$; since it is assumed that the time is discrete, this is the same as the probability of detecting the target during the period [0, T–1]. Then (Pollock, 1970),

$$p(u(\mathfrak{x}_1),T) = \begin{cases} u(\mathfrak{x}_1)(1-\epsilon_1)+(1-u(\mathfrak{x}_1)(1-\epsilon_1))\,p(u(\mathfrak{x}_1\,|\,1),T-1), & \text{first observation of } \mathfrak{x}_1, \\ u(\mathfrak{x}_2)(1-\epsilon_2)+(1-u(\mathfrak{x}_2)(1-\epsilon_2))\,p(u(\mathfrak{x}_1\,|\,2),T-1), & \text{first observation of } \mathfrak{x}_2, \end{cases} \tag{2.69}$$

and for optimal strategy, this probability is the following:

$$p(u(\mathfrak{x}_1),T) = \max\begin{cases} u(\mathfrak{x}_1)(1-\epsilon_1)+(1-u(\mathfrak{x}_1)(1-\epsilon_1))\,p(u(\mathfrak{x}_1\,|\,1),T-1), \\ u(\mathfrak{x}_2)(1-\epsilon_2)+(1-u(\mathfrak{x}_2)(1-\epsilon_2))\,p(u(\mathfrak{x}_1\,|\,2),T-1), \end{cases}$$
$$p(u(\mathfrak{x}_1),0)=0 \tag{2.70}$$

It is clear that using the threshold probability $u^*(\mathfrak{x}_1)$, the probability $p(u(\mathfrak{x}_1),T)$ of detection is defined by the same token as in Equation 2.66. Detailed consideration of the Pollock model with numerical examples is given in the book (Kagan & Ben-Gal, 2013) by Kagan and Ben-Gal.

Unfortunately, the Pollock model of search cannot be directly extended to the search in the domain with more than two points that makes it useless for practical tasks. To overcome this lack, in 1974, Dobbie (1974) suggested another model of search in two cells, where he used continuous time instead of discrete time in Pollock model; as a result, transition and detection probabilities were replaced by transition, and detection rates and observations were replaced by effort density. The following consideration briefly introduces this model.

Similar to the Pollock model, let $\mathfrak{X} = \{\mathfrak{x}_1, \mathfrak{x}_2\}$ and the probability that the target is in the point $\mathfrak{x}_i$ during the period $[t, t+dt)$ is $u(\mathfrak{x}_i,t)dt$, $i = 1, 2$. Given that the target is in the point $\mathfrak{x}_i$, the probability that it will move to the other point $\mathfrak{x}_j$, $i, j = 1, 2$, is defined by the transition rate ς_{ij} such that $\varsigma_{ij}dt$ is a transition probability; it is clear that this probability replaces the transition probability $\hat{\rho}_{ij}$ used in the Pollock model. Regarding the searcher, it is assumed that given that the target is in the point $\mathfrak{x}_i$, the probability that it will be detected by the searcher observing this point during the period $[t, t+dt)$ is $\upsilon(\mathfrak{x}_i,t)dt$; the value $\upsilon(\mathfrak{x}_i,t)$ is called detection rate and determines the velocity $v(\mathfrak{x}_i,t) = d\varphi(\mathfrak{x}_i,\kappa)/dt$ with which detection probability $\varphi(\mathfrak{x}_i,\kappa)$ changes in time. In particular, if $\varphi(\mathfrak{x}_i,\kappa)$ is the Koopman function (see Equation 2.8), then $v(\mathfrak{x}_i,t) = \frac{d\kappa(\mathfrak{x}_i,t)}{dt}\exp(-\kappa(\mathfrak{x}_i,t))$, where, as earlier, $\kappa(\mathfrak{x}_i,t)$ is an amount of search effort applied to the point $\mathfrak{x}_i$, $i = 1, 2$, during the period $[t, t+dt)$.

Finally, notice that in contrast to the static target search, in the search for a moving target, the accumulation of search efforts applied to any point is meaningless. Thus, it is assumed that at each time, the searcher distributes over a domain all available amount of search efforts K. So for each t, it is assumed that $K(t) = K$ and a proportion of search efforts applied to the point $\mathfrak{x}_i$, $i = 1, 2$, during the period $[t, t+dt)$ is defined by the search strategy w and $w(\mathfrak{x}_1,t)+w(\mathfrak{x}_2,t)=1$. Such assumptions slightly differ from the assumptions used in the methods of search for a static target. It is clear that if φ is Koopman function, then also $K = 1$.

Then, the target's location probabilities are governed by the following equations (Dobbie, 1974):

$$\frac{du(\mathfrak{x}_1,t)}{dt} = -u(\mathfrak{x}_1,t)\left[\varsigma_{12} + v(\mathfrak{x}_1,t)w(\mathfrak{x}_1,t)\right] + u(\mathfrak{x}_2,t)\varsigma_{21}, \tag{2.71}$$

$$\frac{du(x_2,t)}{dt} = -u(x_2,t)\left[\varsigma_{21} + v(x_2,t)w(x_2,t)\right] + u(x_1,t)\varsigma_{12}; \tag{2.72}$$

the probability that the target was not detected up to the time t using the strategy w is

$$q(w,t) = u(x_1,t) + u(x_2,t), \tag{2.73}$$

and the expected time required for certain detection of the target using the strategy w given an initial target's location probabilities $u(x_1,0) = u(x_1)$ and, consequently, $u(x_2,0) = u(x_2) = 1 - u(x_1)$ is

$$t(w) = \int_0^\infty q(w,t)\,dt. \tag{2.74}$$

Denote

$$w_1(t) = \frac{1}{\left(v(x_1,t) + v(x_2,t)\right)}\left[\left(v(x_1,t) - v(x_2,t)\right)\sqrt{\frac{\varsigma_{12}\varsigma_{21}}{v(x_1,t)v(x_2,t)}} + \varsigma_{21} - \varsigma_{12} + v(x_2,t)\right]. \tag{2.75}$$

Considering the strategy w as a control function and using the Pontryagin maximum principle (Pontryagin et al., 1961), Doobie demonstrated that (Dobbie, 1974)

1. If $0 < w_1(t) < 1$, then the strategy w can obtain one of three values $w(x_1,t) \in \{0, w_1(t), 1\}$
2. If $w_1(t) < 0$ or $w_1(t) > 1$, then it can obtain only two values $w(x_1,t) \in \{0,1\}$

And the exact value from $\{0, w_1(t), 1\}$ or $\{0, 1\}$ is chosen in such a way that it provides a minimum of the probability $q(w,t)$. The values $w(x_2,t)$ follow from the assumption that $w(x_1,t) + w(x_2,t) = 1$.

As indicated by Doobie (1974), the suggested two-cell model of search in continuous time can be extended to the search in the domain with arbitrary number of points, and in the book (Stone, 1975), Stone presented preliminary studies of this case. The further results in this direction were summarized by Iida (1992).

The considered two-cell models of search for a moving target demonstrate that there rises additional problem regarding the expected time $t(w)$ of search and that the problem regarding the probability $p(w, T) = 1-q(w, T)$ of detecting the target during the period $[0, T]$ requires different approach. For convenience, let us formulate the problems of the moving target search explicitly.

Recall that W stands for a set of all possible search strategies w over the domain $\mathfrak{X}$ given the target's initial probability distribution $u(x)$, $x \in \mathfrak{X}$. Then the problems are formulated as follows.

Problem 2.4: (Moving Target Search)

Find the strategy w^* such that

$$t(w^*) = \min_{w \in W}\left\{t(w) < \infty \,\middle|\, p(w, t(w)) = 1\right\}; \tag{2.76}$$

$$p(w^*, T) = \max_{w \in W}\left\{p(w, T) \,|\, T < \infty\right\}. \tag{2.77}$$

If, similar to the sequential search (see Problem 2.3), the strategy is represented by the sequence $\vec{\mathfrak{d}}$ of decision rules and, in addition, it is assumed that the observed areas $a(t)$ chosen at times $t \in [0, T]$ are pairwise neighboring, then the strategy forms a trajectory of the searcher in the domain $\mathfrak{X}$.

Also notice that in the case of search game (Alpern & Gal, 2003; Gal, 1980), where the target escapes from being caught by the searcher, the solution of the first problem not necessary exists in the sense that the trajectories of the searcher and the target never intersect in time and space and $t(w)$ is infinite for any strategy w.

2.3.2 Moving Target Search in Discrete Domain

Now let us consider a discrete time search for a moving target in the discrete domain $\mathfrak{X} = \{\mathfrak{x}_1, \mathfrak{x}_2, \ldots, \mathfrak{x}_n\}$ with arbitrary number n of points. As earlier, it is assumed that the domain $\mathfrak{X}$ is a square and each point $\mathfrak{x}_i$, $i = 1, 2, \ldots, n$, is defined by a pair (i_x, i_y) of indices $i_x = 1, 2, \ldots, n_x$ and $i_y = 1, 2, \ldots, n_y$ such that $n = n_x \times n_y$ and $i = ((i_x - i)n_x + i_y)$. As earlier, it is assumed that at the beginning of search at $t = 0$, the target's location probabilities $u(\mathfrak{x}_i, 0) = u(\mathfrak{x}_i)$, $i = 1, 2, \ldots, n$, are known, while at each next times, they depend on the searcher's actions. Certainly, $\sum_{i=1}^{n} u(\mathfrak{x}_i, t) = 1$ for each $t = 0, 1, 2, \ldots$.

Similar to the Pollock model (see Equation 2.59), the target's movement is governed by a Markov process with transition probabilities matrix $\hat{\rho} = \|\hat{\rho}_{ij}\|_{n \times n}$, where

$$\hat{\rho}_{ij} = Pr\{\hat{\mathfrak{x}}(t+1) = \mathfrak{x}_j \mid \hat{\mathfrak{x}}(t) = \mathfrak{x}_i\}, \quad i, j = 1, 2, \ldots, n, \quad t = 0, 1, 2, \ldots, \tag{2.78}$$

and $\hat{\mathfrak{x}}(t)$ and $\hat{\mathfrak{x}}(t+1)$ are locations of the target at the times t and $t + 1$, correspondingly. Also, it is assumed that the observations are independent and the search efforts do not cumulate in time; so detection probabilities $\varphi(\mathfrak{x}_i, \kappa(\mathfrak{x}_i, t))$, $t = 1, 2, \ldots$, defined by the strategy w do not depend on the history of observations.

Then the changes of the target's location probabilities it time are governed by the following equation (Ciervo, 1976; Iida, 1992):

$$u(\mathfrak{x}_j, t+1) = \frac{\sum_{i=1}^{n} u(\mathfrak{x}_i, t)\left[1 - \varphi(\mathfrak{x}_i, \kappa(\mathfrak{x}_i, t))\right]\hat{\rho}_{ij}}{\sum_{i=1}^{n} u(\mathfrak{x}_i, t)\left[1 - \varphi(\mathfrak{x}_i, \kappa(\mathfrak{x}_i, t))\right]}, \tag{2.79}$$

where $t = 0, 1, 2, \ldots$ and $u(\mathfrak{x}_i, 0) = u(\mathfrak{x}_i)$, $i = 1, 2, \ldots, n$. Consequently, the probability $p(w, T)$ of target's detection using the strategy w, which specifies the search efforts $\kappa(\mathfrak{x}_i, t)$, $t = 1, 2, \ldots$, is the following (Ciervo, 1976; Iida, 1992):

$$p(w, t) = 1 - \prod_{\tau=0}^{t-1} \sum_{i=1}^{n} u(\mathfrak{x}_i, t)\left[1 - \varphi(\mathfrak{x}_i, \kappa(\mathfrak{x}_i, t))\right], \quad t = 1, 2, \ldots \tag{2.80}$$

This equation specifies the probability of target's detection up to time t for any search strategy w. The dynamics of target's location probabilities is illustrated by the following example.

Example 2.4

Assume that the domain $\mathfrak{X}$ includes $n = 100 \times 100$ points and that the target starts with the initial location probabilities as presented in Figure 2.7a; these probabilities are the same is in the previous examples (see Figure 2.2a). The target's motion is governed by a discrete time Markov process corresponding to a nonbiased diffusion process, where the probabilities of staying at the current location or moving one step to each of four possible directions—north, south, east, and west are equal, while the probabilities of the other steps are zero. A distribution of search efforts is constant and corresponds to the binormal distribution of detection probabilities over a domain, that is,

$$\varphi\left(\mathfrak{x}_i, \kappa\left(\mathfrak{x}_i, t\right)\right) = \frac{K\left(\mathfrak{x}_i\right)}{2\pi\sigma^2}\exp\left(-\frac{(\mathfrak{x}_i - \bar{\mathfrak{x}})^2}{2\sigma^2}\right), \quad t = 0, 1, 2, \ldots \tag{2.81}$$

The center of distribution is in the point $\bar{\mathfrak{x}} = (70,30)$ and deviation in both coordinates is $\sigma = 15$, and the detection probabilities $\varphi\left(\mathfrak{x}_i, \kappa\left(\mathfrak{x}_i, t\right)\right)$, $i = 1, 2, \ldots, n$, are normalized by $K\left(\mathfrak{x}_i\right)$ such that maximum probability is equal to one. Notice that such normalization is possible since detection probabilities at different points are independent. The distribution of detection probabilities is shown in Figure 2.7b.

Figure 2.7c presents distribution of the target's location probabilities at time $t = 0$ without the search, and Figure 2.7d demonstrates these probabilities while the search is conducted.

It is seen that without search (see Figure 2.7c), the distribution of the target's probabilities changes only because of slow diffusion of the probabilities to the boundaries of the domain. In contrast, if the search is conducted (see Figure 2.7d), then, in addition to diffusion, location probabilities are changed correspondingly; in the shown case, they increase down to zero at the points around the point $\bar{\mathfrak{x}} = (70,30)$. ■

General Equations 2.79 and 2.80 specify the dynamics of target's location probabilities and the probability of target's detection following an assumption that the search is conducted by large

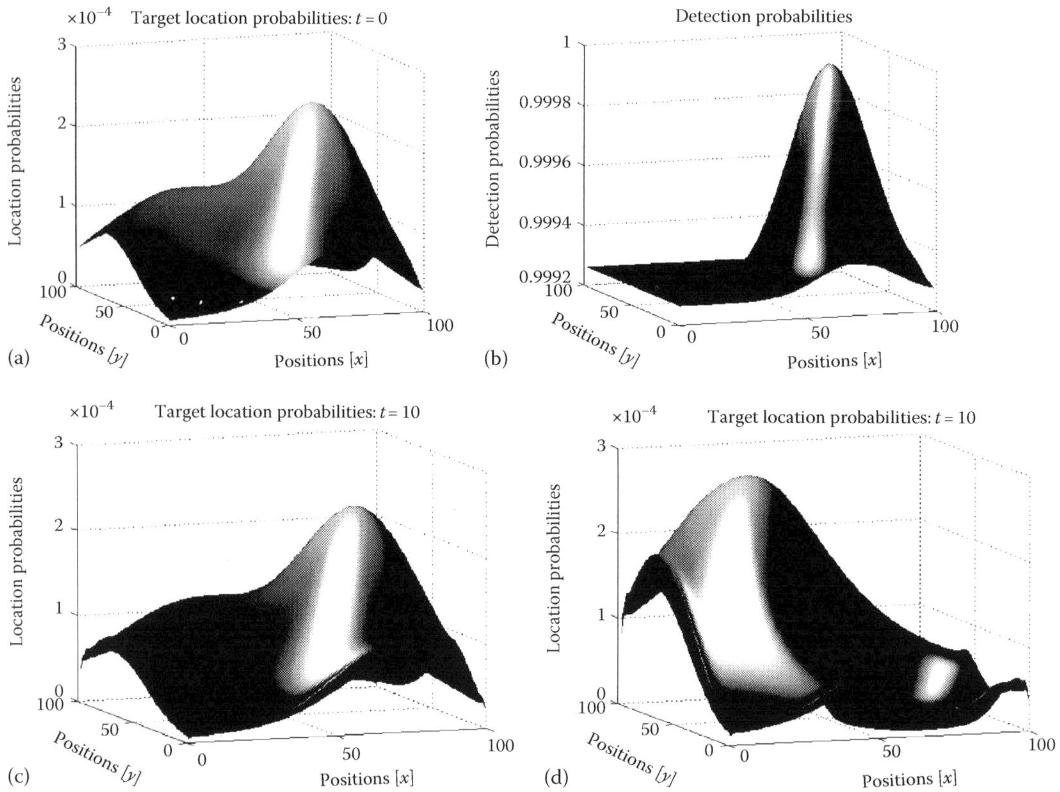

FIGURE 2.7 (a) Initial target's location probabilities at $t = 0$ (as in Figure 2.2a); (b) detection probabilities according to Equation 2.81; (c) target's location probabilities at $t = 10$ without search; (d) target's location probabilities at $t = 10$ while the detections according to the probabilities shown in the figure (a) are applied at each time $t = 1, \ldots, 10$.

search system such that the cost of search efforts' distribution as well as the time required for their distribution over a domain are relatively small and can be omitted.

In opposite, in the case of search by a single searcher, the observed areas or points can be chosen sequentially in such a manner that each next point be in the close neighborhood of the current point. The sequence of the chosen points forms a trajectory of the searcher over a domain. Denote by $\vec{\mathfrak{a}} = \langle a(0), a(1), a(2), \ldots, a(T) \rangle$ a trajectory of the searcher, where $a(t)$, $t = 0, 1, 2, \ldots, T$, stands for the observed area chosen at time t. In terms of online group-testing search with perfect observations, trajectory $\vec{\mathfrak{a}}$ is unambiguously defined by the strategy $\vec{\mathfrak{d}}$, and the areas $a(t)$ are specified by decision rules $\mathfrak{d}(t)$ basing on observation results, that is, $a(t) = \mathfrak{d}\left(a(t-1) \mid z\left(a(t-1)\right)\right)$. However, in the search planning problems, the observations are undefined; so the trajectory $\vec{\mathfrak{a}}$ is considered as a realization of certain stochastic process obtained with some probability. If all areas $a(t)$ include only a single point, in which the searcher is located at time t, then the trajectory $\vec{\mathfrak{a}}$ is directly specified as a sequence of points $\vec{\mathfrak{a}} = \langle \breve{\mathfrak{x}}(0), \breve{\mathfrak{x}}(1), \breve{\mathfrak{x}}(2), \ldots, \breve{\mathfrak{x}}(T) \rangle$. The searcher's trajectory is illustrated in Figure 2.8.

In the shown trajectory, observed area a around location $\mathfrak{x}$ consists of nine points including the point $\mathfrak{x}$ itself. In the other cases, observed areas can consist of arbitrary number of points either including or excluding location $\mathfrak{x}$. In particular, in the previous considerations, location areas included single point, that is, $a = \{\mathfrak{x}\}$; the algorithms presented below in this section implement such single-point areas.

In addition, denote by $\mathfrak{n}(\mathfrak{x}_i) \subset \mathfrak{X}$ a neighborhood of the point $\mathfrak{x}_i$, that is, a set of points, to which the search efforts can be applied at the next step, while at the current step, the searcher is in the point $\mathfrak{x}_i$. In the case of single-point observed areas, $\mathfrak{n}\left(\breve{\mathfrak{x}}(t)\right)$ is a set of points, from which the searcher chooses its next location $\breve{\mathfrak{x}}(t+1)$. In the considered algorithms acting in discrete Euclidian domain, it is assumed that $\mathfrak{n}\left(\breve{\mathfrak{x}}(t)\right)$ includes five points corresponding to the movements to the north, south, east, or west and to staying in the current location; so

$$\mathfrak{n}(\mathfrak{x}_i) = \mathfrak{n}\left(\left(i_x, i_y\right)\right) = \left\{\left(i_x, i_y + 1\right), \left(i_x, i_y - 1\right), \left(i_x - 1, i_y\right), \left(i_x + 1, i_y\right), \left(i_x, i_y\right)\right\} \tag{2.82}$$

for each $i = 1, 2, \ldots, n$ except boundary points.

The indicated possibility of choice of the next location constrains available distributions of search efforts that can be implemented by the searcher, and problem 0 of moving target search should be considered using these constraints.

Similar to the searcher's trajectory, denote by $\vec{\mathfrak{o}} = \left\langle \hat{\mathfrak{x}}(0), \hat{\mathfrak{x}}(1), \hat{\mathfrak{x}}(2), \ldots, \hat{\mathfrak{x}}(T) \right\rangle$ a trajectory of the target over the domain, where $\hat{\mathfrak{x}}(t) \in \mathfrak{X}$ are target locations at the times $t = 0, 1, 2, \ldots, T$. Since it is assumed that at each time t the target can occupy only one point $\hat{\mathfrak{x}}(t)$, trajectory $\vec{\mathfrak{o}}$ is always

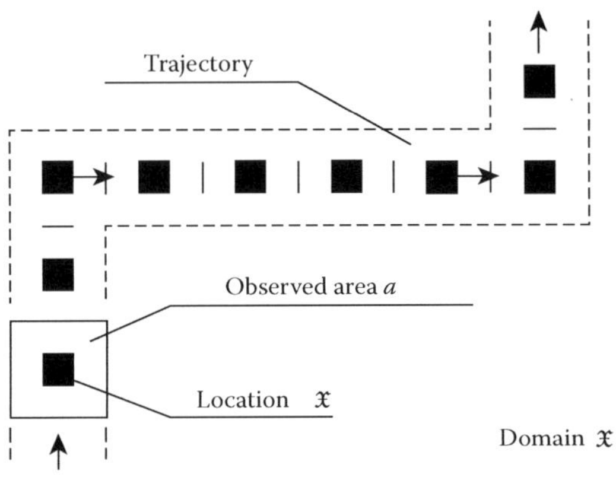

FIGURE 2.8 Discrete time trajectory of the searcher over a discrete domain.

considered as a sequence of points, or if it is needed formally, as a sequence of single-point areas $o(t) = \{\hat{x}(t)\} \subset \mathfrak{X}$.

Finally, following the already used notation of a priori probability $\hat{u}(x,t)$ that the target is located in the point x at time t and was not detected up to t, for each target's trajectory $\vec{o}$, denote by $\hat{u}(\vec{o})$ a priori probability that the target follows this trajectory. Certainly, such probabilities have to be specified prior to the search planning according to the conditions of the specific task.

As earlier, assume that the detection probability $\varphi(x_i, \kappa(x_i,t))$ is defined by the Koopman formula with search parameters $\beta_i = \beta(x_i) > 0$ (see Equations 2.8 and 2.10), that is, for discrete domain,

$$\varphi(x_i, \kappa(x_i,t)) = 1 - \exp(-\beta(x_i)\kappa(x_i,t)), \quad i = 1,2,\ldots,n, \quad t = 0,1,2,\ldots. \tag{2.83}$$

In addition, recall that for discrete time, the search effort $\kappa(x_i,t)$ applied to the point x_i up to time t is

$$\kappa(x_i,t) = \sum_{\tau=0}^{t} w(x_i,\tau), \quad i = 1,2,\ldots,n, \quad t = 0,1,2,\ldots, \tag{2.84}$$

where $w(x_i,\tau)$ is an amount search effort applied to the point x_i at time τ, $\tau = 0, 1, 2,\ldots, t$, according to the search strategy w.

Let $\mathfrak{D}$ be a set of all possible trajectories of the target of the length T. Then, the probability of nondetecting the target up to the time T is defined as follows (Stewart, 1979):

$$q(w,T) = \sum_{\vec{o}\in\mathfrak{D}} u(\vec{o})\exp\left(-\sum_{t=0}^{T}\beta(\hat{x}(t))w(\hat{x}(t),t)\right). \tag{2.85}$$

Consequently, the probability $p(w, T)$ of detecting the target up to the time T is

$$p(w,T) = 1 - q(w,T) = 1 - \sum_{\vec{o}\in\mathfrak{D}} u(\vec{o})\exp\left(-\sum_{t=0}^{T}\beta(\hat{x}(t))w(\hat{x}(t),t)\right), \tag{2.86}$$

and according to the second problem of search for a moving target (see Problems 2.4, Equation 2.77), it is required to find a strategy w^* such that for given T the probability $p(w^*, T)$ is maximal.

There are several methods for solving this problem (Stewart, 1979). Below are considered two of the most popular algorithmic solutions: the Brown algorithm (Brown, 1980) and the Washburn algorithm known as forward and backward (FAB) algorithm (Washburn, 1983), which follows an earlier developed procedure (Washburn, 1980).

Both algorithms follow the same key idea and deal with the search strategy w, which specifies the proportions $w(x,\tau)$ of search efforts applied to the points $x \in \mathfrak{X}$ at times τ = 1, 2,…, T. The algorithms start with $t = 0$ with some initial values $w(x,\tau)$ and, then, by incrementing t sequentially, find the values $w(x,\tau)$ for the times $\tau = t + 1, 2,\ldots, T$. If, in addition, the searcher's starting point $\breve{x}(0)$ is known, then the algorithms can be applied for evaluating the searcher's trajectory $\vec{a}$ (Stewart, 1979). Assume that according to the strategy evaluated for the time τ, the search efforts were placed at the point $x(\tau)$. Then at the next time $\tau + 1$, the search efforts are required be placed at one of the neighboring points of $x(\tau)$, that is, $\breve{x}(\tau+1)(n(\breve{x}(\mathrm{I})))$. As a result, the searcher's trajectory $\vec{a} = \langle \breve{x}(0), \breve{x}(1), \breve{x}(2),\ldots, \breve{x}(T)\rangle$ is specified by the values $w(x,\tau)$ of the strategy w in the points $x \in \mathfrak{X}$ at the times τ = 1, 2,…, T.

Let us start with the Brown algorithm (Brown, 1980). Denote by w_{st}^{t} the strategy of search for a static target calculated for the period $[t, T]$; if $t = 0$, then w_{st}^{0} is exactly the strategy considered in Section 2.2 by the methods of search for a static target in discrete space and time. Recall that $\hat{u}(\mathfrak{x},t)$ stands for the probability that the target is located in the point $\mathfrak{x}$ at time t and was not detected up to this time. Consequently, denote by $\hat{u}_{-t}^{0..T}(\mathfrak{x},t)$ the probability that at time t the target is located in the point $\mathfrak{x}$ and the search at all times except t failed. In other words, $\hat{u}_{-t}^{0..T}(\mathfrak{x},t)$ is a probability of target location in point $\mathfrak{x}$ at time t, while the search during the period $[0, t-1]$ has been failed and the search during the period $[t + 1, T]$, that is,

$$\hat{u}_{-t}^{0..T}(\mathfrak{x}, t) = \sum_{\vec{o}\in\mathfrak{O}:\ \hat{\mathfrak{x}}(t)=\mathfrak{x},\ \hat{\mathfrak{x}}(t)\in\vec{o}} \hat{u}(\vec{o})\exp\left(-\sum_{\tau=0,\ \tau\neq t}^{T} \beta(\hat{\mathfrak{x}}(\tau))w(\hat{\mathfrak{x}}(\tau),\tau)\right), \quad \mathfrak{x}\in\mathfrak{X}. \tag{2.87}$$

Notice that in this equation, in contrast to Equation 2.85, the summation is restricted by the trajectories, which pass the point $\mathfrak{x}$ at time t. Then, the probability of nondetecting the target at time t is defined by the use of the strategy w_{st}^{t} of search for a static target. As it follows form Equation 2.17, this probability is

$$q(w_{st}^{t},T) = \sum_{i=1}^{n} \hat{u}_{-t}^{0..T}(\mathfrak{x}_i,t)\exp(-\beta(\mathfrak{x}_i)w_{st}^{t}(\mathfrak{x}_i,t)). \tag{2.88}$$

By the use of these values, the Brown algorithm is outlined as follows. Slightly different formulations of the Brown algorithm are given in the paper (Stewart, 1979) by Stewart and the book (Kagan & Ben-Gal, 2013) by Kagan and Ben-Gal.

Algorithm 2.3: (Brown, 1980)

Given discrete domain $\mathfrak{X} = \{\mathfrak{x}_1,\mathfrak{x}_2,\ldots,\mathfrak{x}_n\}$, *prior* probabilities $\hat{u}(\vec{o})$ of target's trajectories $\vec{o}\in\mathfrak{O}$ of the length T, search parameters $\beta(\mathfrak{x}_i)$, $i = 1, 2,\ldots, n$, and accuracy value ε, $0 < \varepsilon \ll 1$, do:

1. Start with $t = 0$ and some initial search strategy w. For example, if the searcher's starting point is unknown, then w either arbitrary or $w(\mathfrak{x},t) = 0$, $\mathfrak{x}\in\mathfrak{X}$; otherwise, $w(\breve{\mathfrak{x}}(0),t) = 1$ and $w(\mathfrak{x},t) = 0$, $\mathfrak{x}\in\mathfrak{X}\setminus\{\breve{\mathfrak{x}}(0)\}$, $t = 0, 1, 2,\ldots, T$.
2. For $t = 0, 1, 2, \ldots, T$ do:
3. Set $w' = w$.
4. For $\tau = t + 1, \ldots, T - 1$ do:
5. Find the strategy w_{st}^{τ} for which the probability $q(w_{st}^{\tau},T)$ is minimal.
6. For $i = 1, 2, \ldots, n$ set $w'(\mathfrak{x}_i,\tau) = w_{st}^{\tau}(\mathfrak{x}_i,T)$ (set new values of the search strategy w' by the values of the strategy w_{st}^{τ} of search for a static target at time τ).
7. End for (over τ).
8. Set $w = w'$.
9. If $|q(w',T)-q(w,T)| < \varepsilon$, then break (that is, go to the line 0).
10. End for (over t).
11. Return w.

The key operation in the algorithm is the finding of the strategy w_{st}^{τ} of search for a static target over the period $[\tau, T]$ (line 0). An existence of such strategy was proven by Stone (1975), but its calculation is a nontrivial task and depends on the conditions of search and assumptions regarding the target's motion. If the search is conducted by the large search system, where the goal is to find an immediate distribution of the search efforts, then the strategy w_{st}^{τ} can be specified using the Stone Algorithm 2.1; such a strategy was implemented in the book (Kagan & Ben-Gal, 2013) and earlier in the paper (Kagan, Goren, & Ben-Gal, 2012). The actions of the Brown algorithm in such a case are illustrated by the following example.

Example 2.5

Similar to the previous examples, let $\mathfrak{X}$ be a square domain containing $n = 100 \times 100$ points, and assume that the target starts with the same initial location probabilities and moves according to the same discrete time Markov process as in Example 2.4 (see distributions (a) and (c) in Figure 2.7); so all possible trajectories of the target have equal probabilities. As indicated previously, calculation of the strategy w_{st}^{τ} of search for a static target in line 5 of the Brown Algorithm 2.3 is conducted using the Stone Algorithm 2.1. While calling this algorithm, it is assumed that the search efforts are cumulative, that is, $\kappa(x_i, t) = \sum_{\tau=0}^{t} \beta(x_i) w(x_i, \tau)$ that equal coefficients $\beta(x_i) = 1000$, $i = 1, 2, \ldots, n$; then $K(t) = \sum_{i=1}^{n} \kappa(x_i, t)$ and $K = \sum_{t=0}^{T} K(t)$. Results of the algorithm's actions for zero initial search strategy are shown in Figure 2.9.

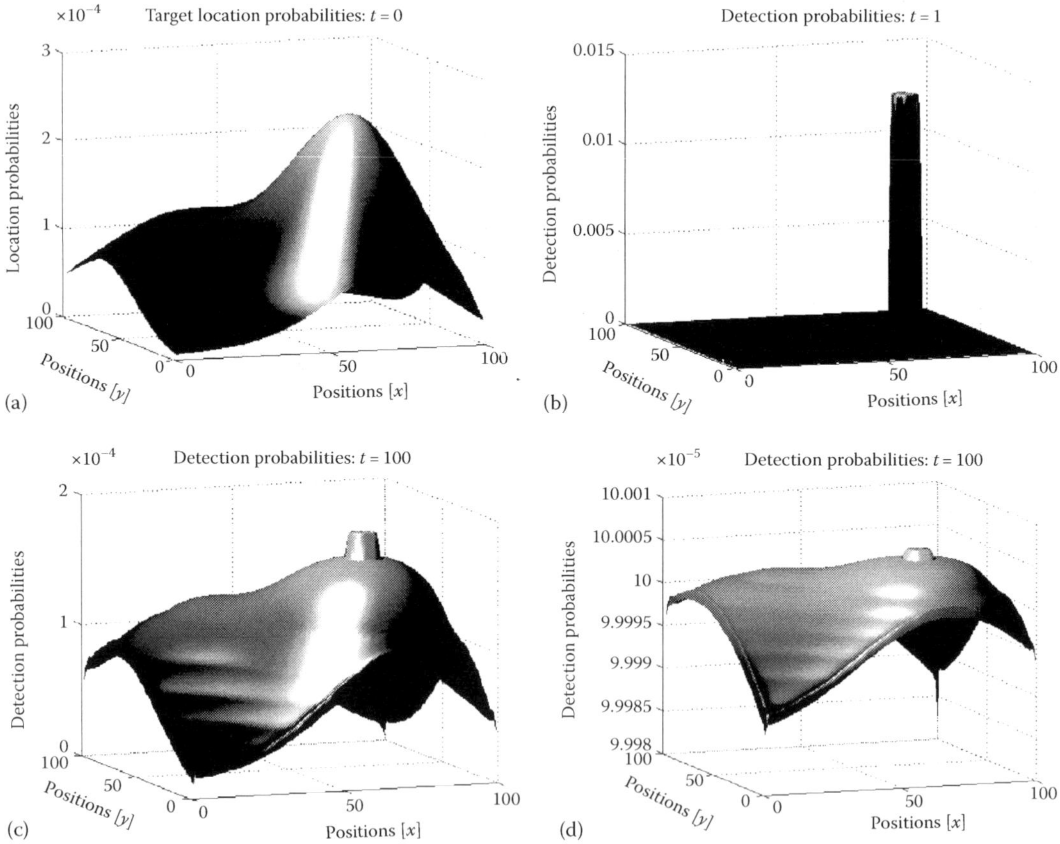

FIGURE 2.9 (a) Initial target's location probabilities at $t = 0$ (as in Figure 2.7a); (b) search strategy obtained by the Stone Algorithm 2.1 at time $t = 1$ with total search effort $K = 1.001 \times 10^3$; (c) and (d) search strategies obtained by the Brown Algorithm 2.3 at the times $t = 10$ with $K = 5.5 \times 10^4$ and $t = 100$ with $K = 5.05 \times 10^6$, respectively.

It is seen that at the beginning of search, the most search efforts are applied to the small area with maximal target's location probability (Figure 2.9b). The continuation of search and consequent increasing of the total search effort leads to distributing the efforts according to the target's location probabilities at that time. However, in contrast to the static target search, the proportion of the efforts applied to the area with maximal initial target's location probability is preserved essentially greater than the proportion of efforts applied to the other points (Figure 2.9c). The further continuation of search and increasing of the total search changes such a relation (Figure 2.9d), and distribution of search efforts follows the target's location probabilities and underlying discrete time Markov process. ■

Now assume that the search is conducted by a single searcher. In this case, as it follows from Equations 2.87 and 2.88, direct calculation of the strategy w_{st}^{τ} requires consideration of all target's trajectories that for practical tasks are certainly impossible. To overcome this problem in the search for a Markovian target, Brown suggested the following techniques (Brown, 1980).

Let $\hat{\rho} = \|\hat{\rho}_{ij}\|_{n\times n}$ be a transition probability matrix, which defines the probabilities of target's movements over the points $\mathfrak{x}_i$ of the domain $\mathfrak{X}$, $i = 1, 2, \ldots, n$, as it is specified by Equation 2.78; for convenience, denote $\hat{\rho}_{ij} = \hat{\rho}(\mathfrak{x}_i, \mathfrak{x}_j)$, that is, the probability of target's movement form the point $\mathfrak{x}_i$ to the point $\mathfrak{x}_j$, $i, j = 1, 2, \ldots, n$. Recall that $\hat{u}(\mathfrak{x}_i, t)$ stands for the probability that the target is located in the point $\mathfrak{x}_i$ at time t and was not detected up to this time (see Equations 2.2 and 2.32). In discrete time search with the detection probability given by Equation 2.83, this probability is defined as follows:

$$\hat{u}(\mathfrak{x}_i,t) = u(\mathfrak{x}_i,t)\exp\left(-\sum_{\tau=0}^{t}\beta(\mathfrak{x}_i)w(\mathfrak{x}_i,\tau)\right), \quad i = 1,2,\ldots,n, \quad t = 0,1,2,\ldots,T, \tag{2.89}$$

where $\hat{u}(\mathfrak{x}_i,0) = u(\mathfrak{x}_i,0) = u(\mathfrak{x}_i)$ is an initial probability of target's location in the point $\mathfrak{x}_i$ at the beginning of the search and $\hat{u}(\mathfrak{x}_i,T)$ is the probability that the target is located in the point $\mathfrak{x}_i$ at the end of the search and survived during all search period. Then, the probability $\hat{u}(\vec{o})$ that the target follows the trajectory $\vec{o} = \hat{\mathfrak{x}}(0), \hat{\mathfrak{x}}(1), \hat{\mathfrak{x}}(2), \ldots, \hat{\mathfrak{x}}(T)$ and escapes from being caught by the searcher is

$$\hat{u}(\vec{o}) = \hat{u}(\hat{\mathfrak{x}}(0),0)\times\hat{\rho}(\hat{\mathfrak{x}}(0),\hat{\mathfrak{x}}(1))\times\hat{\rho}(\hat{\mathfrak{x}}(1),\hat{\mathfrak{x}}(2))\times\cdots\times\hat{\rho}(\hat{\mathfrak{x}}(T-1),\hat{\mathfrak{x}}(T))\times\hat{u}(\hat{\mathfrak{x}}(T),T). \tag{2.90}$$

Let us fix a certain time t, $0 < t < T$ and consider a point $\mathfrak{x}_j$. According to Equations 2.89 and 2.90, the probability that the target reaches this point at time t without being detected is

$$\hat{u}_{-t}^{0\ldots t}(\mathfrak{x}_j,t) = \sum_{i_0,\ldots,i_{t-1}} u(\mathfrak{x}_{i0})\hat{\rho}(\mathfrak{x}_{i0},\mathfrak{x}_{i1})\hat{\rho}(\mathfrak{x}_{i1},\mathfrak{x}_{i2})\cdots\hat{\rho}(\mathfrak{x}_{i_{t-1}},\mathfrak{x}_j)\exp\left(-\sum_{\tau=0}^{t-1}\beta(\mathfrak{x}_{i_\tau})w(\mathfrak{x}_{i_\tau},\tau)\right). \tag{2.91}$$

Here, $u(\mathfrak{x}_{i0})\hat{\rho}(\mathfrak{x}_{i0},\mathfrak{x}_{i1})\hat{\rho}(\mathfrak{x}_{i1},\mathfrak{x}_{i2})\cdots\hat{\rho}(\mathfrak{x}_{i_{t-1}},\mathfrak{x}_j)$ is the probability that the target follows the trajectory $\mathfrak{x}_{i0}, \mathfrak{x}_{i1}, \mathfrak{x}_{i2}, \ldots, \mathfrak{x}_j$, which starts in the point $\mathfrak{x}_{i0}$ and ends in the point $\mathfrak{x}_j$; $\exp\left(-\sum_{\tau=0}^{t-1}\beta(\mathfrak{x}_{i_\tau})w(\mathfrak{x}_{i_\tau},\tau)\right)$ is the probability that the target was not detected before reaching the point $\mathfrak{x}_j$; and the summation is conducted over all possible trajectories, which end in the point $\mathfrak{x}_j$. In other words, the probability $reach(\mathfrak{x}_j, t)$ is a sum of probabilities of all possible trajectories of the length t ending in the point $\mathfrak{x}_j$ with respect to the probability that the target was not detected while it follows these trajectories.

Similarly, the probability that the target leaves the point $\mathfrak{x}_j$ and continues its motion up to the end of the search period at time T without being detected is:

$$\hat{u}_{-t}^{t..T}(\mathfrak{x}_j,t) = \sum_{i_{t+1},\dots,i_T} \hat{\rho}(\mathfrak{x}_j,\mathfrak{x}_{i_{t+1}})\hat{\rho}(\mathfrak{x}_{i_{t+1}},\mathfrak{x}_{i_{t+2}})\cdots\hat{\rho}(\mathfrak{x}_{i_{T-1}},\mathfrak{x}_{i_T})u(\mathfrak{x}_{i_T},T)\exp\left(-\sum_{\tau=t+1}^{T}\beta(\mathfrak{x}_{i_\tau})w(\mathfrak{x}_{i_\tau},\tau)\right). \quad (2.92)$$

Using the probabilities $\hat{u}_{-t}^{0..t}(\mathfrak{x}_j,t)$ and $\hat{u}_{-t}^{t..T}(\mathfrak{x}_j,t)$, $j = 1, 2,\dots, n$, $t = 0, 1, 2,\dots, T$, the overall probability $\hat{u}_{-t}^{0..T}(\mathfrak{x}_j,t)$ that at time t the target is located in the point $\mathfrak{x}$ and the search at all times except t failed is defined by the following recursion (Brown, 1980):

$$\hat{u}_{-t}^{0..T}(\mathfrak{x}_j,t) = \hat{u}_{-t}^{0..t}(\mathfrak{x}_j,t)\times\hat{u}_{-t}^{t..T}(\mathfrak{x}_j,t), \quad j = 1,2,\dots,n, \quad (2.93)$$

$$\hat{u}_{-t}^{0..t}(\mathfrak{x}_j,0) = \hat{u}(\mathfrak{x}_j,0), \quad (2.94)$$

$$\hat{u}_{-(t+1)}^{0..t+1}(\mathfrak{x}_j,t+1) = \sum_{i=1}^{n}\hat{u}_{-t}^{0..t}(\mathfrak{x}_i,t)\exp\left(-\beta(\mathfrak{x}_i)w(\mathfrak{x}_i,t)\right)\hat{\rho}(\mathfrak{x}_i,\mathfrak{x}_j), \quad (2.95)$$

$$\hat{u}_{-t}^{t..T}(\mathfrak{x}_j,T) = \hat{u}(\mathfrak{x}_j,T), \quad (2.96)$$

$$\hat{u}_{-(t-1)}^{t-1..T}(\mathfrak{x}_j,t-1) = \sum_{i=1}^{n}\hat{\rho}(\mathfrak{x}_j,\mathfrak{x}_i)\exp\left(-\beta(\mathfrak{x}_i)w(\mathfrak{x}_i,t)\right)\hat{u}_{-t}^{t..T}(\mathfrak{x}_i,t). \quad (2.97)$$

These equations provide the probabilities $\hat{u}_{-t}^{0..T}(\mathfrak{x}_j,t)$ and can be used instead of Equation 2.87 in the calculations of the probability $q(w_{\text{st}}^t,T)$ according to Equation 2.88, which appears in the line 5 of the Brown Algorithm 2.3. It is clear, that recursion (2.93–2.97) also requires huge computations; however, if it is known that the target can start its motion in a small number of points and moves without jumps, then it can be conducted straightforwardly even for reasonable domains.

At the same time, a slightly different method of search for a moving target was suggested by Washburn (1980). The complete algorithm developed by Washburn also implements recursive procedures and is known as FAB algorithm (Washburn, 1983).

Let us introduce the following characteristic (cf. Equation 2.45):

$$h(w,t) = \mathcal{E}\left\{\sum_{\tau=0}^{t-1}C(w,\tau)\prod_{k=0}^{\tau}\left(1-\varphi\left(\hat{\mathfrak{x}}(k),\kappa\left(\hat{\mathfrak{x}}(k),k\right)\right)\right)\right\}$$

$$+\sum_{i=1}^{n}\hat{u}_{\text{f}}(\mathfrak{x}_i,t)\left[1-\varphi\left(\mathfrak{x}_i,\kappa(\mathfrak{x}_i,t)\right)\right]\hat{u}_{\text{b}}(\mathfrak{x}_i,t), \quad (2.98)$$

where $\hat{u}_{\text{f}}(\mathfrak{x}_i,t)$, $i = 1, 2,\dots, n$, satisfy forward equations for $t = 0, 1,\dots, T-1$ (cf. Equation 2.79),

$$\hat{u}_{\text{f}}(\mathfrak{x}_i,0) = u(\mathfrak{x}_i), \quad (2.99)$$

$$\hat{u}_{\text{f}}(\mathfrak{x}_i,t+1) = \sum_{j-1}^{n}\hat{u}_{\text{f}}(\mathfrak{x}_j,t)\left[1-\varphi\left(\mathfrak{x}_j,\kappa(\mathfrak{x}_j,t)\right)\right]\hat{\rho}(\mathfrak{x}_j,\mathfrak{x}_i), \quad (2.100)$$

and $\hat{u}_b(\mathfrak{x}_i, t)$, $i = 1, 2, \ldots, n$, satisfy backward equations for $t = T, \ldots, 2, 1$:

$$\hat{u}_b(\mathfrak{x}_i, T) = \mathcal{E}\{C(w, T) \mid \hat{\mathfrak{x}}(T) = \mathfrak{x}_i\}, \tag{2.101}$$

$$\hat{u}_b(\mathfrak{x}_i, t-1) = \mathcal{E}\{C(w, t-1) \mid \hat{\mathfrak{x}}(t-1) = \mathfrak{x}_i\} + \sum_{j=1}^{n} \left[1 - \varphi(\mathfrak{x}_j, \kappa(\mathfrak{x}_j, t))\right] \hat{\rho}(\mathfrak{x}_j, \mathfrak{x}_i) \hat{u}_b(\mathfrak{x}_j, t). \tag{2.102}$$

In these equations, $\mathcal{E}\{\cdot\}$ stands for the expected value of random variable and $C(w, t)$, as earlier, denotes the cost of application of the strategy w at time t, which is independent on the target's locations $\hat{\mathfrak{x}}(\tau)$ at previous times $\tau = 0, 1, \ldots, t$.

If the cost $C(w, t)$ represents some probabilistic value and $C(w, t) = 0$ for all times $t = 0, 1, 2, \ldots, T - 1$ and $C(w, T) = 1$ at the end of the search, then $h(w, t) = q(w, t)$, and $p(w, T) = 1 - h(w, T)$ is the probability of detecting the target up to the time T. Thus, according to the Washburn scheme, the goal of the searcher is to find such strategy w^* that minimizes the value $h(w^*, T)$ at the end of the search at time T.

Similar to the aforementioned, let w_{st}^t stand for a strategy of search for a static target calculated at time t, and denote by $\kappa_{st}(\mathfrak{x}_i, t)$, $i = 1, 2, \ldots, n$, the search efforts defined by this strategy, $t = 0, 1, \ldots, T$. Then, following the requirement of minimization of the characteristic $h(w, T)$ and using Equations 2.99 through 2.102, the Washburn algorithm is outlined as follows.

Algorithm 2.4: (Washburn, 1983)

Given discrete domain $\mathcal{X} = \{\mathfrak{x}_1, \mathfrak{x}_2, \ldots, \mathfrak{x}_n\}$, initial target's location probabilities $u(\mathfrak{x}_i)$, $i = 1, 2, \ldots, n$, target's transition probabilities matrix $\hat{\rho} = \|\hat{\rho}_{ij}\|_{n \times n}$ and random costs $C(w,t)$, $t = 0, 1, \ldots, T$, do:

1. Start with $t = 0$ and some initial search strategy w.
2. For $i = 1, 2, \ldots, n$ do:
3. Set initial $\hat{u}_f(\mathfrak{x}_i, t)$ according to forward Equation 2.99.
4. For $\tau = t, 2, \ldots, T$ calculate $\hat{u}_b(\mathfrak{x}_i, \tau)$ according to the backward equations 0 and 0.
5. End for (over i).
6. While $t < T$ do:
7. Set $t = t + 1$.
8. For $i = 1, 2, \ldots, n$ calculate $\hat{u}_f(\mathfrak{x}_i, t)$ according to forward equation 2.100.
9. Find the strategy w_{st}^t for which the characteristic $h(w_{st}^t, t)$ is minimal.
10. For $i = 1, 2, \ldots, n$ set $w(\mathfrak{x}_i, t) = w_{st}^t(\mathfrak{x}_i, T)$ (Notice that that w_{st}^t is calculated for the period $[t, T]$ and its resulting values are $w_{st}^t(\mathfrak{x}_i, T)$).
11. End while.
12. Return w.

As it was demonstrated by Washburn (1983), this algorithm converges for any detection function φ, and for regular detection function (see Section 2.2.1) especially for the Koopman function defined by Equation 2.83, the obtained strategy w provides global optimum. In the aforementioned case with $C(w, t) = 0$ for $t = 0, 1, \ldots, T - 1$, and $C(w, T) = 1$, this strategy results in minimum of the probability q and maximum of the probability $p(w, T) = 1 - q(w, T)$ of detecting the target up to the time T. The actions of the Washburn Algorithm 2.4 are illustrated by the following example.

Example 2.6

Assume that the search is conducted in the same as aforementioned square domain $\mathfrak{X}$ of the size $n = 100 \times 100$, and, as earlier, define the detection function φ by the use of the Koopman formula (2.83). In addition, assume that the costs in such a way that $h(w, T) = q(w, T)$ and $p(w, T) = 1 - h(w, T)$. Similar to Example 2.5, the target starts with the initial target's location probabilities shown in Figure 2.9a and moves according to the same discrete time Markov process. The strategy w_{st}^t of search for a static target required in the line 9 of the Washburn Algorithm 2.4 is specified by the use of the Stone Algorithm 2.1. Then, the searcher starts with the same detection probabilities as shown in Figure 2.9b. In addition, in contrast to Example 2.5, assume that $\hat{u}_f(\mathfrak{x}_i, 0) = u(\mathfrak{x}_i)$, $\hat{u}_b(\mathfrak{x}_i, T) = 1$ if $\mathfrak{x}_i = \hat{\mathfrak{x}}(T)$, and $\hat{u}_b(\mathfrak{x}_i, T) = 0$ otherwise, $i = 1, 2, \ldots, n$. The resulting target's location probabilities at time $t = 10$ and the detection probabilities at this time are shown in Figure 2.10.

It is seen that in this case, the search efforts are concentrated around the point $\hat{\mathfrak{x}}(T) = (70, 30)$, while in the other points, they are essentially smaller. With the continuation of search, the area around the point $\hat{\mathfrak{x}}(T) = (70, 30)$, becomes greater with respect to the target's location probabilities. As it follows from further simulations, after a large number of steps, the distribution of search efforts converges to the uniform distribution, and the efforts cover all points of the domain in the same proportion. ■

The considered Brown and Washburn algorithms follow a general scheme of the branch and bound algorithms of search in discreet time and space. The initial consideration of this scheme was published by Stewart (1979), and an overview of the family of such algorithms was presented by Washburn (1998). The other methods of solving the optimization problem in the search for a moving target, while detection function φ is not necessary regular, implement the methods of partially observable Markov decision models. In this direction, in 1984, Eagle (1984) found the solution in the form of optimal search policy $\vec{\mathfrak{d}}$ (see Section 2.2.2) for finite horizon, and nearly 20 years later, Singh and Krishnamurthy (2003) succeeded to solve this problem for infinite horizon. A detailed description of these and similar methods appears in the book (Kagan & Ben-Gal, 2013) by Kagan and Ben-Gal.

2.3.3 Moving Target Search in Continuous Domain

Finally, let us briefly consider the search for a moving target in continuous time and space. Since such a search extends the search for a static target in continuous time and space, the following discourse follows the same notation as in Section 2.2.1. As usual, it is assumed that $\mathfrak{X} \subset \mathbb{R}^2$ and each

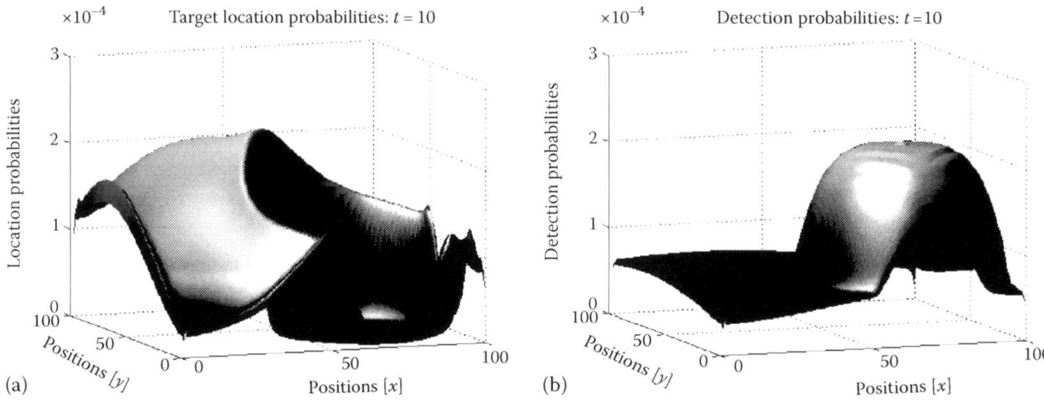

FIGURE 2.10 (a) Target's location probabilities at $t = 10$ (for initial location probabilities shown in Figure 2.9a) while the searcher applies the search strategy shown in the figure (b); (b) the search strategy obtained by the Washburn algorithm with $b_b(\mathfrak{x}, T) = 1$ if $\mathfrak{x} = \hat{\mathfrak{x}}(T) = (70, 30)$, and $b_b(\mathfrak{x}, T) = 0$ otherwise.

point $\mathfrak{x} = (x, y) \in \mathcal{X}$ is defined by its Cartesian coordinates x and y and that the target's location is defined by its probability density function $u : \mathcal{X} \times [0, T] \to \mathbb{R}^+$, which defines conditional target's location probabilities $u(\mathfrak{x}, t)$ in the points $\mathfrak{x} \in \mathcal{X}$ given that the search up to time $t \in [0, T]$ was unsuccessful. Trajectories of the searcher and of the target in the continuous time and space are defined by the functions $\breve{\mathfrak{x}} : \mathcal{X} \times [0, T] \to \mathcal{X}$ and $\hat{\mathfrak{x}} : \mathcal{X} \times [0, T] \to \mathcal{X}$, correspondingly. Usually, it is assumed that the functions $\breve{\mathfrak{x}}$ and $\hat{\mathfrak{x}}$ are smooth for both arguments; such property means that the searcher and the targets move over a domain without jumps. Trajectory of the searcher in the continuous time and domain is illustrated in Figure 2.11.

In the shown trajectory, the observed area a is defined by a disc of radius r around the searcher's location $\breve{\mathfrak{x}}(t)$. A width of the search track, in which the shown area is equal to the diameter $2r$ of the observed area, is often called sweep width (Stone, 1975).

Denote by $\hat{u} : \mathcal{X} \times [0, T] \to \mathbb{R}^+$ the probability density function, which defines the target's movement without the search. In general, this function is governed by the following equation:

$$\begin{aligned}
\frac{\partial}{\partial t}\hat{u} &= \frac{\partial}{\partial x}\left(\hat{s}_{11}(\mathfrak{x}, t, \hat{u})\frac{\partial}{\partial x} + \hat{s}_{12}(\mathfrak{x}, t, \hat{u})\frac{\partial}{\partial y}\right)\hat{u} + \frac{\partial}{\partial y}\left(\hat{s}_{21}(\mathfrak{x}, t, \hat{u})\frac{\partial}{\partial x} + \hat{s}_{22}(\mathfrak{x}, t, \hat{u})\frac{\partial}{\partial y}\right)\hat{u} \\
&\quad - \frac{\partial}{\partial x}\left(\hat{m}_1(\mathfrak{x}, t, \hat{u})\hat{u}\right) - \frac{\partial}{\partial y}\left(\hat{m}_2(\mathfrak{x}, t, \hat{u})\hat{u}\right) \\
&= U\left(\hat{u}; \|\hat{s}\|, \|\hat{m}\|\right). \qquad (2.103)
\end{aligned}$$

This equation is called the Fokker–Planck equation (Frank, 2005; Risken, 1989) (also known as forward Kolmogorov equation), which describes the motion of the particles in different environmental conditions. In this equation, two first terms specify the diffusion with a rate defined by the functions $\hat{s}_{ij}$, $i, j = 1, 2$, and two last terms specify the drift with a trend defined by the functions $\hat{m}_i$, $i = 1, 2$. In particular, these functions are constant such that

$$\hat{s}_{11}(\mathfrak{x}, t, \hat{u}) = \frac{1}{2}\hat{\sigma}_1^2, \quad \hat{s}_{22}(\mathfrak{x}, t, \hat{u}) = \frac{1}{2}\hat{\sigma}_2^2, \quad \hat{s}_{12}(\mathfrak{x}, t, \hat{u}) = \hat{s}_{21}(\mathfrak{x}, t, \hat{u}) = 0,$$

$$\hat{m}_1(\mathfrak{x}, t, \hat{u}) = \hat{\mu}_1, \quad \hat{m}_1(\mathfrak{x}, t, \hat{u}) = \hat{\mu}_2,$$

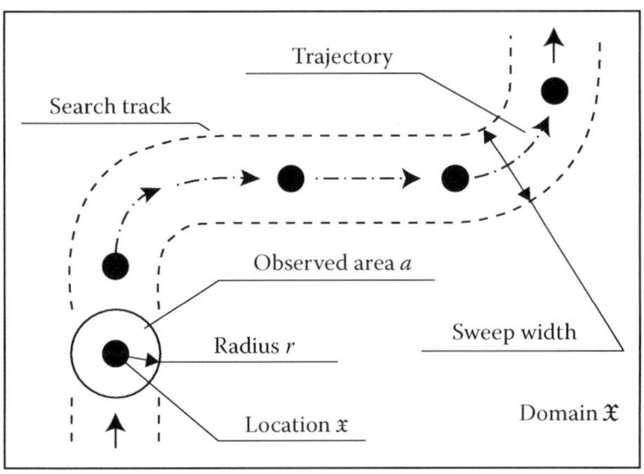

FIGURE 2.11 Continuous-time trajectory of the searcher over a continuous domain.

and then Equation 2.103 is reduced to the diffusion equation

$$\frac{\partial}{\partial t}\hat{u}(\mathfrak{x},t) = U(\hat{u};\hat{\sigma}_1,\hat{\sigma}_2,\hat{\mu}_1,\hat{\mu}_2) = \frac{1}{2}\left(\hat{\sigma}_1^2\frac{\partial^2}{\partial x^2} + \hat{\sigma}_2^2\frac{\partial^2}{\partial y^2}\right)\hat{u}(\mathfrak{x},t) - \left(\hat{\mu}_1\frac{\partial}{\partial x} + \hat{\mu}_2\frac{\partial}{\partial y}\right)\hat{u}(\mathfrak{x},t), \tag{2.104}$$

which describes a drifted (or biased) Brownian motion in 2D domain. Solution of Equation 2.104 is given by 2D Gaussian distribution

$$\hat{u}(\mathfrak{x},t) = u(\mathfrak{x},0)\frac{1}{2\pi\hat{\sigma}_1\hat{\sigma}_2 t}\exp\left(-\frac{(x-\hat{\mu}_1 t)^2}{2\hat{\sigma}_1^2 t} - \frac{(y-\hat{\mu}_2 t)^2}{2\hat{\sigma}_2^2 t}\right), \tag{2.105}$$

where $u(\mathfrak{x},0) = u(\mathfrak{x})$, $\mathfrak{x} \in \mathfrak{X}$, are initial target's location probabilities. For detailed consideration of Brownian motion, see Section 4.2, where it is implemented for the movements of the searcher.

Then, using Equation 2.103, conditional target's location density u is defined as follows (Hellman, 1972):

$$\frac{\partial}{\partial t}u(\mathfrak{x},t) = U(u;\|\hat{s}\|,\|\hat{m}\|) - \left(w(\mathfrak{x},t) - \int_{\mathfrak{X}} w(\mathfrak{x},t)u(\mathfrak{x},t)d\mathfrak{x}\right)u(\mathfrak{x},t). \tag{2.106}$$

This equation has the same form as Equation 2.1 with additional operator $U(u;\|\hat{s}\|,\|\hat{m}\|)$, which defines the target's movement over a domain $\mathfrak{X}$. It was formulated in 1972 by Hellman (1972). Later, Hellman (1985) demonstrated an existence of the solutions of this equation for a wide class of operators U and derived necessary conditions for the optimality of search. However, strict solutions based on Bessel functions were obtained only for a special case of nonbiased diffusion process with $U(u;\hat{\sigma},\hat{\sigma},0,0)$.

Starting from Equation 2.106, the problem of finding optimal trajectories of the searcher is formulated as follows (Lukka, 1977; Mangel, 1982; Ohsumi, 1991). Similar to the search density function w, let $\psi : \mathfrak{X}\times\mathfrak{X} \to \mathbb{R}^+$ be a function such that the value $\psi(\hat{\mathfrak{x}}(t),\breve{\mathfrak{x}}(t))dt$ is a probability of detecting the target during the period $[t,t+dt)$, while the target's location is $\hat{\mathfrak{x}}(t)$ and the searcher's location is $\breve{\mathfrak{x}}(t)$. Then the function u is governed by the following equation:

$$\frac{\partial}{\partial t}u(\mathfrak{x},t) = U(u;\|\hat{s}\|,\|\hat{m}\|) - \left(\psi(\mathfrak{x}(t),\breve{\mathfrak{x}}(t))dt - \int_{\mathfrak{X}} \psi(\mathfrak{x},\breve{\mathfrak{x}}(t))u(\mathfrak{x},t)d\mathfrak{x}\right)u(\mathfrak{x},t). \tag{2.107}$$

For the static target search and for the search by a large search system, the probability $\psi(\hat{\mathfrak{x}}(t),\breve{\mathfrak{x}}(t))dt$ is equivalent to the probability $w(\mathfrak{x},t)dt$ of detecting the target in the point $\mathfrak{x}$ during the period $[t,t+dt)$ given that the target is located in $\mathfrak{x}$, and Equation 2.107 is reduced to Equation 2.106.

To formulate the optimization problem, let us notice that the value $\left(\int_{\mathfrak{X}} \psi(\mathfrak{x},\breve{\mathfrak{x}}(t))u(\mathfrak{x},t)d\mathfrak{x}\right)dt$ is the probability of detecting the target during the period $[t,t+dt)$. Denote by $p(\breve{\mathfrak{x}},t)$ the probability of detecting the target during the period $[0,t)$. Then, the probability $dp(\breve{\mathfrak{x}},t)$ that the target will be detected at the period $[t,t+dt)$ with certainty is defined as follows:

$$\begin{aligned}
dp(\breve{\mathfrak{x}},t) &= Pr\{\text{the target was not detected during the period } [0,t)\} \\
&\quad \times Pr\{\text{the target will be detected during the period } [t,t+dt)\} \\
&= (1-p(\breve{\mathfrak{x}},t))\left(\int_{\mathfrak{X}} \psi(\mathfrak{x},\breve{\mathfrak{x}}(t))u(\mathfrak{x},t)d\mathfrak{x}\right)dt.
\end{aligned} \tag{2.108}$$

Thus,

$$\frac{d}{dt}p(\breve{\mathfrak{x}},t)=\left(1-p(\breve{\mathfrak{x}},t)\right)\int_{\mathfrak{X}}\psi\left(\mathfrak{x},\breve{\mathfrak{x}}(t)\right)u(\mathfrak{x},t)\,d\mathfrak{x}, \tag{2.109}$$

and so

$$p(\breve{\mathfrak{x}},t)=1-\exp\left(-\int_0^t\int_{\mathfrak{X}}\psi\left(\mathfrak{x},\breve{\mathfrak{x}}(\tau)\right)u(\mathfrak{x},\tau)\,d\mathfrak{x}\,d\tau\right). \tag{2.110}$$

Definition of the probability $p(\breve{\mathfrak{x}},t)$ in the form given by Equation 2.110 or in the form provided by complimentary equation for the probability $q(\breve{\mathfrak{x}},t)=1-p(\breve{\mathfrak{x}},t)$ appears in the papers by Lukka (1977) and by Mangel (1981, 1982) and then used by Ohsumi (1991) and by Israel, Khmelnitsky, and Kagan (2012). The aforementioned consideration follows Mangel (1981, 1982) and Israel et al. (2012).

Using the probability defined earlier $p(\breve{\mathfrak{x}},t)$, the first problem of search for a moving target (see Problems 2.4, Equation 2.76, and the first Pollock problem in page 33) requires to define the searcher's motion such that it results in certain detection of the target in minimal expected time $t(\breve{\mathfrak{x}})$. In other words, it is required to find a trajectory $\breve{\mathfrak{x}}^*$ of the searcher such that

$$t(\breve{\mathfrak{x}}^*)=\min_{\breve{\mathfrak{x}}}\left\{t(\breve{\mathfrak{x}})<\infty \mid p\left(\breve{\mathfrak{x}},t(\breve{\mathfrak{x}})\right)=1\right\}. \tag{2.111}$$

Since $t(\breve{\mathfrak{x}})>0$, the expected time is defined as follows (Israel et al., 2012):

$$t(\breve{\mathfrak{x}})=\int_0^\infty Pr\left\{\text{the target was not detected during the period }[0,t)\right\}dt. \tag{2.112}$$

In terms of probability $p(\breve{\mathfrak{x}},t)$ and by the use of Equation 2.110, one obtains

$$t(\breve{\mathfrak{x}})=\int_0^\infty\left(1-p(\breve{\mathfrak{x}},t)\right)dt=\int_0^\infty\exp\left(-\int_0^t\int_{\mathfrak{X}}\psi\left(\mathfrak{x},\breve{\mathfrak{x}}(\tau)\right)u(\mathfrak{x},\tau)\,d\mathfrak{x}\,d\tau\right)dt. \tag{2.113}$$

Thus, the minimization problem (2.111) is formulated as follows:

$$\breve{\mathfrak{x}}^*=\operatorname{argmin}_{\breve{\mathfrak{x}}}\int_0^\infty\exp\left(-\int_0^t\int_{\mathfrak{X}}\psi\left(\mathfrak{x},\breve{\mathfrak{x}}(\tau)\right)u(\mathfrak{x},\tau)\,d\mathfrak{x}\,d\tau\right)dt, \tag{2.114}$$

subject to Equation 2.107, which governs the target's location density u.

Problem (2.114) provides general optimization of search with unconstrained in time search paths and in general cannot be solved analytically. Restricted variants of this problem and of the underlying Equation 2.107 were studied by Hellman (1972), Lukka (1977), Mangel (1981), and Ohsumi (1991). In 1985, Hellman presented a detailed review of the results in this direction and corresponding existence theorems in the book (Hellman, 1985); additional information regarding this problem as well as its consideration in the terms of minimization of risk can be found in the book (Iida, 1992) by Iida.

2.4 SUMMARY

The chapter presents classical theory of optimal search, in which it is assumed that the search is conducted by a large number of seekers that share information and are coordinated and controlled from a definite central unit. The actions of the seekers are described by the joint search density function that is used for specification of the distribution of search efforts over a domain or of the sequence of areas, which have to be observed by the searcher.

The problem of search for a static target (Section 2.2) is considered following three main approaches:

1. Constrained search (Section 2.2.1) addresses the distribution of the limited search efforts such that the probability of finding the target in certain period reaches its maximum. The most popular algorithm, which provides near-optimal solution of this problem, is the Stone Algorithm 2.1.
2. Group-testing search (Section 2.2.2) deals with the optimal search trees, which provide the sequence of nested areas that have to be observed up to finding the target. The optimal search tree is provided by the Zimmermann Algorithm 2.2, which is also known as the Huffman coding algorithm.
3. Sequential search (Sections 2.2.3 and 2.2.4) combines the methods of constrained and group-testing searches and considers the sequence of the observed areas, which not necessary nested. The main results in this direction were obtained by Chew and by Leipata, and the recursive procedure, which defines the number of observations, was created by Hellman (Section 2.2.4).

The problem search for a moving target (Section 2.3) is more complex than the problem of search for a static target, and its discrete and continuous versions are considered separately.

1. Search in discrete domain (Section 2.3.2) extends the methods of sequential search to the case of moving target and provides the trajectory of the searcher in the form of the sequence of areas such that the next area can be reached from the current one in one step. The most effective off-line procedures, which solve this problem, are the Brown Algorithm 2.3 and the Washburn Algorithm 2.4.
2. Search in continuous domain (Section 2.3.3) essentially differs from the search for moving target in discrete domain and deals with probabilistic dynamics of the searcher specified by corresponding Fokker–Planck equation (2.107). The main results in this field were obtained by Lukka, Mangel, Ohsumi, and Hellman (for references, see Section 2.3.3), but general analytic solution of the problem was not yet reported.

The considered methods and algorithms of optimal search for static and moving target are the basic methods used in the modern search and screening theory. These methods provide optimal or near-optimal solutions of the search planning problems and are applied in many systems, which are based on the off-line solutions of the problem. However, notice that because of the complexity of the problem of search, especially of the search for moving target, the implemented optimization methods often cannot provide online solutions of the problem and, as a result, cannot be used for navigation of the searchers in real time.

3 Methods of Optimal Foraging

In contrast to the search and screening theory, the theory of foraging considers an observed behavior of the forager and assumes that this behavior is optimal tries to develop its formal model (MacArthur & Pianka, 1966). This chapter includes general formulations of the problems addressed by the theory of foraging and describes classical probabilistic approaches of their solution.

3.1 PREYING AND FORAGING BY PATCHES

Assume that there is a predator hunting for a prey. In general, the hunting consists of three main processes: the first process is a search for a prey; the second is making a decision whether to terminate a search and start a chase for a prey or to omit the chase and continue searching; and the third process is a pursuit for the prey, if such decision was made. In the theory of optimal foraging, the studies are concentrated on the first two processes (Stephens & Krebs, 1986).

In contrast to the theory of search, where the studies start with the formulation of the optimization problem subject to the given characteristics of search agent or search system and result in the solution, which prescribes the searcher's behavior, in the foraging theory, the consideration follows an opposite direction. It starts with the observed or expected behavior of the forager and then, with respect to certain meaningful assumptions regarding forager's goals and abilities, attempts to formulate an optimization problem such that its solution corresponds to the observed motion of the forager.

The basic model of the forager's behavior, which represents an observed movement and a decision-making process, follows an assumption that the prey items are distributed differently in different regions or patches in the environment and the predators hunt in the patch during certain period and then pass it for the other patch (MacArthur & Pianka, 1966). Such model is known as a model of foraging in a patchy environment. Regarding the patches, it is assumed that each patch is characterized by the availability of the prey items or, in the simplest case, by the number of items, which is known for the forager. The model is illustrated by Figure 3.1.

In the figure, the observed trajectory (Figure 3.1a) is modeled by the process (Figure 3.1b) in which the forager starts with the patch of type 5, hunts there during certain period, then moves to the patch of type 2, and, after hunting in it, continues search in the patch of type 3 and so on.

The optimization problem regarding the forager's behavior follows general assumption that the forager acts as economical as possible. Usually, the problem is formulated either as a problem of minimization of the time spent for capturing the prey item, including the time of search for a patch and the time of hunting in the patch (MacArthur & Pianka, 1966), as a problem of maximization of the utilized energy per prey item (Charnov, 1976). Then, the prey model deals with the decision making whether to stay in the patch or to continue search, while the patch model addresses a question: how long the forager has to stay and hunt in a certain patch, or when should the forager leave the current patch and continue searching (Stephens & Krebs, 1986)?

In the deterministic setup, the solution of the patch problem was found in the form of the marginal value theorem (Charnov, 1976). In this theorem, it is assumed that the resources in the patches are not renewed and that the times of movements between the patches are proportional to the distances. Then, following the requirement that the forager maximizes the net rate of the energy intake, the theorem states that the forager should leave the patch if the marginal rate of gain in the patch becomes equivalent to the long-term average rate of energy intake in the habitat (Charnov, 1976). Notice that since the overall energy rate depends on the rates in the

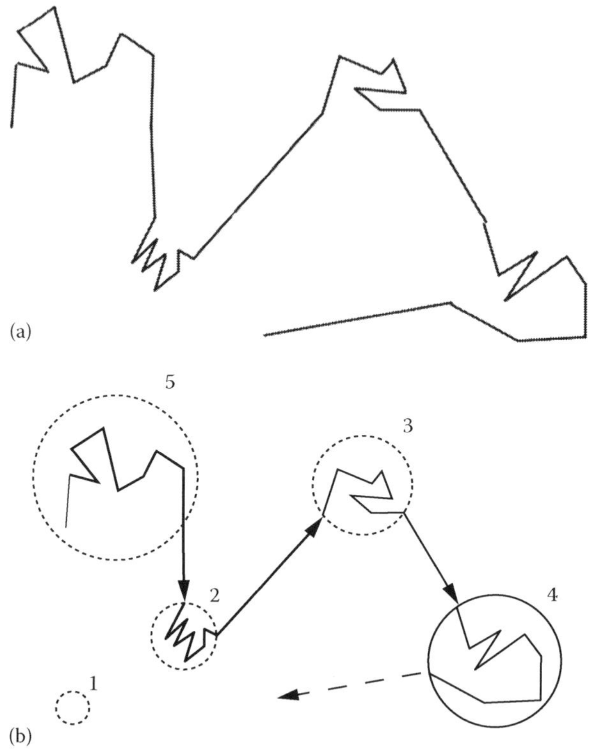

FIGURE 3.1 Foraging in a patchy environment: (a) observed trajectory and (b) model of foraging by patches.

patches, according to the model, the forager will return to the already-explored patches up to their complete depletion (Stephens & Krebs, 1986).

It is clear that the requirements of the deterministic setup of the problem in practice cannot be satisfied in their complete form: the forager cannot know an exact rate of the energy intake in the patch and certainly cannot know the long-term rate of energy intake over the habitat. To overcome these difficulties, the problem is formulated in more realistic probabilistic setup (Oaten, 1977). In such setup, it is assumed that the captures of the prey in the patches are random events and that the forager does not inform about the number of prey items in the patches. However, the forager keeps a distribution of the prey items over the patches such that it defines the probability that the chosen patch includes a certain amount of prey. Then, the strategy of the forager is specified in the terms of probabilities of gain and energy rates, but the resulting solution is not necessarily optimal (Oaten, 1977), and the forager may terminate hunting and leave a patch before reaching the threshold value of the energy rate.

Such forager's behavior can be represented by patch sampling and assessing potential gain in this patch (Stephens & Krebs, 1986) that implies certain predictive abilities of the forager. In terms of foraging in random environment, such abilities are represented by the potential function (McNamara, 1982), which specifies relative advantages of continuing hunting in the patch given that the future behavior of the forager is optimal. An optimal policy is also defined by the use of potential function and prescribes to stay in the patch while the value of potential is positive and to leave the patch when it drops to zero.

The indicated models, as well as their modifications (Green, 1980; Stephens, Brown & Ydenberd, 2007; Stephens & Charnov, 1982; Stephens & Krebs, 1986), follow a common idea: given a distribution of the prey items in the patches, they provide motion policies and finally result in the forager's trajectories over a certain domain. In the deterministic setup, the obtained trajectory is unique for each distribution of the prey, while in the stochastic case, each distribution results in a variety of trajectories. In other words, given a distribution of the prey, the models result in the probabilities of the forager location such that the forager's income reaches its maximum. It is clear that such result is literally the same as the result of the optimal search procedures indicated in the previous chapter. The following discourse is aimed to demonstrate this similarity and to show how the methods of the theory of search can contribute to the theory of foraging and vice versa.

3.2 SPATIAL DYNAMICS OF POPULATIONS

Let us start with general dynamics of probability distributions of theoretical animal's populations over a habitat. In contrast to a well-known population dynamics, which specify the changes in time of the population size (see, e.g., the chapter by Holt & Kimbrell in the book [Stephens et al., 2007]; for detailed review and recent results, see, e.g., the book by [Turchin, 2003]), the considered equations define the probabilities of finding at least one individual in certain points of the domain. On one hand, such equations demonstrate the close relation between the models of foraging and the models of search, especially in continuous domain (see Section 2.3.3) and, on the other hand, form a basis for further consideration of swarm dynamics (see Section 3.5).

A general equation, which is used in population dynamics for defining an evolution of the density function $f : \mathfrak{X} \times [0, T] \rightarrow \mathbb{R}^+$, is specified as follows:

$$\frac{\partial}{\partial t} f(\mathfrak{x}, t) = F(f; \|s\|, \|m\|), \tag{3.1}$$

where

$F(f; \|s\|, \|m\|)$ is an operator defined similarly to the operator U in Equation 2.103

$\|s\|$ is a matrix of functions s_{ij}, $i, j = 1, 2$, which define a diffusion rate

$\|m\|$ is a vector of functions m_i; and $i = 1, 2$, which defines a drift

Formally, Equation 3.1 defines a probability that the particle is located in certain point $\mathfrak{x} \in \mathfrak{X}$ of the domain $\mathfrak{X}$ at time t. Also especially in population dynamics, instead of considering the location probabilities of a single particle, it is addressed a large number of moving particles, so the density of the particles specifies the indicated probability (Patlak, 1953). In other words, Equation 3.1 defines the dynamics of population of constant size, where density function f specifies a population density on the domain $\mathfrak{X}$ at time t (Holmes, Lewis, Banks, & Veit, 1994; Okubo, 1986; Patlak, 1953; Skellam, 1951). Then the value $f(\mathfrak{x}, t)d\mathfrak{x}dt$ is a probability of finding at least one individual in the neighborhood $[\mathfrak{x}, \mathfrak{x} + d\mathfrak{x})$ of the point x during the period $[t, t + dt)$.

In the general case of population dynamics, Equation 3.1 includes additional term, which defines an evolution of the population size. Usually, there are considered two main types of populations:

1. Malthusian population with exponential growth:

$$\frac{\partial}{\partial t} f(\mathfrak{x}, t) = F(f; \|s\|, \|m\|) + \gamma f(\mathfrak{x}, t), \tag{3.2}$$

2. Logistic population, where the size of population satisfies logistic function:

$$\frac{\partial}{\partial t} f(\mathfrak{x}, t) = F(f; \|s\|, \|m\|) + \gamma(1 - f(\mathfrak{x}, t)) f(\mathfrak{x}, t). \tag{3.3}$$

In 1951, Skellam (1951) studied these equations with $F(f; \langle\sigma, \sigma\rangle, \langle 0, 0\rangle)$ that corresponds to the homogeneous diffusion (cf. Equation 2.104) and different boundary conditions and provided both

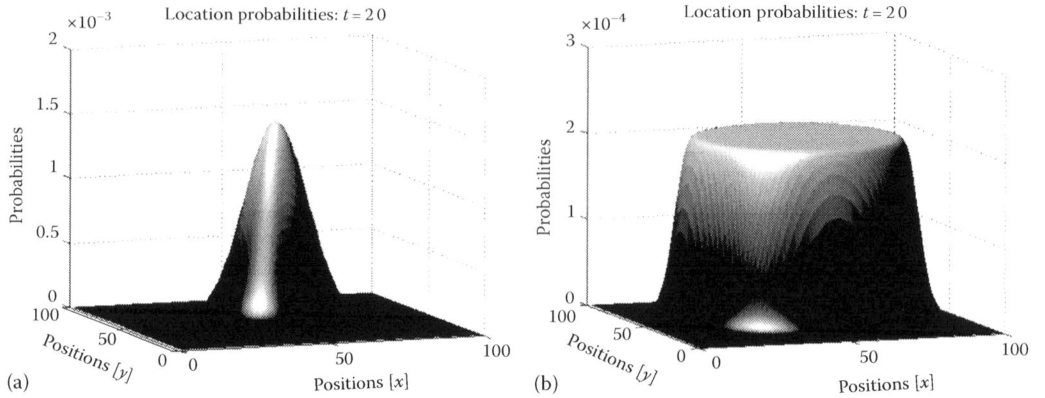

FIGURE 3.2 Probability distribution for (a) Malthusian and (b) logistic populations.

analytical (using Bessel functions) and numerical solutions. Solutions of Equations 3.2 and 3.3 with the parameters $\sigma = \frac{1}{2}$ and $\gamma = 1$ are illustrated in Figure 3.2. Initial distribution is Gaussian with $\sigma_x = \sigma_y = 10$ concentrated in the central point $\mathfrak{x} = (50, 50)$.

It is seen that since for the Malthusian population the growth of population is greater than its diffusion, the form of distribution is remained Gaussian. For the logistic population, in contrast, while the speed of growth decreases, the population is distributed by diffusion.

In 1953, Patlak (1953) considered Equation 3.1 with $\gamma = 0$ and nonzero drift for analysis of the motion of theoretical populations of constant size and clarified it formal relation with the equations appearing in statistical physics. These results and similar considerations of spatial dynamics in biological and physical systems allowed to formulate general equations of motion and to analyze different regimes of spatial evolution (see, e.g., Romanovsky, Stepanova, & Chernavsky, 1984; Okubo, 1986). Here, for illustration, let us consider the simplest 1D equations suggested by Okubo (1986); more general equations and their analysis are presented in Section 3.5.

Assume that the animal moves in 1D domain $\mathfrak{X} = \mathbb{R}$ starting from the point $x = 0$. At time t, the animal can move dx units left or right from its current location or can stay in it for the time dt. Following Equation 2.103, an equation, which defines the animal's location probabilities in 1D domain $\mathfrak{X}$, is the following (Okubo, 1986):

$$\frac{\partial}{\partial t} f(x,t) = \frac{1}{2}\sigma^2\left((1-2\theta)\frac{\partial}{\partial x}\left(\frac{\partial}{\partial x}(1-\rho_0(x,t))f(x,t)\right) + \frac{\partial}{\partial x}\left((1-\rho_0(x,t))\frac{\partial}{\partial x}f(x,t)\right)\right)$$
$$-\frac{\partial}{\partial x}\left(m(x,t)f(x,t)\right), \qquad (3.4)$$

where

$\theta \in [0, 1]$ is a parameter, which defines a regime of probabilities' evolution
$\rho_0(x, t)$ is a probability that at time t the animal stays in its current location x

If $\rho_0(x, t) \equiv 0$, then the motion corresponds to the Brownian walk; in addition, if the function s is defined by the use of transition probabilities function ρ as $s(x,t) = \frac{1}{2}\sigma^2(1-\rho(x,t))$, then the Equation 3.4 is rewritten as follows:

$$\frac{\partial}{\partial t} f(x,t) = \frac{\partial}{\partial x}\left(s(x,t)\frac{\partial}{\partial x}\right)f(x,t) - \frac{\partial}{\partial x}\left(m(x,t) - (1-2\theta)\frac{\partial}{\partial x}s(x,t)\right)f(x,t). \qquad (3.5)$$

This equation is interpreted as follows (Okubo, 1986). The anisotropy in the transition probabilities results in a drift $m(x, t)$ and in a spatially heterogeneous diffusion governed by the values $s(x, t)$. In addition, the diffusion depends on the behavior of the animal, which is defined by the parameter θ. In particular (Okubo, 1986):

1. If $\theta < \frac{1}{2}$, then the animal tends to move toward the direction, in which the diffusion decreases and, consequently, decreases the degree of the motion randomness. In terms of the number of animals, that means that the animals tend to aggregate in the area, where the diffusivity is small. If $\theta = 0$, such behavior is called repulsive.
2. If $\theta > \frac{1}{2}$, then the animal tends to move toward the direction, in which the diffusion increases and, consequently, increases the degree of the motion randomness. That is, the animals tend to aggregate in the area, where the diffusivity is large. If $\theta = 1$, such behavior is called attractive.
3. If $\theta = \frac{1}{2}$, then the diffusive yawing vanishes and the animal follows its direction according to the parameter $m(x, t)$ only. In this case, the animals tend to distribute uniformly over a domain; such a behavior is called neutral.

The evolution of probabilities defined by Equation 3.5 is illustrated in Figure 3.3 (see also Figure 1 in the paper by Okubo [1986]). To stress the dependence of the evolution on the values of the parameter θ and to avoid oscillating and nonlinear effects, in the shown graphs, it is assumed that $m(x, t) \equiv 0$ and $s(x,t) \equiv \frac{1}{2}\sigma^2(1 - rx)$, $\hat{r} > 0$, and the values of the parameters are $\sigma = 1$ and $r = 1$.

In the figure, it is seen that if $\theta = \frac{1}{2}$ (Figure 3.3b), then the probabilities are governed by the diffusion and are flatting up to uniform distribution. If $\theta = 0$ (Figure 3.3c), then the probabilities tend to the left side of the domain, and if $\theta = 1$ (Figure 3.3d), they tend to the right side of the domain.

It is clear that Equation 3.1 and, consequently, Equations 3.2 through 3.5, can be implemented for the evolution of any theoretical populations or, equivalently, for the dynamics of mobile agents or particles. In particular, Equation 3.1 has the same form as Equation 2.103, which defines the target's movement without search, and Equation 3.2, which defines Malthusian population, by appropriate definition of the parameter γ, is transformed to Equation 2.106 or 2.107, which specifies the dynamics of target's location probabilities obtained during the search. Now, define the predator's or searcher's dynamics; since the dynamics of both is defined by the same equations, the following terms are used interchangeably. By the same token, the terms "prey" and "target" are used to sign the agent, which is governed by Equations 2.103, 2.106, and 2.107.

Similar to the definitions given in Section 2.2.1, let $v : \mathfrak{X} \times [0, T] \rightarrow \mathbb{R}^+$ be a probability density function, which defines the movement of searcher over a domain $\mathfrak{X} \subset \mathbb{R}^2$. Then $v(\mathfrak{x}, t)d\mathfrak{x}$ is a probability that the searcher (predator) is located in the neighborhood $[\mathfrak{x}, \mathfrak{x} + d\mathfrak{x})$ of the point x at time $t \in [0, T]$, and $v(\mathfrak{x}, t)dt$ is a probability that the searcher is located in the point x during the period $[t, t + dt)$; both probabilities are assumed conditional given that up to the time t, the search was unsuccessful. Similar to the target's distribution, it is assumed that at the beginning of search, the values $v(\mathfrak{x}, 0) = v(\mathfrak{x})$, $\mathfrak{x} \in \mathfrak{X}$, are specified a priori by a known function $v : \mathfrak{X} \rightarrow \mathbb{R}^+$, and $v(\mathfrak{x})d\mathfrak{x}$ is a probability of the searcher's location in the neighborhood $[\mathfrak{x}, \mathfrak{x} + d\mathfrak{x})$ of the point x at time $t = 0$. Similar to the target's density function u, for the searcher's density function v, it holds true that $\int_{\mathfrak{X}} v(\mathfrak{x}, t)d\mathfrak{x} = 1$ for each $t \in [0, T]$. Notice that in spite of close relation between the density function v and the search density function w (see Section 2.2.1), these functions have different nature: the function w defines the distribution of search efforts, while the function v defines the movement of the searcher.

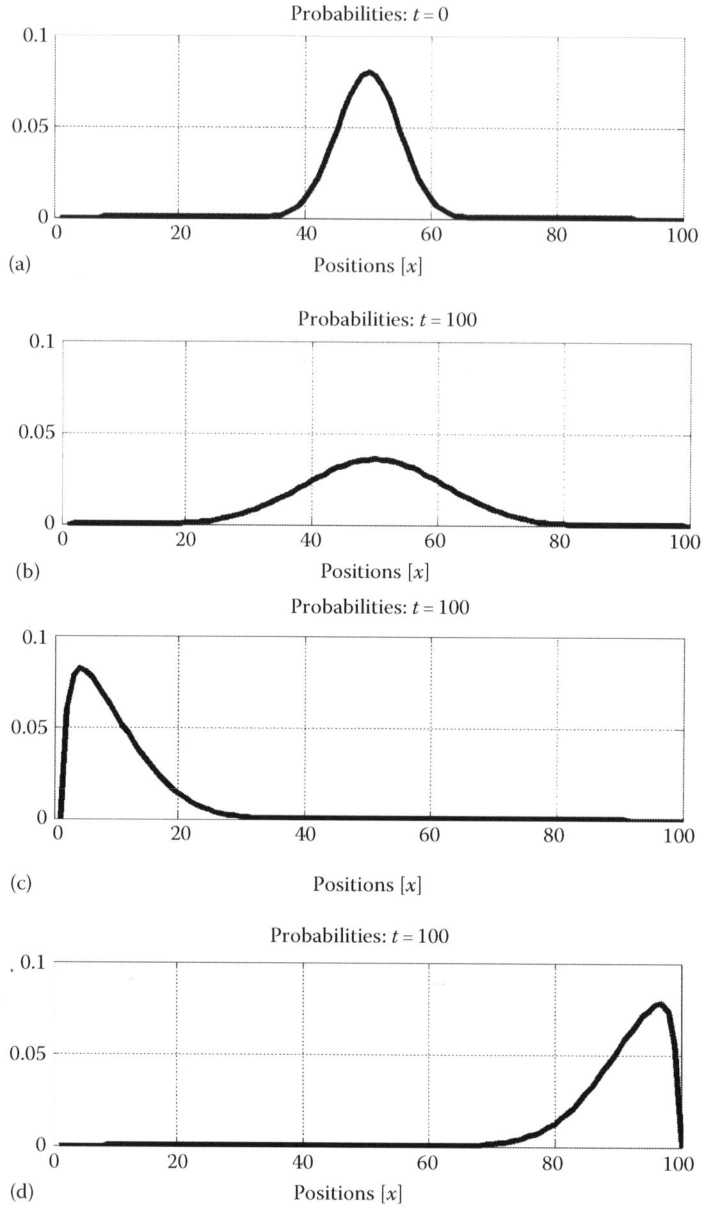

FIGURE 3.3 Evolution of location probabilities in 1D domain defined by Equation 3.5 with different parameter θ: (a) initial distribution with the center $x = 50$, (b) distribution for $\theta = \frac{1}{2}$ (neutral), (c) distribution for $\theta = 0$ (repulsive), and (d) distribution for $\theta = 1$ (attractive).

Finally, recall that distribution of the location probabilities and its drift are defined by the functions s and m, which appear in the operator $F\left(f;\|s\|,\|m\|\right)$ in Equation 3.1 (cf. Equation 2.103), as it is illustrated by Equations 3.4 and 3.5 with parameter θ (see Figure 3.3), and notice that the searcher (predator) acts according to the target's location probabilities. Then, the Fokker–Planck equation, which defines the movement of the searcher, can be written as follows:

$$\frac{\partial}{\partial t} v\left(\breve{x},t\right) = V\left(v;\|\breve{s}\|,\|\breve{m}\|\right), \tag{3.6}$$

where operator $V\left(v;\left\|\breve{s}\right\|,\left\|\breve{m}\right\|\right)$ has the same form as operator U in Equation 2.103, that is,

$$V\left(v;\left\|\breve{s}\right\|,\left\|\breve{m}\right\|\right)=\frac{\partial}{\partial x}\left(\breve{s}_{11}\frac{\partial}{\partial x}+\breve{s}_{12}\frac{\partial}{\partial y}\right)v+\frac{\partial}{\partial y}\left(\breve{s}_{21}\frac{\partial}{\partial x}+\breve{s}_{22}\frac{\partial}{\partial y}\right)v-\frac{\partial}{\partial x}\left(\breve{m}_1 v\right)-\frac{\partial}{\partial y}\left(\breve{m}_2 v\right),$$

However, in contrast to Equations 2.103, 2.106, and 2.107, here, the functions $\breve{s}_{ij}$, which define diffusion rate, and the functions $\breve{m}_i$, which define a drift, depend both on the target's density u and on the searcher's density v, that is,

$$\breve{s}_{ij}=\breve{s}_{ij}\left(\mathfrak{x},t,u,v\right) \quad \text{and} \quad \breve{m}_i=\breve{m}_i\left(\mathfrak{x},t,u,v\right) \quad i,j=1,2$$

Thus, in the most general form, the equations, which govern the process of search and foraging in the system "target–searcher," or in the terms of foraging theory, in the system "prey–predator," are defined as follows:

$$\begin{cases}\dfrac{\partial}{\partial t}u=U\left(u;\left\|\hat{s}\left(\mathfrak{x},t,u\right)\right\|,\left\|\hat{m}\left(\mathfrak{x},t,u\right)\right\|\right)-\left(\psi\left(\hat{\mathfrak{x}},\breve{\mathfrak{x}}\right)dt-\displaystyle\int_X \psi\left(\mathfrak{x},\breve{\mathfrak{x}}\right)u\,d\mathfrak{x}\right)u,\\[2ex] \dfrac{\partial}{\partial t}v=V\left(v;\left\|\breve{s}\left(\mathfrak{x},t,u,v\right)\right\|,\left\|\breve{m}\left(\mathfrak{x},t,u,v\right)\right\|\right),\end{cases} \tag{3.7}$$

where, as in Equation 2.107, function ψ defines the probabilities $\psi\left(\hat{\mathfrak{x}}(t),\breve{\mathfrak{x}}(t)\right)dt$ of detecting the target during the period $\left[t,t+dt\right)$ while the target's location is $\hat{\mathfrak{x}}(t)$ and the searcher's location is $\breve{\mathfrak{x}}(t)$.

Notice that with respect to the considered dynamics, the system (3.7) implies that the population size of the predator does not change, while the size of the prey population can change only because of the searcher's actions. In the considerations of general population dynamics, the equations certainly have to include additional terms that define the population growth like in the aforementioned Malthusian or logistic populations (see Figure 3.2).

In the search and screening theory, which was considered in Chapter 2, the main attention is concentrated on the first equation of the system (3.7), and given the dynamics of the target, the optimization addresses the search strategy w and, consequently, the function ψ. The theory of foraging, in contrast, deals with the second equation of the system (3.7) and given the target's location probabilities considers the trajectories of the searcher.

3.3 METHODS OF OPTIMAL FORAGING

As indicated earlier, the theory of optimal foraging deals with two general models: the prey model deals with the problem "whether a forager should attack the item it has just encountered or pass it over" and the patch model asks "how long the forager should stay in a patch" (Stephens & Krebs, 1986). The following discourse addresses the second problem and describes the optimization techniques used for specification of the forager's motion in the patchy environment. In terms of the system (3.7), such techniques result in certain solutions of the second equation with zero diffusion given the target's location probabilities.

3.3.1 General Assumptions and Deterministic Foraging

Probably, the first approach that explicitly separates the problems addressed by the prey and patch models was suggested in 1966 by MacArthur and Pianka (1966) and Emlen (1966). Following the critical review by Pyke (1984), the underlying assumptions of this approach are the following (for contemporary analysis of these assumptions and their consequences, see recently published book [Stephens et al., 2007]):

1. An individual fitness that is a contribution of the animal to the next generation depends on its foraging activity and can be measured
2. A functional or quantitative relationship between foraging behavior and individual fitness also called the currency of fitness is known

These two assumptions provide a general possibility of quantitative analysis of foraging. Furthermore, it is assumed that

3. The main patterns of foraging behavior are heritable and are transmitted from the parent generation to the child generation
4. The evolution of foraging behavior is not prevented by genetic constraints
5. The evolution of foraging behavior follows definite and determined functional constraints, like morphologies and physical properties of the animals
6. The foraging activity is more rapid than any changes of functional properties of the foragers

From these last four assumptions, it follows that (Pyke, 1984, p. 524)

> The evolution and adaptation of foraging behavior should approximately reach completion with individuals foraging in ways close to (i.e. statistically indistinguishable from) those that maximize their expected fitness, subject to any functional constraints. In this sense, it is hypothesized that animals forage "optimally."

This hypothesis of optimality of the animals foraging together with the possibility of quantitative analysis of foraging activity forms a basis for optimization techniques used in optimal foraging theory.

Notice again that in contrast to the search and screening theory, which at first determines an optimal behavior of the searcher and then the search agent is constructed or controlled according to the determined behavior, the foraging theory starts with the hypothesis that the forager behavior is optimal and then addresses the models, which can describe such behavior.

Let us consider the Charnov model of optimal foraging (Charnov, 1976); this fundamental deterministic model was suggested in 1976 and formed a terminological and methodological basis for further studies of foraging both in deterministic and in probabilistic setup. The following consideration follows the original paper (Charnov, 1976) by Charnov and the book (Stephens & Krebs, 1986) by Stephens and Krebs.

Let, as earlier, $\mathcal{X} \subset \mathbb{R}^2$ be a domain representing a prey's habitat where the forager acts. It is assumed that the points of the domain $\mathcal{X}$ are combined into the areas or patches, $A^j \subset \mathcal{X}$, each of which is labeled by its type $j = 1, 2, \ldots, k$ (see Figure 3.1b). The type of the patch is a formal term, which is used for distinguishing between the patches or for indicating additional information regarding the patch, for example, potential reward or potential payoff; it is assumed that the forager has a priori knowledge about patches' types and instantaneously recognizes the type of the patch. Formally, a division of the domain $\mathcal{X}$ to patches is specified by a partition $\mathcal{A}_{1\ldots m}^{0\ldots k} = \left\{A^0, A_1^{j_1}, A_2^{j_2}, \ldots, A_m^{j_m}\right\}$, where $j_i \in \{1, 2, \ldots, k\}$, $i = 1, 2, \ldots, m$, and $A^0 = \mathcal{X} \backslash \left(\bigcup_{i=1}^{m} A_i^{j_i} \right)$. Such a partition represents a patchy environment and allows a formulation of the foraging problem as a problem of sequential search considered in Section 2.2.3.

As indicated in Section 3.1, the goal of the forager is to maximize the net energy intake rate by choosing which patches to hunt and when to leave the patch. Similar to the group-testing and sequential search (see Sections 2.2.2 and 2.2.3), denote by $\vec{\mathfrak{d}} = \mathfrak{d}(0), \mathfrak{d}(1), \mathfrak{d}(2), \ldots$ a strategy that is a sequence of decision rules $\mathfrak{d}\left(t \mid A^j\right)$ that specify the patch, which has to be chosen at time t given that the forager is hunting in the patch A^j of type j. Notice that here, in contrast to sequential search, the strategy $\vec{\mathfrak{d}}$ is not limited by certain time T.

Denote by $r_e(\vec{\eth})$ the net energy intake rate determined by the strategy $\vec{\eth}$. Then, the problem of foraging in patchy environment $\mathcal{A}_{1..m}^{0..k}$ is formulated as the problem of unconstrained optimization as follows.

Problem 3.1: (Foraging in a Patchy Environment: Deterministic Setup)

In the set $\mathfrak{D}$ of all possible strategies $\vec{\eth}$ of foraging in the environment $\mathcal{A}_{1..m}^{0..k}$, find the strategy $\vec{\eth}^*$ such that

$$r_e(\vec{\eth}^*) = \max_{\vec{\eth} \in \mathfrak{D}} \left\{ r_e(\vec{\eth}) \mid 0 \le t_j < \infty \right\}, \quad (3.8)$$

where
$t_j = t(A^j)$ stands for a residence time, that is, a time spent in the patch of type j
$t_0 = t(A^0)$ denotes the transit time, that is, the time of moving between the patches

Denote by $\eta_j = \eta(A^j) \in [0, 1]$ an encounter rate that is a proportion of the patches of type j, $j = 1, 2, \ldots, k$, and by $\xi_j(t_j) \ge 0$ an expected net energy gained from the patch A^j of type j given that the forager spent t_j time units hunting in A^j. In addition to the transit time t_0, let $C_0 = C(A^0) \ge 0$ be an energy cost of such movement. Then, for some strategy $\vec{\eth}$, the net energy intake rate $r_e(\vec{\eth})$ is defined as follows (Charnov, 1976):

$$r_e(\vec{\eth}) = \frac{\left(\sum_{j=1}^{k} \eta_j \xi_j(t_j) - t_0 C_0\right)}{\left(t_0 + \sum_{j=1}^{k} \eta_j t_j\right)}, \quad (3.9)$$

where numerator specifies an overall energy gained from the patches without the energy spend for the movement between the patches and denominator specifies overall time of the foraging, that is, a time of the movement between the patches and the time of hunting in the patches.

The central role in the optimization of the net energy intake rate $r_e(\vec{\eth})$ plays the gain function $\xi : \mathbb{R}^+ \to \mathbb{R}^+$. In Problem 3.1, it is assumed that for each type $j = 1, 2, \ldots, k$, this function satisfies the following conditions (Stephens & Krebs, 1986):

1. $\xi_j(0) = 0$.
2. $\frac{d\xi_j}{dt}(0) > 0$, that is, the function ξ is at least initially increasing.
3. There exists such t that $\frac{d^2\xi_j}{dt^2}(\tau) < 0$ for all $\tau \ge t$, that is, starting from a certain time, the function ξ is concave.

Notice that instead of condition 3, in the initial setup (Charnov, 1976), the concave on all its domain gain function was used. Examples of the gain function (with normalized values) are shown in Figure 3.4.

In the figure, the exponential gain function (Figure 3.4a) is increasing and concave for all $t \in [0, 100]$; quadratic gain function (Figure 3.4b) is concave on all interval [0, 100], increases on the interval [0.70), reaches its maximum at the point $t = 70$, and then decreases on the interval (70,100]. In contrast, logistic gain function (Figure 3.4c) is increasing on [0, 100], is convex on the interval [0, 50), and is concave on the interval (50, 100] with inflection point at $t = 50$.

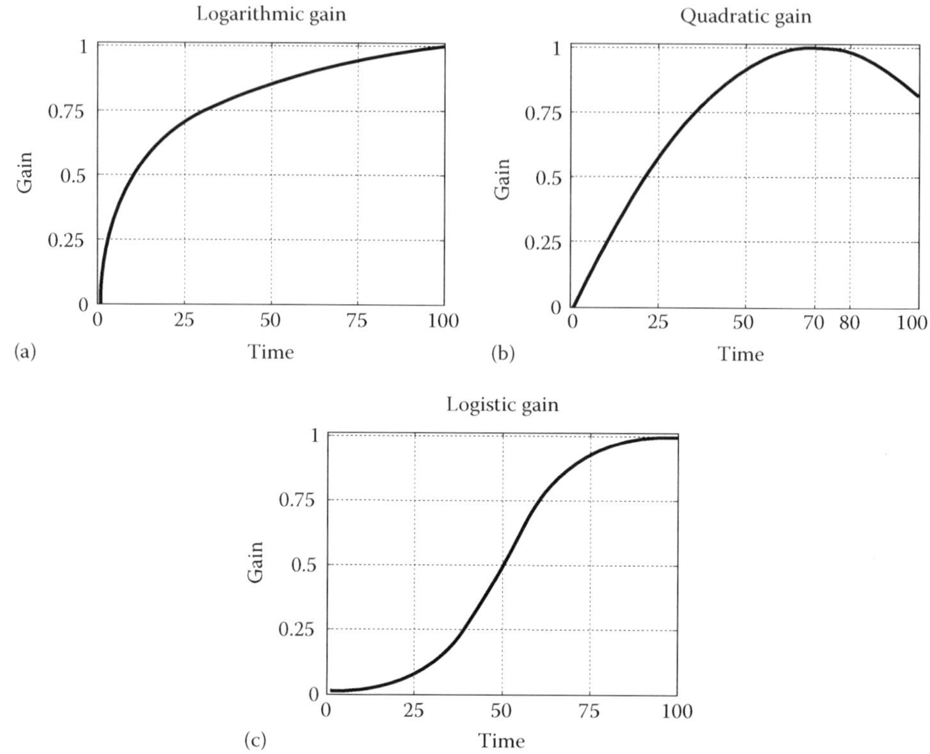

FIGURE 3.4 Examples of the gain function ξ: (a) logarithmic gain, (b) quadratic gain, and (c) logistic gain.

As indicated earlier, Problem 3.1 of optimal foraging requires to find the decision rules $\eth\left(t \mid A^j\right)$ that specify the patch, which has to be chosen at time t given that the forager is in the patch A^j of type j. Thus, each decision rule includes two decisions: first, the forager has to decide whether to leave the current patch A^j or to choose A^j and to continue hunting, and, second, if the forager has decided to leave A^j, then it has to decide to which patch to move from the patch A^j. However, recall that the movement between the patches follows in the patch A^0; hence, the first decision is to choose when to leave the current patch A^j for the patch A^0, and the second decision is to choose to which patch to move while the current patch is A^0.

Let us consider the first decision regarding leaving the current patch. Assume that the times t_j of hunting in the patches of the types $j = 1, 2, \ldots, k$ as well as the transit time t_0 are mutually independent and that these times do not depend on the encounter rates η_j. Then, according to Equation 3.9, the net energy intake rate $r_c\left(\vec{\eth} \mid A^j\right)$ given that the forager is in the patch A^j is defined as follows (Charnov, 1976; Stephens & Krebs, 1986):

$$r_c\left(\vec{\eth} \mid A^j\right) = \frac{\left(\eta_j \xi_j\left(t_j\right) + \alpha_{-j}\right)}{\left(\eta_j t_j + \beta_{-j}\right)}, \tag{3.10}$$

where

$$\alpha_{-j} = \sum_{i=1, i \neq j}^{k} \eta_i \xi_i\left(t_i\right) - t_0 C_0 \quad \text{and} \quad \beta_{-j} = \sum_{i=1, i \neq j}^{k} \eta_i t_i + t_0$$

combine the terms, which do not depend on t_j; since the times t_0 and t_j, $j = 1, 2, \ldots, k$, are mutually independent, while the forager is in A^j, these values remain constant.

Under the indicated assumptions regarding the gain function ξ, the rate $r_e\left(\vec{\partial} \mid A^j\right)$ reaches its maximum if

$$\frac{\partial}{\partial t_j} r_e\left(\vec{\partial} \mid A^j\right) = \frac{\eta_j\left(\eta_j t_j + \beta_{-j}\right)\frac{d}{dt_j}\xi_j\left(t_j\right) - \eta_j\left(\eta_j \xi_j\left(t_j\right) + \alpha_{-j}\right)}{\left(\eta_j t_j + \beta_{-j}\right)^2} = 0, \tag{3.11}$$

that in terms of the net energy, gain ξ is

$$\frac{d}{dt_j}\xi_j\left(t_j\right) = \frac{\left(\eta_j \xi_j\left(t_j\right) + \alpha_{-j}\right)}{\left(\eta_j t_j + \beta_{-j}\right)}, \quad j = 1, 2, \ldots, k. \tag{3.12}$$

Since in this equation, the left side represents an instantaneous rate of the net energy gained from the patch A^j and the right side is an average rate of the energy gained from all the patches except A^j, it leads to the following condition of leaving the current patch:

> The forager should leave the current patch A^j if the instantaneous rate $\frac{d}{dt_j}\xi_j\left(t_j\right)$ of the net energy gained in A^j drops to the average rate of the energy gained from all the patches except A^j.

Notice that if the gain function ξ has both intervals of concavity and convexity (see, e.g., Figure 3.4c), then the equivalence specified by Equation 3.12 can be reached several times, and additional consideration is required. In particular, in the case of logistic gain, the last drop of the instantaneous rate should be used (see Example 3.1).

The left side value in Equation 3.12 is also known as the marginal rate, and the condition defined by this equation is called the marginal value theorem. The formulated condition was derived by Charnov (1976) in 1976 and formed a basis for the most studies in the theory of optimal foraging (Stephens & Krebs, 1986; Stephens et al., 2007). For a discrete time case, the application of the marginal value theorem is summarized in the form of the following algorithm.

Algorithm 3.1: (The Charnov Foraging in Discrete Time)

Given a domain $\mathcal{X} \subset \mathbb{R}^2$, patchy environment $\mathcal{A}^{0\ldots k}_{1\ldots m} = \left\{A^0, A_1^{j_1}, A_2^{j_2}, \ldots, A_m^{j_m}\right\}$, transit time t_0, cost C_0, encounter rates η_j, residence times t_j, $j = 1, 2, \ldots, k$, and total foraging time T do:

1. Start with initial forager's location $\breve{x}(0)$.
2. Set $t = 1$.
3. Choose patch $A^j = A(t)$ with respect to the forager location $\breve{x}(t-1)$, time t_0 and cost C_0.
4. Move to the patch A^j and set $\breve{x}(t)$ at the entrance to A^j.
5. Set $\tau = 0$.
6. For $t = 2, 3, \ldots, T$ do:
7. Move one step in the patch A^j and set $\breve{x}(t)$.
8. Mark patch A^j as visited.
9. Set $\tau = \tau + 1$.
10. If $\tau > t_j$ (residence time is over) then go to 14 (to the choice of the next patch).
11. Calculate $\alpha_{-j} = \sum_{i=1, i\neq j}^{k} \eta_i \xi_i\left(t_i\right) - t_0 C_0$ and $\beta_{-j} = \sum_{i=1, i\neq j}^{k} \eta_i t_i + t_0$.
12. Calculate $f(\tau) = \frac{d\xi_j}{dt_j}(\tau)$ and $g(\tau) = \frac{\eta_j \xi_j(\tau) + \alpha_{-j}}{\eta_j t_j + \beta_{-j}}$ (left and right sides of Equation 3.12).

13. If $f(\tau) = g(\tau)$ (Equation 3.12) and $\frac{df}{d\tau} < 0\left(\frac{d\xi_j}{dt_j}\text{ decreases in }\tau\right)$ (leaving condition holds) then
14. If not-visited patches remain then
15. Choose next patch $A^j = A(t)$ with respect to the forager location $\breve{x}(t)$, time t_0 and cost C_0.
16. Move to the chosen patch A^j and set $\breve{x}(t)$ at the entrance to A^j.
17. Set $\tau = 0$.
18. Else (all patches were visited)
19. Break (terminate foraging).
20. End if (choice of the next not-visited patch).
21. End if (leaving condition)
22. End for (over t).

In this algorithm, the global time t corresponds to the movement of the forager in the patches and between the patches, while the local time τ is used for checking whether to leave the current patch or to continue the movement in this patch. The actions of the Charnov foraging Algorithm 3.1 in discrete time are illustrated by the following example.

Example 3.1

Let, as earlier, $\mathfrak{X}$ be a square domain of the size $n = 100 \times 100$, and assume that there are $k = 5$ circle patches $A^1,\ldots,A^5$ with the following geometric characteristics:

Patch	A^1	A^2	A^3	A^4	A^5
Center coordinates	(20, 20)	(45, 55)	(15, 70)	(70, 25)	(80, 80)
Radius r_j	5	7	10	12	15

The patches are shown by the circles in Figure 3.5b.

Let the total foraging time be $T = 1000$. The residence times t_j for the patches are defined as

$$t_j = \left(r_j / \sum_{i=1}^{5} r_i\right) \times T, \quad j = 1,\ldots,5,$$

and the encounter rates η_j are defined by the probability

$$\eta_j = \sum_{i=1}^{5} |A^i| / |\mathfrak{X}| = \sum_{i=1}^{5} \pi r_i^2 / n, \quad j = 1,\ldots,5,$$

where $|\cdot|$ stands for the size of the set.

The gain function is assumed logistic (see Figure 3.4c) that given the residence times t_j results in

$$\xi_j(\tau) = 1/\left(1 + \exp\left[-a\left(\tau - t_j/2\right)\right]\right), \quad j = 1,\ldots,5.$$

The example of the gain function for the third patch A^3 and corresponding instantaneous (marginal) and average gain rates is shown in Figure 3.4a.

The forager starts in the point $\breve{x}(0) = (40,5)$ (see white point in Figure 3.5b). Between the patches, it moves by immediate flights so that the transit time $t_0 = 0$, and in the patches, it moves according to the Brownian walk. The trajectory of the forager, which moves according to the criterion given by Equation 3.12 (the Charnov condition), is shown in Figure 3.5c. For comparison, the trajectory, obtained for a foraging that uses all available residence times, is shown in Figure 3.5d.

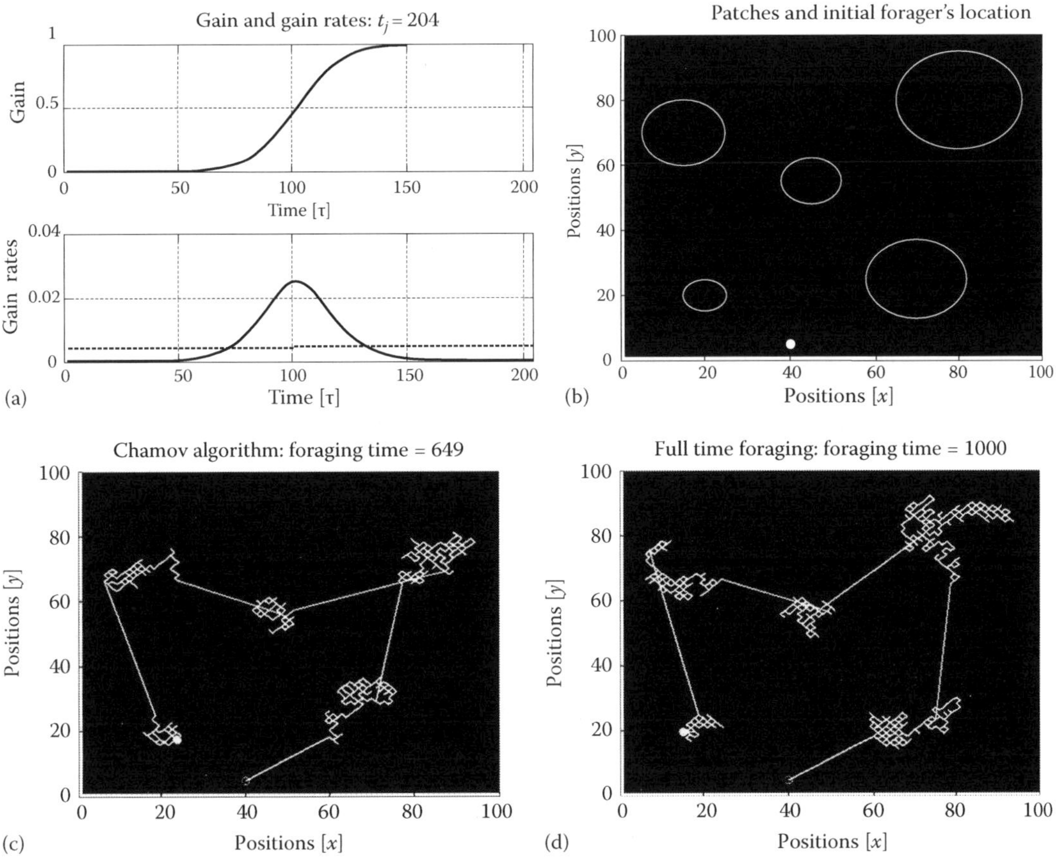

FIGURE 3.5 (a) Gain function and gain rates for the patch of type $j = 3$: solid line depicts instantaneous rate, dashed line average rate; (b) patches and initial forager's location $\breve{x}(0) = (40,5)$ (white circle); (c) forager's trajectory by the Charnov algorithm; and (d) forager's trajectory according to full residence times.

In the figure, it is seen that according to the Charnov condition, the forager leaves the patches at the times before the end of available residence times (e.g., according to Figure 3.4a, it leaves the patch A^3 at $\tau = 132$, while the residence time is $t_3 = 204$). Consequently, the density of the trajectories in the patches for the Charnov foraging (see Figure 3.4c) is less than the density of the trajectories for the foraging using the full residence times (see Figure 3.4d). As a result, the Charnov foraging terminates before ending the total foraging time; see foraging time $t = 649$ in Figure 3.4c and foraging time $t = 1000$ in Figure 3.4d. ■

The condition of leaving the patch defined by Equation 3.12 (the Charnov marginal value theorem) implements a deterministic setup of the problem of optimal foraging. It implies that the forager has complete information about all the patches in the domain that is certainly not satisfied for real-world foragers like animals and insects. However, the considered setup and foraging algorithm form a basis for the following stochastic models of foraging in patchy environment and for the algorithms of individual search and foraging, which are considered in the next chapter.

3.3.2 Foraging in Stochastic Environment

As it follows from Equation 3.12, the Charnov foraging considered earlier is essentially based on the assumption that, in addition to the knowledge regarding the patches' number, their locations, and transit time t_0 and cost C_0, the forager has complete information about the potential of all available patches represented by residence times t_j, $j = 1, \ldots, k$. Such an assumption can hold in the tasks of path planning and navigation of artificial mobile agents acting according to predefined data or communicating with certain supervising agent. However, in the case of foraging by natural agents as insects and animals that can use only information obtained from their local neighborhood, this assumption in general does not hold.

3.3.2.1 Oaten Model

The model of foraging that extends the Charnov model to the stochastic case was suggested in 1977 by Oaten (1977). In particular, given a domain $\mathcal{X} \subset \mathbb{R}^2$ divided to the patches $A_1, A_2, \ldots, A_m$, the Oaten model implements the following assumptions (cf. Charnov model considered in Section 3.3.1):

1. The forager is not informed about the exact fitness of the patches, but it is informed about the probabilities $b_j(A) = Pr\{\text{the chosen patch } A \text{ containes } j \text{ units of prey}\}$. In other words, in contrast to the Charnov model, the forager is not informed about the types of the patches but can expect that the chosen patch A is of the type j with probability $b_j(A)$.
2. Let initially, when the forager enters the patch, the patch contains j units of prey, and let τ_i, $i = 1, 2, \ldots, j$, be an expected time of capturing the prey i given that the preys $1, 2, \ldots, i$ are already captured. Then, it is assumed that the forager is informed about joint probability density $\mathfrak{f}(\tau_1, \tau_2, \ldots, \tau_i \mid j)$ and this function is the same for all patches $A_1, A_2, \ldots, A_m$. In the terms of the Charnov model, it means that the unknown residence time depends on the amount of prey in the patch and the forager can expect this time. Notice that the density $\mathfrak{f}(\tau_1, \tau_2, \ldots, \tau_i \mid j)$ represents the results of foraging in the current patch. In addition, let τ_{j+1} be an expected time of foraging in the patch after capturing the last prey.
3. The found prey is not replaced by a new unit. In other words, it is assumed that the prey population in the considered domain $\mathcal{X}$ grows essentially slower than it is eaten by the foraging predator, so as in the Charnov model, such a growth can be omitted.
4. Similar to the Charnov model, the transit time $t_0 = t_0$ of traveling between the patches, that is, the time of movement in the patch $A^0 = \mathcal{X} \backslash \left(\bigcup_{k=1}^{m} A_k \right)$, is constant and, in general, is greater than zero. If the times of movement between the patches are random, then t_0 is an average of these times.

As earlier, denote by $\vec{\partial} = \partial(0), \partial(1), \partial(2), \ldots$ a strategy of the forager acting in the domain $\mathcal{X}$. Since the forager is not informed about the patches types, the decision rules $\partial(t \mid A(t))$ of the strategy $\vec{\partial}$ specify the patch $A \in \{A^0, A_1, A_2, \ldots, A_m\}$, which has to be chosen at time t given that at this time the forager is hunting in the patch $A(t) \in \{A_1, A_2, \ldots, A_m\}$. The choice of the patch A^0 is interpreted as a choice to leave the current patch.

In the considered Oaten model (1977), the decision is obtained basing on the transit time t_0; probabilities $b_j(A_k)$, $k = 1, 2, \ldots, m$; and the probability densities $\mathfrak{f}(\tau_1, \tau_2, \ldots, \tau_i \mid j)$. Notice again that if the forager acts in the patch A_k, which initially contained j units of prey, and consequently captures the preys $1, 2, \ldots, i$, $i \le j$, then the density $\mathfrak{f}(\tau_1, \tau_2, \ldots, \tau_i \mid j)$ represents the results of foraging in A_k.

The optimization problem is formulated in the terms of the expected number of prey captured in the chosen patch with respect to the time of foraging to this prey. Since an amount of eaten prey unambiguously defines the energy intake in the patch, the average rate of prey capture is proportional to the energy intake rate $r_c(\vec{\partial})$ as it is used in the Charnov model (see Section 3.3.1) and vice versa.

Similar to the Charnov model (see Equation 3.10), denote by $r_c(\vec{\partial} \mid A)$ the ratio of the expected number $n(A)$ of prey captured at the chosen patch to the expected time of foraging according to the strategy $\vec{\partial}$, that is,

$$r_c(\vec{\partial} \mid A) = \frac{n(A)}{(t_0 + t(A))}, \tag{3.13}$$

where $t(A)$ is an expected time of foraging in the patch A; following the Charnov model, this time is also referred as expected residence time. Consequently, for the strategy $\vec{\eth}$, the capture rate $r_c(\vec{\eth})$ is defined as an average rate over all patches chosen according to the strategy $\vec{\eth}$; formally, it is given by the following equation:

$$r_c(\vec{\eth}) = \frac{n(\vec{\eth})}{(t_0 + t(\vec{\eth}))}. \tag{3.14}$$

Then the problem of foraging in the environment $\mathcal{A}^0_{1...m} = \{A^0, A_1, A_2, \ldots, A_m\}$ requires to find a strategy that maximizes the rate $r_c(\vec{\eth})$ and is formulated as follows (cf. Problem 3.1).

Problem 3.2: (Foraging in a Patchy Environment: Stochastic Setup)

In the set D of all possible strategies $\vec{\eth}$ of foraging in the environment $\mathcal{A}^0_{1...m}$, find the strategy $\vec{\eth}^*$ such that

$$r_c(\vec{\eth}^*) = \max_{\vec{\eth}\in\mathfrak{D}} \{ r_c(\vec{\eth}) \,|\, \tau_1, \tau_2, \ldots, \tau_{jk} > 0 \quad \text{for each } A_k, k = 1, 2, \ldots, m \}. \tag{3.15}$$

In this equation, let us consider the decision either to leave the current patch or to continue searching in it. Following Oaten (1977), in this case, the strategy of the forager is represented by a sequence of "giving up" times t_1, $t_2(\cdot)$, $t_3(\cdot,\cdot)$,... such that t_1 is a constant and each next time $t_i(\cdot,\cdot,\ldots)$ depends on $i - 1$ times τ of previous captures. Following this strategy, the forager will leave the patch if no prey has been caught up to time $t_2(\tau_1)$; otherwise, it will continue foraging and will leave the patch if the next prey has not been caught up to time $t_2(\tau_1)$ and so on up to the time $t_{j+1}(\tau_1,\ldots,\tau_j)$. In the algorithmic form, this process is outlined as follows.

Algorithm 3.2: (The Oaten Foraging)

Given a domain $\mathfrak{X} \subset \mathbb{R}^2$, patchy environment $\mathcal{A}^0_{1...m} = \{A^0, A_1, A_2, \ldots, A_m\}$, transit time t_0, giving up times t_1, t_2, t_3,... for each patch $A \in \mathcal{A}^0_{1...m}$ and total foraging time T do:

1. Start with initial forager's location.
2. While the foraging time T is not over do:
3. Choose a patch A from $\mathcal{A}^0_{1...m}$ with respect to the current forager's location and transit time t_0.
4. Move to the chosen patch A and enter it.
5. Set $i = 1$ and $t = 0$.
6. While $t \le t_i$ do (hunting in the patch A)
7. Hunt in the patch A and increase actual foraging time t accordingly.
8. If the prey is captured then set $i = i + 1$ (hunt for the next prey).
9. End while (over hunting time t).
10. Set current forager's location at the exit of the patch A.
11. End while (over total foraging time T).

Given the giving up times t_1, t_2, t_3,..., the actions of the algorithm are obvious, and, following Problem 3.2, the goal is to find such times $t^* = \langle t_1^*, t_2^*(\cdot), t_3^*(\cdot,\cdot), \ldots \rangle$ that the values defined by Equations 3.13 and 3.14 reach their maxima.

Assume that the forager is in a patch A, which initially contains j units of prey. Then, the expected number $n(A)$ of captured prey and the expected residence time $t(A)$ for the patch A are defined as follows (Oaten, 1977):

$$n(A)=\sum_{i=0}^{\infty}Pr\{\text{at least } i \text{ units of prey are captured}\}$$
$$=\sum_{i=0}^{\infty}\sum_{j=0}^{\infty}b_j(A)Pr\{\text{at least } i \text{ units of prey were captured} \mid \text{initially there were } j \text{ units of prey}\} \tag{3.16}$$

$$t(A)=\sum_{i=0}^{\infty}\mathcal{E}\left\{\min(t_i,t_i)\prod_{j=0}^{i-1}\text{ind}(t_j\le t_i)\right\}$$
$$=\sum_{i=0}^{\infty}\left[b_{i-1}(A)\mathcal{E}\left\{\min(t_i,t_i)\prod_{j=0}^{i-1}\text{ind}(t_j\le t_i)\mid\text{initially there were } i-1 \text{ units of prey}\right\}\right.$$
$$\left.+\sum_{k=i}^{\infty}b_k(A)\mathcal{E}\left\{\min(t_i,t_i)\prod_{j=0}^{i-1}\text{ind}(t_j\le t_i)\mid\text{initially there were } k \text{ units of prey}\right\}\right], \tag{3.17}$$

where

$\mathcal{E}\{\cdot\}$ stands for the expected value of random variable

$\text{ind}(t_j \le t_i)$ is an indicator value such that $\text{ind}(t_j\le t_i)=\begin{cases}1 & \text{if } t_j\le t_i,\\ 0 & \text{otherwise,}\end{cases}$

t_j are actual times of hunting in the patch up to the capture with the jth prey, $j = 0, 1, 2,\ldots,$ (see lines 6–9 in Algorithm 3.2)

In the terms of density functions $\mathfrak{f}(\tau_1,\tau_2,\ldots,\tau_i \mid j)$, $i = 1, 2,\ldots, j$, $j = 1, 2, \ldots,$ the values $n(A)$ and $t(A)$ are specified by the following equations (Oaten, 1977):

$$n(A)=\sum_{i=0}^{\infty}\sum_{j=i}^{\infty}b_j(A)\int_0^{t_1}\int_0^{t_2}\ldots\int_0^{t_i}\mathfrak{f}(\tau_1,\tau_2,\ldots,\tau_i \mid j)\,d\tau_1 d\tau_2,\ldots,d\tau_i, \tag{3.18}$$

$$t(A)=b_0(A)t_1+\sum_{j=1}^{\infty}b_j(A)\int_0^{t_1}\int_0^{\infty}\min(\tau_1,t_1)\mathfrak{f}(\tau_1 \mid j)\,d\tau_1 d\tau_{j+1}$$
$$+\sum_{j=2}^{\infty}\left[b_{j-1}(A)\int_0^{t_1}\ldots\int_0^{t_{j-1}}t_j\mathfrak{f}(\tau_1,\ldots,\tau_{j-1} \mid j-1)\,d\tau_1,\ldots,d\tau_{j-1}\right.$$
$$\left.+\sum_{k=j}^{\infty}b_k(A)\int_0^{t_1}\ldots\int_0^{\infty}\min(\tau_j,t_j)\mathfrak{f}(\tau_1,\ldots,\tau_j \mid k)\,d\tau_1,\ldots,d\tau_j d\tau_{j+1}\right]. \tag{3.19}$$

In Equation 3.19, the first term stands for the expected time up to capturing the first prey, the second term stand for the expected time of capturing the second prey given that the first prey was captured, and, consequently, the third term defines the expected time of capturing the next preys given that the previous preys were captured.

Similar to the search and screening problems (see Section 2.3.3), a strict solution of Equations 3.18 and 3.19 with respect to the expected giving up times $t_1, t_2, \ldots, t_j$ in general case is rather problematic. However, a necessary condition for the optimality of the forager's strategy can be found in the terms of the capture rate, and for any patch A, it is defined as follows (Oaten, 1977).

Denote by $q(t_i > t_i | j, \tau_1, \tau_2, \ldots, \tau_{i-1}) > 0$ the conditional probability that no prey will be captured up to time t_i given that initially there were j units of prey and expected capture times are $\tau_1, \tau_2, \ldots, \tau_{i-1}$. Then the conditional probability $b\left(j \middle| \tau_1, \tau_2, \ldots, \tau_{i-1}, t_i \rangle t_i\right)$ that initially in the patch A were j units of prey given that the expected capture times are $\tau_1, \tau_2, \ldots, \tau_{i-1}$ and no prey will be captured up to time t_i is defined as follows:

$$b\left(j \middle| \tau_1, \tau_2, \ldots, \tau_{i-1}, t_i \rangle t_i\right) = \frac{b_j(A)\mathfrak{f}\left(\tau_1, \ldots, \tau_{i-1} | j\right) q\left(t_i > t_i | j, \tau_1, \ldots, \tau_{i-1}\right)}{\sum_{j=i-1}^{\infty} b_j(A)\mathfrak{f}\left(\tau_1, \ldots, \tau_{i-1} | j\right) q\left(t_i > t_i | j, \tau_1, \ldots, \tau_{i-1}\right)}. \tag{3.20}$$

In addition, denote by $\mathfrak{f}\left(t_i | j, \tau_1, \ldots, \tau_{i-1}\right)$ the conditional probability density of giving up time, that is,

$$\mathfrak{f}\left(t_i | j, \tau_1, \ldots, \tau_{i-1}\right) = \frac{\mathfrak{f}\left(\tau_1, \ldots, \tau_i | j\right)}{\mathfrak{f}\left(\tau_1, \ldots, \tau_{i-1} | j\right)}, \tag{3.21}$$

where it is assumed that if $i = 1$, then $\mathfrak{f}\left(\tau_1, \ldots, \tau_{i-1} | j\right) = \mathfrak{f}\left(\tau_0 | j\right) = 1$ for any j. Then, the conditional capture rate for the giving up time t_i given that initially in the patch A were j units of prey and the expected capture times are $\tau_1, \tau_2, \ldots, \tau_{i-1}$ is

$$r_c\left(t_i | j, \tau_1, \ldots, \tau_{i-1}\right) = \frac{\mathfrak{f}\left(t_i | j, \tau_1, \ldots, \tau_{i-1}\right)}{q\left(t_i > t_i | j, \tau_1, \tau_2, \ldots, \tau_{i-1}\right)}; \tag{3.22}$$

notice again that $q(t_i > t_i | j, \tau_1, \tau_2, \ldots, \tau_{i-1}) > 0$ for any $i > 0$ and $j > 0$.

Finally, the conditional capture rate $r_c(t_i | j)$ for the giving up time t_i given that initially in the patch A were j units of prey is specified by the sum over the expected future captures

$$r_c\left(t_i | j\right) = \sum_{j=i}^{\infty} b\left(j \middle| \tau_1, \tau_2, \ldots, \tau_{i-1}, t_i > t_i\right) r_c\left(t_i | j, \tau_1, \ldots, \tau_{i-1}\right); \tag{3.23}$$

so the capture rate $r_c(t_i)$ for the giving up time t_i is defined by the following recursion (Oaten, 1977):

$$r_c\left(t_i\right) = r_c\left(t_i | j\right) + \sum_{j=i}^{\infty} b\left(j \middle| \tau_1, \tau_2, \ldots, \tau_{i-1}, t_i > t_i\right) r_c\left(t_i | j, \tau_1, \ldots, \tau_{i-1}\right) \mathcal{E}\left\{j_{i+}^{*} - r_c\left(t_i\right) t_{i+}^{*} | j, \tau_1^{*}, \ldots, \tau_i^{*}\right\}, \tag{3.24}$$

where

the expected value $\mathcal{E}\left\{j_{i+}^{*} - r_c\left(t_i\right) t_i^{*} | j, \tau_1^{*}, \ldots, \tau_i^{*}\right\}$ represents the future foraging in the patch A after capturing the ith prey

here, j_{i+}^{*} stands for additional captures

t_{i+}^{*} stands for additional time after capturing the ith prey while the forager acts according to optimal strategy

$\tau_1^{*}, \ldots, \tau_i^{*}$ are optimal capture times.

The obtained Equation 3.24 allows a verification of the expected giving up times $t_1, t_2,\ldots, t_j$ based on the capture rates $r_c(t_i)$, $i = 1, 2,\ldots, j$. Namely, if some means of optimal capture rates $r_c^*(t_i)$ are known, then the optimal giving up times $t_1^*, t_2^*,\ldots,t_j^*$ will with necessity result in $r_c^*(t_i)$ by the means of Equation 3.24. However, as indicated earlier, in general case, the sufficient conditions for the optimality of the giving up times $t_1, t_2,\ldots, t_j$ are still unknown.

Let us consider an implementation of the Oaten model for the random search with definite number of prey units and corresponding probabilities. Following Oaten (1977), assume that each patch $A \in \{A_1, A_2,\ldots, A_m\}$ is either empty with probability $b_0(A)$ or includes exactly j units of prey with probability $b_j(A) = 1-b_0(A)$. In addition, assume that the forager searches randomly with a constant speed υ, that is, the proportion of the patch area checked per time unit (cf. the detection rate used in the Pollock model in Section 2.3.1).

Then, given that the patch contains j units of prey, the probability $q(t > t \mid j)$ that the next prey will be captured after a giving up time t (i.e., will not be captured up to giving up time t) and, correspondingly, the probability $p(t \le t \mid j) = 1-q(t > t \mid j)$ that the prey will be captured up to the time t are defined as follows:

$$q(t > t \mid j) = \exp(-j\upsilon t) \quad \text{and} \quad p(t \le t \mid j) = 1-\exp(-j\upsilon t). \tag{3.25}$$

Consequently, Equations 3.16 and 3.17 for the expected number $n(A)$ of captures and the expected time $t(A)$ of foraging in the patch A obtain the following form (Oaten, 1977):

$$\begin{aligned} n(A) &= b_j(A)\,p(t_1 \le t_1 \mid j)\sum_{k=0}^{j} Pr\{t_i \le t_i, 2 \le i \le k \mid j, t_1 \le t_1\} \\ &= b_j(A)\left[1-\exp(-j\upsilon t_1)\right]\sum_{k=1}^{j} p_k, \end{aligned} \tag{3.26}$$

where $p_k = Pr\{t_i \le t_i, 2 \le i \le k \mid j, t_1 \le t_1\}$ is denoted for shortness:

$$t(A) = b_0(A)t_1 + b_j(A)\left[1-\exp(-j\upsilon t_1)\right]\left(\sum_{k=1}^{j}\frac{p_k}{\upsilon(j-k+1)} + p_j\mathcal{E}\left\{t_{j+1} \mid \prod_{k=1}^{j}\text{ind}(t_k \le t_k) = 1\right\}\right), \tag{3.27}$$

where it is taken into account that $\mathcal{E}\{\min(t_1, t_k)\} = p(t_k \le t_k \mid j)/(\upsilon(j-k+1))$.

Assume that the forager leaves the patch immediately after capturing the last prey or if the time t_j is over. Then, $t_{j+1} = 0$ with certainty and $\mathcal{E}\left\{t_{j+1} \mid \prod_{k=1}^{j}\text{ind}(t_k \le t_k) = 1\right\} = 0$. In addition, notice that for optimal giving up times, it holds true that $p_k = 1$ for each $k = 0, 1, 2,\ldots, j$. Then,

$$n(A) = b_j(A)\,j\left[1-\exp(-j\upsilon t_1)\right], \tag{3.28}$$

$$t(A) = b_0(A)t_1 + b_j(A)\left[1-\exp(-j\upsilon t_1)\right]\sum_{k=1}^{j}\frac{1}{\upsilon(j-k+1)}, \tag{3.29}$$

and Equation 3.13 for a capture rate $r_c(\vec{\eth} \mid A)$ is

$$r_c(\vec{\eth} \mid A) = \frac{b_j(A)\,j\left[1-\exp(-j\upsilon t_1)\right]}{t_0 + b_0(A)t_1 + b_j(A)\left[1-\exp(-j\upsilon t_1)\right]\sum_{k=1}^{j} 1/\upsilon(j-k+1)}, \tag{3.30}$$

t_1	t_2	t_3	t_4	t_5
8.917	14.159	19.046	23.858	28.704
t_6	t_7	t_8	t_9	t_{10}
33.655	38.772	44.134	49.850	56.091
t_{11}	t_{12}	t_{13}	t_{14}	t_{15}
63.149	71.572	82.531	99.112	133.750

(a)

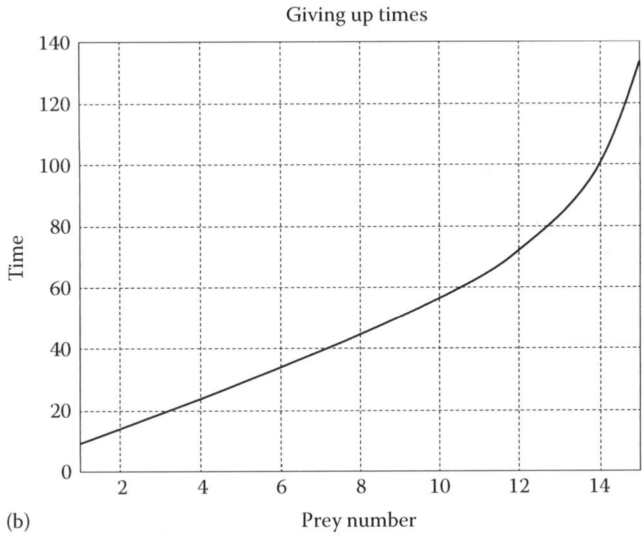

(b)

FIGURE 3.6 (a) Values of giving up times and (b) dependence of giving up time on the prey number.

where the giving up time t_1 is a solution of the following equation:

$$b_0(A)\exp(j\upsilon t_1) - b_0(A)\, j\upsilon t_1 = j\upsilon t_0 + b_0(A) \quad \text{with } t_0 = t_0. \tag{3.31}$$

The next giving up time t_2 is then defined by the same equation:

$$b_0(A)\exp((j-1)\upsilon t_2) - b_0(A)(j-1)\upsilon t_2 = 2(j-1)\upsilon t_0 + b_0(A), \tag{3.32}$$

and so on up to the time t_j. For example, if $b_0(A) = 0.25$, $j = 15$, $\upsilon = 0.01$, and $t_0 = 1$, then numerical computation of the giving up times t_i, $i = 1, 2, \ldots, j$, results in the values shown in Figure 3.6.

As it is expected, the giving up times increase exponentially with i and prescribe the forager to leave a newly chosen patch after a short time if no prey has been captured. However, if the prey was captured, then the forager should continue searching as long as the number of prey units becomes greater. For more detailed description of the optimization techniques and corresponding reasoning, see the original paper (Oaten, 1977) by Oaten. The actions of Algorithm 3.2 of the Oaten foraging with the giving up times defined by Equations 3.31 and 3.32 are illustrated by the following example.

Example 3.2

Let the domain $\mathfrak{X}$ and the patches $A_1,\ldots,A_5$ be the same as in Example 3.1, and, as earlier, assume that the patch $A \in \{A_1,\ldots, A_5\}$ is empty with probability $b_0(A) = 0.25$ or contains $j = 15$ units of prey with probability $b_{10}(A) = 1 - b_0(A) = 0.75$. The patches with distributed prey are shown in Figure 3.7b. Also, the transit time is $t_0 = 1$.

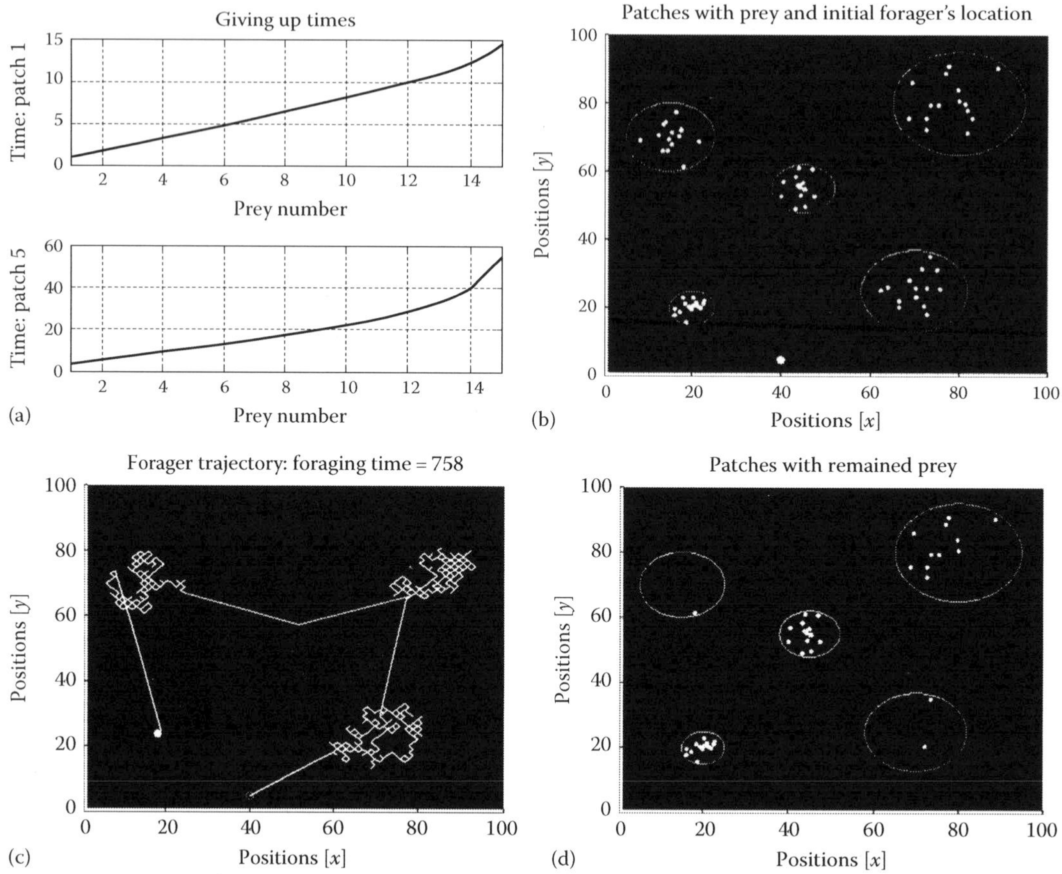

FIGURE 3.7 (a) Giving up times for the patches A_1 and A_5 (the total number of prey in each patch is j = 15); (b) patches with initially distributed prey (white points) and initial forager's location $\breve{x}(0) = (40, 5)$ (white circle); (c) forager's trajectory by the Oaten algorithm; (d) patches with the prey that remained after foraging in all patches.

The speed υ, which defines a proportion of the patch area checked per time unit, depends on the area of the patch; for the circle patches with radiuses r_k, it is $\upsilon_k \sim 1/\pi r_k^2$, $k = 1, 2, \ldots, j$. The corresponding giving up times for the smallest patch A_1 and the largest patch A_5 are shown in Figure 3.7a. Notice that all times are normalized so that the global foraging time is T = 1000 and the residence times are proportional to the patches' radiuses (see Example 3.1).

Also, as in Example 3.1, the forager starts in the point $\breve{x}(0) = (40,5)$ (see Figure 3.7b). The choice of the next patch follows the same rule as in Example 3.1 that is a minimum of the distance to the patch divided by its residence time. The resulting trajectory of the forager is shown in Figure 3.7c. It is assumed that the forager checks its closest neighborhood (its location and four neighboring points) and eats the prey located in these points; the patches with the prey remained after the forager actions are shown in Figure 3.7d.

It is seen that for a smaller patch, the giving up times are also smaller, and, as a result, after foraging, there remains more prey than in the larger patches. Because of stochastic setup of the Oaten foraging, the obtained total foraging time is greater than this time for the Charnov foraging with the same time setup (see Example 3.1). However, notice that the Oaten foraging does not require to be terminated after visiting all the patches so that the forager can return to already-visited patch. ■

The considered implementation of the Oaten model for the random search provides a strong relation between optimal foraging and the methods of search and screening considered in Section 2.2. In fact, as it follows from the random search assumption, the probability $p(t \leq t|j)$ that the prey will be captured up to the time t is defined by the same equation as the Koopman detection probability for random search (see Equation 2.8) with the search effort is $\kappa(\mathfrak{x}, t) = j\upsilon t$. That means that the considered setup of the Oaten model is a variant of the constrained search for a static target presented in Section 2.2, in which the search efforts are defined by the time of search. Thus, corresponding algorithms and methods of search and screening applied to the Oaten model will result in the distribution of the giving up times with respect to the number of prey j and forager's speed υ and can provide additional useful technique for the studies in optimal foraging. Similarly, an application of the Oaten and similar models in the framework of the theory of search and screening can help in specification of the searcher's trajectories and for the consideration of search for several targets.

3.3.2.2 McNamara Model

The Oaten model directly addresses the giving up times and provides the necessary conditions for optimality of these times. In 1982, McNamara (1982) considered the same process of foraging in stochastic environment from the other point of view. Namely, having in mind the initial Charnov model with the patchy environment $\mathcal{A}_{1...m}^{0...k}$ (see Section 3.3.1), McNamara suggested to use a potential reward and an expected reward rate to define the giving up times, in which the forager has to leave the patches in order to maximize its income. As a result, the forager's strategy is formed by the decision rules, which are defined alternatively by one of the following conditions (McNamara, 1982) (cf. the Charnov condition in Section 3.3.1):

1. Continue searching in the current patch A while the potential reward that can be obtained from the patch A is greater than zero, and leave the current patch A when the potential reward drops to zero.
2. Continue searching in the current patch A while the expected reward rate for the patch A is greater than the maximal average reward rate for the environment, and leave the current patch A when the expected reward rate drops to the maximal average reward rate.

Let us consider these conditions given a patchy environment $\mathcal{A}_{1...m}^{0...k} = \left\{A^0, A_1^{j_1}, A_2^{j_2}, \ldots, A_m^{j_m}\right\}$ defined over a domain $\mathfrak{X}$, where $j_i \in \{1, 2, \ldots, k\}$, $i = 1, 2, \ldots, m$, stand for the types of the patches and $A^0 = \mathfrak{X} \setminus \left(\bigcup_{i=1}^{m} A_i^{j_i}\right)$ denotes the part of the domain between the patches. A proportion of the patches of a certain type j is defined by the encounter rate $\eta_j = \eta(A^j) \in [0, 1]$, $j = 1, 2, \ldots, k$. As indicated earlier (see Section 3.3.1), an average transit time, that is, an average time of movement between the patches, is denoted by $t_0 = t(A^0)$, and an average cost of such movement is denoted by $c_0 = c(A^0)$. Similar to the Charnov model, both t_0 and c_0 can be considered as constants or as means of the appropriate random variables; however, notice that here c_0 is the overall cost, while in the Charnov model C_0 is the transit cost per time unit.

Let $t = t(A)$ be a time spent so far in the patch $A \in \mathcal{A}_{1...m}^{0...k} \setminus A^0$, and denote by $i(t)$ the informational state of the forager, which represents the knowledge of the forager; in particular, it includes forager's experience in this patch and also can include its location $\mathfrak{x}(t)$ at t and time t itself. At any time $\tau < t$, the state $i(t)$ is considered as a random variable, which at time t obtains its actual value $\mathbb{I}(t)$. Denote by $R^{0...t}(\mathbb{I}(t))$ a reward obtained by the forager up to and including the time t so that the forager's state is $\mathbb{I}(t)$, where the notation $0...t$ stresses the fact that $\mathbb{I}(t)$ includes the foragers experience from entering the patch A and during the period $[0, t]$.

By the same token, if $t = t(A)$ is an expected giving up time for the patch A, then $\mathcal{R}_{-t}^{t...t}\{i(t) \mid i(t) = \mathbb{I}(t)\} = \mathcal{R}_{-t}^{t...t}(\mathbb{I}(t))$ denotes an expected reward, which will be obtained by the forager in the patch A up to the giving up time t given that at time t the state of the forager is $i(t) = \mathbb{I}(t)$. It is clear that at the entrance of the patch, the reward $R^{0...0}(\mathbb{I}(0)) = 0$ and the expected reward rate is $\mathcal{R}_{-0}^{0...t}\{i(t) \mid i(0) = \mathbb{I}(0)\} = \mathcal{R}_{-0}^{0...t}(\mathbb{I}(0))$. Then

$$\mathcal{R}_{-t}^{t...t}(\mathbb{I}(t)) = \mathcal{R}_{-0}^{0...t}(\mathbb{I}(t)) - R^{0...t}(\mathbb{I}(t)), \tag{3.33}$$

where the first term at the right side is the expected reward, which will be obtained up to the time t given that at the time $t < t$ the forager's state is $\mathbb{I}(t)$, and the second term is the reward obtained during the period $[0, t]$.

Recall that the time t is an expected giving up time, which stands for the expected time that the forager will spend from entering the patch and up to leaving it. Then, given that at time t the forager's state is $\mathbb{I}(t)$, the expected residence time $t_{-t}^{t\ldots\infty}\left(\mathbb{I}(t)\right)$ that will be spent in the patch after t is

$$t_{-t}^{t\ldots\infty}\left(\mathbb{I}(t)\right) = t - t. \tag{3.34}$$

Then, the potential reward $\mathcal{R}\left(\mathbb{I}(t)\right)$ that can be obtained from the patch A given that at time t the forager's state is $\mathbb{I}(t)$ (see the McNamara condition 1) is defined as follows (McNamara, 1982):

$$\mathcal{R}\left(\mathbb{I}(t)\right) = \mathcal{R}_{-t}^{t\ldots t}\left(\mathbb{I}(t)\right) - r_{\mathrm{c}}^{*}\left(\mathcal{A}_{1\ldots m}^{0\ldots k}\right) t_{-t}^{t\ldots\infty}\left(\mathbb{I}(t)\right), \tag{3.35}$$

where $r_{\mathrm{c}}^{*}\left(\mathcal{A}_{1\ldots m}^{0\ldots k}\right)$ is the maximal average reward rate over the patchy environment $\mathcal{A}_{1\ldots m}^{0\ldots k}$ that is indicated in the McNamara condition 2. In other words, the potential reward in the patch is the reward that is expected to be obtained from the patch starting from time t, excluding the expected average reward, which could be obtained by foraging in other patches at this time.

To define $r_{\mathrm{c}}^{*}\left(\mathcal{A}_{1\ldots m}^{0\ldots k}\right)$, recall that t_0 is an average transit time and c_0 is an average transit cost in $\mathcal{A}_{1\ldots m}^{0\ldots k}$. The average reward rate over the environment $\mathcal{A}_{1\ldots m}^{0\ldots k}$ provided by the foraging strategy $\vec{\mathfrak{d}}$ is defined as follows (McNamara, 1982) (cf. Charnov Equations 3.9 and 3.10 and Oaten Equations 3.13 and 3.14):

$$r_{\mathrm{c}}\left(\vec{\mathfrak{d}} \mid \mathcal{A}_{1\ldots m}^{0\ldots k}\right) = \frac{\left(\mathcal{R}_{-0}^{0\ldots t}\left(\mathbb{I}(0)\right) - c_0\right)}{\left(t_{-0}^{0\ldots\infty}\left(\mathbb{I}(0)\right) + t_0\right)}. \tag{3.36}$$

Then, the maximal average reward rate over the patchy environment $\mathcal{A}_{1\ldots m}^{0\ldots k}$ is an average reward rate, which is provided by the optimal strategy $\vec{\mathfrak{d}}^{*}$, that is,

$$r_{\mathrm{c}}^{*}\left(\mathcal{A}_{1\ldots m}^{0\ldots k}\right) = r_{\mathrm{c}}\left(\vec{\mathfrak{d}}^{*} \mid \mathcal{A}_{1\ldots m}^{0\ldots k}\right) = \max_{\vec{\mathfrak{d}}\in\mathfrak{D}} r_{\mathrm{c}}\left(\vec{\mathfrak{d}} \mid \mathcal{A}_{1\ldots m}^{0\ldots k}\right), \tag{3.37}$$

where, as usual, $\mathfrak{D}$ is the set of all possible strategies $\vec{\mathfrak{d}}$, which specify the giving up times t for the patches of the environment $\mathcal{A}_{1\ldots m}^{0\ldots k}$.

Finally, given the forager's state $\mathbb{I}(t)$, the instantaneous expected reward rate $r_{\mathrm{c}}\left(\mathbb{I}(t)\right)$ in the patch A is

$$r_{\mathrm{c}}\left(\mathbb{I}(t)\right) = \lim_{\varepsilon\to 0} \frac{\mathcal{R}_{-t}^{t\ldots t}\left\{i(t+\varepsilon) \mid i(t) = \mathbb{I}(t)\right\} - R^{0\ldots t}\left(\mathbb{I}(t)\right)}{\varepsilon}. \tag{3.38}$$

Using the defined values, the McNamara conditions indicated earlier are formulated as follows:

1. Continue searching in the current patch A while $\mathcal{R}\left(\mathbb{I}(t)\right) > 0$ and leave A otherwise.
2. Continue searching in the current patch A while $r_{\mathrm{c}}\left(\mathbb{I}(t)\right) > r_{\mathrm{c}}^{*}\left(\mathcal{A}_{1\ldots m}^{0\ldots k}\right)$ and leave A otherwise.

These conditions extend the Charnov model (see Section 3.3.1) to the foraging in stochastic environment. The model is illustrated by the following example, which appears in the McNamara paper (McNamara, 1982).

Example 3.3: (McNamara, 1982)

Assume that, in addition to the area A^0, the environment $\mathcal{A}_{1,2}^{0,1} = \{A^0, A_1^0, A_2^1\}$ includes two patches A_1^0 and A_2^1 such that the patch A_1^0 contains 0 item of prey and the patch A_2^1 contains 1 item of prey, and let the time required for finding prey in the patch A_2^1 be exponential with mean 1 time unit. Let the transit time be $t_0 = 1$ and transit cost be $c_0 = 0$. Define the state of the forager as $\mathbb{I}(t) = (t, n)$, where n is a number of prey items captured during the period $[0, t]$.

The forager starts somewhere in the area A^0 between the patches and enter a patch A. Let the initial probability that the entered patch is A_2^1 be $\frac{1}{2}$, that is, $Pr\{A = A_2^1 \mid t = 0\} = \frac{1}{2}$. If by foraging t time in the patch A the prey was not found, then the probability that this patch is A_2^1 is $Pr\{A = A_2^1 \mid t\} = (e^t + 1)^{-1}$.

Assume that one item of prey provides one reward unit. Then, since there is only one item of prey in the environment, it is optimal to leave the patch immediately after capturing the prey; otherwise, the forager should leave the patch A at the giving up time t. The initial probability that the patch contains the prey is $Pr\{A = A_2^1 \mid t = 0\} = \frac{1}{2}$, and, as earlier, for the random search, the probability of capturing the prey in the patch A_2^1 by t time of foraging is $(1 - e^{-t})$. Thus, $Pr\{t < t\} = \frac{1}{2}(1 - e^{-t})$.

Finally, since capturing of the prey is equivalent to obtaining one reward unit, if follows that

$$\mathcal{R}_{-0}^{0\ldots t}(\mathbb{I}(0)) = \frac{1}{2}(1 - e^{-t}), \tag{3.39}$$

$$t_{-0}^{0\ldots\infty}(\mathbb{I}(0)) = \frac{1}{2}\int_0^t te^{-t}dt + \frac{1}{2}t(1 + e^{-t}) = \frac{1}{2}(1 + t - e^{-t}). \tag{3.40}$$

Substitution of the obtained values into Equation 3.36 results in the following:

$$r_e(\vec{\mathfrak{d}} \mid \mathcal{A}_{1,2}^{0,1}) = \frac{(1 - e^{-t})}{(2 + t - e^{-t})}. \tag{3.41}$$

where it was taken into account $t_0 = 1$ and $c_0 = 0$. The rate $r_e(\vec{\mathfrak{d}} \mid \mathcal{A}_{1,2}^{0,1})$ reaches its maximum if

$$\frac{d}{dt} r_e(\vec{\mathfrak{d}} \mid \mathcal{A}_{1,2}^{0,1}) = \frac{e^t(2 + t - e^t)}{(e^t(2 + t) - 1)^2} = 0. \tag{3.42}$$

Since $t > 0$, this equation is equivalent to $e^t = 2 + t$, and its approximate solution is $t = 1.146$. Consequently, in the terms of giving up time, the optimal strategy of the forager is to leave the patch immediately after capturing the patch or if the patch was not found, to leave after 1.146 time units. In addition, using the value $t = 1.146$, from Equations 3.39 and 3.41, one obtains that the expected reward is $\mathcal{R}_{-0}^{0\ldots t}(t(0)) = 0.341$ and maximal average reward rate is $r_e^*(\mathcal{A}_{1,2}^{0,1}) = 0.241$. ■

The considered models by Oaten and by McNamara imply that the foraging is conducted in continuous space and time. Certainly, both models follow the random search approach and, in essence, implement the techniques used in the models of search by areas presented in Section 2.2.4. In addition, notice that in the McNamara model, the optimization technique is based on the expected reward up to the end of foraging in the patch that is the same as used in the Washburn Algorithm 4 of search for a moving target.

3.3.2.3 Foraging in Discrete Space and Time

Finally, let us consider the foraging in discrete domain $\mathfrak{X} = \{\mathfrak{x}_1, \mathfrak{x}_2, \ldots, \mathfrak{x}_n\}$ and discrete time $t = 0, 1, 2, \ldots$ The basic model for such foraging was suggested by Green (1980) (see also his recent paper [Green, 2006]). The model applies the same assumptions as the Oaten model (see assumptions 1–4 at the beginning of Section 3.3.2) and addresses the same Problem 3.2, which requires to find a strategy such that the capture rate reaches its maximum. The following consideration follows key ideas of the Green model, but implements additional assumption that the number of patches is known that makes the consideration simpler.

Let $\mathcal{A}^0_{1\ldots m} = \{A^0, A_1, A_2, \ldots, A_m\}$ be an environment defined over $\mathfrak{X}$, and, following Green (1980), assume that the patches $A = \{\mathfrak{x}_1, \mathfrak{x}_2, \ldots, \mathfrak{x}_k\} \in \mathcal{A}^0_{1\ldots m}$ are of the same size k. As earlier, the expected time of movement between the patches that is the residence time in the patch A^0 is denoted by t_0.

Assume that the forager does not search in the points twice; this, probably, holds for the short-term behavior of certain natural predators like flying insects and birds. Then, the residence time t that the forager acts in the patch is equal to the number of checked points, and maximal residence time is equal to the number of points k in the patch.

Denote by $j(t|A)$ the number of prey units captured in the patch A during the period $[0, t]$, and denote by $u(t, j|A)$ the probability that the next point of A will contain a prey given that during the period $[0, t]$, there was captured $j = j(t|A)$ units of prey. At the beginning of forging at $t = 0$, the probabilities $u = u(0,0 \mid A) = u(\mathfrak{x},0)$ of finding the prey in the point $x \in A$ of the patch A are specified by beta distribution with the parameters $\alpha > 0$ and $\beta > 0$ (cf. the parameters used in Example 1 to specify the conditions of search); so the probability density function $\mathfrak{f}(u)$ is (Papoulis & Pillai, 2002)

$$\mathfrak{f}(u) = \frac{1}{\mathrm{B}(\alpha,\beta)} u^{\alpha-1}(1-u)^{\beta-1}, \quad B(\alpha,\beta) = \int_0^1 u^{\alpha-1}(1-u)^{\beta-1}\,du, \quad 0 \le u \le 1; \tag{3.43}$$

the fact that $u = u(0, 0|A)$ is specified by the beta distribution is denoted by $u \sim \beta(\alpha, \beta)$.

Assume that the number of prey units in each patch A satisfies the binomial distribution with the parameters k and u. That is, if $u = u(t, j|A)$ is the probability of finding the prey in the point $\mathfrak{x} \in A$, then for each $t = 0, 1, 2, \ldots, k$, the probability $b_j(t|A)$ that the patch A of the size k contains j units of prey is defined as follows (Papoulis & Pillai, 2002):

$$b_j(t \mid A) = \binom{k}{j} u(t, j \mid A)^j \left[1 - u(t, j \mid A)\right]^{k-j}, \quad j = 0, 1, 2, \ldots, k, \tag{3.44}$$

where (see, e.g., DeGroot [1970, p. 160])

$$u(t, j \mid A) \sim \beta(\alpha + j, \beta + t - j). \tag{3.45}$$

The fact that the number of prey is defined by the binomial distribution is denoted by

$$n(t, j \mid A) \sim \mathrm{B}(k, u(t, j \mid A)). \tag{3.46}$$

Similar to the Oaten model, the goal of the forager is defined by Problem 3.2, and it is required to define when the forager has to leave the patch so that the expected capture rate $r_c(\vec{\mathfrak{d}})$ defined by Equation 3.14 reaches its maximum. Also, recall that for a given patch A, the capture rate $r_c(\vec{\mathfrak{d}} \mid A)$ for A is defined by Equation 3.13.

Denote by $n_{-t}^{t\ldots k}(t, j \mid A)$ the expected number of prey units captured after the time t and by $t_{-t}^{t\ldots k}(t, j \mid A)$ the expected residence time that will be spent in the patch after time t, both given that during the period [0, t] there were found $j(t|A)$ prey units and at time t the forager decides to continue in the patch A. Similarly, denote by $n_{-t}^{t\ldots\infty}(t, j \mid \mathfrak{X})$ the expected number of prey units captured after the time t over all patches and by $t_{-t}^{t\ldots\infty}(t, j \mid \mathfrak{X})$ the expected time of foraging in $\mathfrak{X}$, both given that overall during the period [0, t] there were found $j(t \mid \mathfrak{X})$ prey units.

Assume that the forager does not return to the patch. Then the expected capture numbers $n_{-t}^{t\ldots k}(t, j \mid A)$ and $n_{-t}^{t\ldots\infty}(t, j \mid \mathfrak{X})$ are defined as follows:

$$n_{-t}^{t\ldots k}(t, j \mid A) \sim B\left(k, u(t, j \mid A)\right), \tag{3.47}$$

$$n_{-t}^{t\ldots\infty}(t, j \mid \mathfrak{X}) \sim B\left(mk, u(t, j \mid A)\right), \tag{3.48}$$

where it is taken into account that since each patch is visited only once, the maximal time of foraging over all patches is mk. Given these values, the maximal times of foraging in the patches are equal to the size k of the patches, and the expected times $t_{-t}^{t\ldots k}(t, j \mid A)$ and $t_{-t}^{t\ldots\infty}(t, j \mid \mathfrak{X})$ are

$$t_{-t}^{t\ldots k}(t, j \mid A) = k - n_{-t}^{t\ldots k}(t, j \mid A), \tag{3.49}$$

$$t_{-t}^{t\ldots\infty}(t, j \mid \mathfrak{X}) = mk - n_{-t}^{t\ldots k}(t, j \mid \mathfrak{X}) + t_0. \tag{3.50}$$

Equations 3.47 through 3.50 follow the same approach as the equations suggested by Green (1980). However, in contrast to the Green recursive equations, here, the number m of patches is used, and the required values are drawn directly from the appropriate binomial distributions.

Using these equations, the strategy $\vec{\mathfrak{d}}$ of the forager can be defined according to the Charnov approach by the decision rules $\mathfrak{d}(t)$, which at each time $t = 0, 1, 2, \ldots, k$ implement the following conditions:

1. If $n_{-t}^{t\ldots k}(t, j \mid A) / t_{-t}^{t\ldots k}(t, j \mid A) > n_{-t}^{t\ldots\infty}(t, j \mid \mathfrak{X}) / t_{-t}^{t\ldots\infty}(t, j \mid \mathfrak{X})$, then check the next point in the current patch A.
2. If $n_{-t}^{t\ldots k}(t, j \mid A) / t_{-t}^{t\ldots k}(t, j \mid A) < n_{-t}^{t\ldots\infty}(t, j \mid \mathfrak{X}) / t_{-t}^{t\ldots\infty}(t, j \mid \mathfrak{X})$, then leave the patch A.
3. If $n_{-t}^{t\ldots k}(t, j \mid A) / t_{-t}^{t\ldots k}(t, j \mid A) = n_{-t}^{t\ldots\infty}(t, j \mid \mathfrak{X}) / t_{-t}^{t\ldots\infty}(t, j \mid \mathfrak{X})$, then choose arbitrarily either to leave A or to check the next point in A.

These conditions have the same form as the original Green conditions (Green, 1980) with the right sides calculated for each time t. For convenience, the Green method is outlined in the form of the following algorithm.

Algorithm 3.3: (The Green Foraging in Discrete Space and Time)

Given a domain $\mathfrak{X} \subset \mathbb{R}^2$ and a patchy environment $\mathcal{A}_{1\ldots m}^0 = \{A^0, A_1, A_2, \ldots, A_m\}$ with the patches' size k and beta-binomial prey distribution with the parameters α and β, transit time t_0 and zero transit cost do:

1. Mark all patches in the environment $\mathcal{A}_{1\ldots m}^0 \backslash A^0$ as not-visited and all points x in each patch $A \in \mathcal{A}_{1\ldots m}^0 \backslash A^0$ as not-checked.
2. Start with initial forager's location.

3. While not-visited patches remain do:
4. Choose a patch A from $\mathcal{A}^{0}_{1\dots m}\backslash A^{0}$ with respect to the current forager's location and transit time t_0.
5. Move to the chosen patch A and enter it.
6. Mark the patch A as visited.
7. Set $j = 0$.
8. For $t = 0, 1, 2, \dots, k$ do:
9. If all points in the patch A are checked, then break (leave the patch and continue with line 3).
10. Draw $u(t,j|A)\sim\beta(\alpha + j, \beta + t - j)$.
11. Draw $n_{-t}^{t\dots k}(t,j\,|\,A)\sim B\big(k,u(t,j\,|\,A)\big)$ and $n_{-t}^{t\dots\infty}(t,j\,|\,\mathfrak{X})\sim B\big(mk,u(t,j\,|\,A)\big)$
12. Calculate expected times $t_{-t}^{t\dots k}(t,j\,|\,A)$ and $t_{-t}^{t\dots\infty}(t,j\,|\,\mathfrak{X})$ using Equations 3.49 and 3.50.
13. If $n_{-t}^{t\dots k}(t,j\,|\,A)/t_{-t}^{t\dots k}(t,j\,|\,A) < n_{-t}^{t\dots\infty}(t,j\,|\,\mathfrak{X})/t_{-t}^{t\dots\infty}(t,j\,|\,\mathfrak{X})$ (condition 2), then break (leave the patch A and continue with line 3).
14. If $n_{-t}^{t\dots k}(t,j\,|\,A)/t_{-t}^{t\dots k}(t,j\,|\,A) = n_{-t}^{t\dots\infty}(t,j\,|\,\mathfrak{X})/t_{-t}^{t\dots\infty}(t,j\,|\,\mathfrak{X})$ (condition 3), then choose randomly either to break (leave the patch A and continue with line 3) or to continue foraging in the patch A (that is, to continue with the next line 15).
15. Choose not-checked point $\breve{\mathfrak{x}}(t)\in A$ and move to it (jumps are allowed).
16. Check the point $\breve{\mathfrak{x}}(t)$ and mark it as checked.
17. If the prey unit is found in $\breve{\mathfrak{x}}(t)$, then set $j = j + 1$;
18. End for (over t).
19. End while (over the not-visited patches).

The actions of the Green algorithm of foraging in discrete space and time are illustrated by the following example.

Example 3.4

Assume that similar to the previous examples, the forager acts in the square domain $\mathfrak{X}$ of the size $n = 100 \times 100$ and that there are five circle patches $A_1,\dots,A_5$ with the same aforementioned coordinates of the centers. However, here, following the requirements of the algorithm, it is assumed that all patches have the same radius 10. Thus, the size of each patch is $k = 314$. Let the parameters of the beta distribution be $\alpha = \beta = 0.5$ and the expected transit time be $t_0 = 0$. As earlier, the starting point of the forager is $\breve{\mathfrak{x}}(0) = (40,5)$. The next patch is chosen by maximum number of the remained prey number in the patch divided by the distance from the current forager's location to the patch. The results of the simulation are shown in Figure 3.8.

The forager starts in the initial point $\breve{\mathfrak{x}}(0) = (40,5)$, chooses the patch A_1, and flies to it. In this patch, the forager checks four points and after finding one unit of prey leaves the patch for the patch A_4. After checking six points and finding three units of prey, the forager flies to the patch A_2 and then immediately to the patch A_5. In this patch, the forager checks two points and since no prey was found, it leaves the patch for the last not-visited patch A_3, where the forager checks eight points and finds one unit of prey. Then, the forager decides to leave this patch, and since all five patches were already visited, the foraging terminates. ■

Notice that the Green Algorithm 3.3 of foraging follows the assumptions, which are widely used in the group-testing methods (Sobel & Groll, 1958); for more information and a review of the group-testing algorithms, see the books of Du and Hwang (1993) and Kagan and Ben-Gal (2013). However, in contrast to the group-testing methods, this algorithm specifies the leaving times and results in the trajectory of the forager. In 1982, Stephens and Charnov in their paper (Stephens & Charnov, 1982) summarized the considered methods and suggested additional stochastic models of optimal foraging; in the general framework of optimal foraging, such models are discussed in the book by Stephens and Krebs (1986). These results formed a basis for the attempts to develop unified approach for modeling optimal foraging based on Markov decision processes

At the beginning of foraging

Patch	A_1	A_2	A_3	A_4	A_5
Prey units	63	48	7	82	63

At the end of foraging

Patch	A_1	A_2	A_3	A_4	A_5
Checking points	4	1	8	6	2
Prey found	1	0	1	3	0

(a)

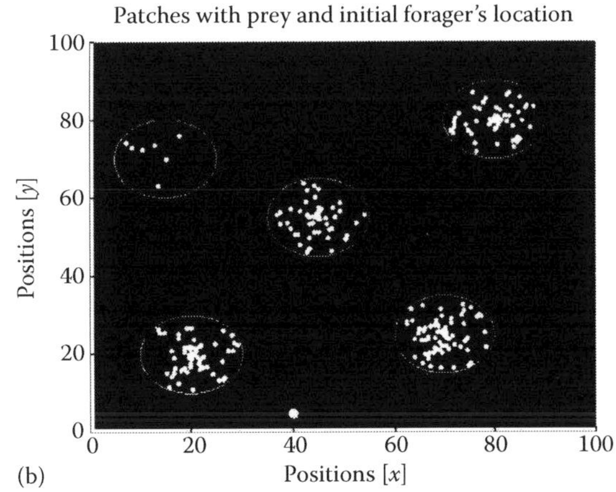

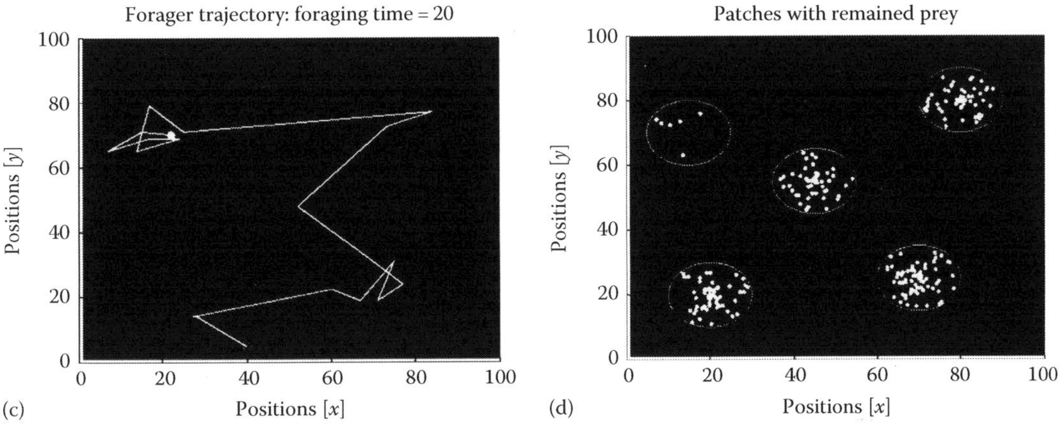

FIGURE 3.8 (a) Numbers of prey units in the patches at the beginning and at the end of foraging; (b) patches with initially distributed prey (white points) and initial forager's location $\breve{x}(0) = (40, 5)$ (white circle); (c) forager's trajectory by the Green algorithm; (d) patches with the prey that remained after foraging in all patches.

(Mangel & Clark, 1986) and allowed to consider optimal foraging using general optimization techniques. The further studies of optimal foraging methods in both deterministic and stochastic setups resulted in unforeseen applications like information foraging (Pirolli, 2007) and algorithms of search planning and navigation of autonomous mobile agents (Chernikhovsky, Kagan, Goren, & Ben-Gal, 2012; Kagan, Goren, & Ben-Gal, 2010, 2012). In Section 4.4, such algorithms will be considered in details. Additional applications of the optimal foraging techniques are indicated in the recent books of Stephens et al. (2007) and Viswanathan, da Luz, Raposo, and Stanley (2011).

3.3.3 Preying Model and Pursuit-Evasion Game

The considered algorithms of foraging by patches follow an assumption that the predator catches the prey immediately after its finding. However, in many cases, such behavior not exactly corresponds to the observed behavior of natural foragers; by different reasons, the predator can avoid attacking the found prey and can continue foraging in the current patch or move to the other patch. This section briefly describes the reasoning, which is used in the considerations of such decision making, and demonstrates its relation with the methods of search games. Since the decisions regarding chasing or omitting the found prey are not in the scope of the book, the following discourse is informal and is aimed to give some key ideas rather than to present a complete theory.

The description of the prey model follows the book by Stephens and Krebs (1986). Assume that the predator hunts to the prey items of k types $i = 1, 2, \ldots, k$. Notice the type i of prey differs from the type patch, which is out of consideration of the prey model.

Denote by $\eth_i = \eth(i) \in [0,1]$ the decision rules, which are associated with the probabilities of attacking the encountered prey item of the type i or leaving it and continuing search. Let $\eta_i = \eta(i)$ be the rate at which the predator encounters the prey of type i during the foraging (i.e., before starting to decide whether to attack the prey or not). Also, let $t_i = t(i)$ be an expected time of chasing the prey of type i if the predator decided to attack it and $\xi_i = \xi(i)$ be an expected net energy, which will be gained from the prey of type i if it will be caught.

Then the problem of preying is to find a policy $\vec{\eth} = \eth_1, \eth_2, \ldots, \eth_k$ such that, similar to the Charnov model (see Section 3.3.1), it maximizes the net energy intake rate (Stephens & Krebs, 1986):

$$r_e(\vec{\eth}) = \left(\sum_{i=1}^{k} \eth_i \eta_i \xi_i\right) \Bigg/ \left(1 + \sum_{i=1}^{k} \eth_i \eta_i t_i\right). \tag{3.51}$$

The model is evaluated by the same manner as the Charnov model. Given the type i, the net energy intake rate $r_e(\vec{\eth} \mid i)$ is defined as follows:

$$r_e(\vec{\eth} \mid i) = \frac{(\eth_i \eta_i \xi_i + \alpha_{-i})}{(\eth \eta_i t_i + \beta_{-i})}, \tag{3.52}$$

where

$$\alpha_{-i} = \sum_{j=1, j \neq i}^{k} \eth_j \eta_j \xi_j \quad \text{and} \quad \beta_{-i} = 1 + \sum_{j=1, j \neq i}^{k} \eth_j \eta_j t_j$$

combine the terms corresponding to the types different from i.

Consider the dependence of the expected rate $r_e(\vec{\eth} \mid i)$ on the probability $\eth_i$. By differentiation of $r_e(\vec{\eth} \mid i)$ with respect to $\eth_i$, one obtains

$$\frac{\partial}{\partial \eth_i} r_e(\vec{\eth} \mid i) = \frac{\eta_i}{(\eth_i \eta_i t_i + \beta_{-i})^2} (\xi_i \beta_{-i} - t_i \alpha_{-i}). \tag{3.53}$$

The sign of the obtained derivative does not depend on the value of $\eth_i$, and it maximizes the expected rate $r_e(\vec{\eth} \mid i)$ if it obtains either the largest feasible value $\eth_i = 1$ or the smallest feasible value $\eth_i = 0$. Such property of the probabilities $\eth_i$ is known as the zero–one rule. The choice between these values is specified according to Equation 3.53 as follows (Stephens & Krebs, 1986):

1. If $\xi_i \beta_{-i} - t_i \alpha_{-i} < 0$, then set $\eth_i = 0$, that is, do not attack the prey item of the type i and continue searching for the prey of the other type.
2. If $\xi_i \beta_{-i} - t_i \alpha_{-i} > 0$, then set $\eth_i = 1$, that is, start chasing the prey item of the type i.
3. If $\xi_i \beta_{-i} - t_i \alpha_{-i} = 0$, then choose arbitrarily either to attack the prey of the type i or to continue search.

Finally, notice that the types of the prey items are assumed to be known to the predator. Thus, its diet can be defined before starting the foraging. Then the algorithm, which specifies this diet, is the following (Stephens & Krebs, 1986):

1. Order the k prey types such that for the expected net energy ξ_i and the expected chasing time t_i, it holds true that $\xi_1/t_1 \geq \xi_2/t_2 \geq \cdots \geq \xi_k/t_k$.
2. Add the types to the diet until $\left(\sum_{j=1}^{i} \eta_j \xi_j\right) \Big/ \left(1 + \sum_{j=1}^{i} \eta_j t_j\right) > \xi_{i+1}/t_{i+1}$, $i = 1, 2, \ldots, k-1$.

Notice that this algorithm follows the same general approach as the methods of search for a static target presented in Section 2.2.1, especially the Stone Algorithm 1.

Now, let us consider this problem from the other point of view. Assume that the predator encounters the prey item of the type i. Then, given the predator's abilities and potential benefit ξ_i of catching the prey, it is required to estimate the time t_i of chasing the prey and a risk of its escape with respect to the potential benefits and risks of omitting the chase and continuing foraging. In other words, the predator is in the situation of the pursuit–evasion game against the prey (Hajek, 1975; Petrosjan, 1993), and the value of this game, that is, $\xi_i\beta_{-i} - t_i\alpha_{-i}$, forms a basis for the predator decision. Some ideas of such games in the form of the search games (Alpern & Gal, 2003; Gal, 1980; Garnaev, 2000) were already introduced in Section 1.3, and a simple one-step game of search was considered in Example 1.

As earlier, assume that the foraging is conducted in the 2D domain $\mathfrak{X} \subset \mathbb{R}^2$ and the points $\mathfrak{x} \in \mathfrak{X}$ are defined by the pairs (x,y) of coordinates. The predator (or searcher) location at time t is denoted by $\breve{\mathfrak{x}}(t) \in \mathfrak{X}$ and the prey (or target) location at time t is denoted by $\hat{\mathfrak{x}}(t) \in \mathfrak{X}$. Then, $d\breve{\mathfrak{x}}/dt$ and $d\hat{\mathfrak{x}}/dt$ are, correspondingly, the velocity vectors of the predator and the prey, and $|d\breve{\mathfrak{x}}/dt|$ and $|d\hat{\mathfrak{x}}/dt|$ are their speeds, which are nonnegative and are upper bounded as $|d\breve{\mathfrak{x}}/dt| \leq \breve{v}$ and $|d\hat{\mathfrak{x}}/dt| \leq \hat{v}$, where $\breve{v} \geq 0$ and $\hat{v} \geq 0$ are constants. Since the pursuit–evasion game begins when the predators encounters the prey, the time $t = 0$ is considered as a starting time of the game. The locations of the predator and the prey at this time are $\breve{\mathfrak{x}}(0)$ and $\hat{\mathfrak{x}}(0)$, respectively.

Assume that the predator recognizes the prey before capturing it, that is, at the time $t = 0$, the distance between the predator and the prey is $\|\breve{\mathfrak{x}}(0) - \hat{\mathfrak{x}}(0)\| > 0$, while the capture holds if for some t it holds true that $\|\breve{\mathfrak{x}}(t) - \hat{\mathfrak{x}}(t)\| = 0$. It is clear that if the prey is able to move faster than the predator, that is, $\breve{v} < \hat{v}$, then for any initial positions $\breve{\mathfrak{x}}(0) \neq \hat{\mathfrak{x}}(0)$, it can avoid the capture, and if the predator's possible speed is greater than the possible speed of the prey, that is, $\breve{v} > \hat{v}$, then the strategy of the predator, which leads to capturing the prey, always exists.

Let $\breve{v} > \hat{v}$, and denote by $\breve{f} : \mathbb{R} \to \mathbb{R}^2$ and $\hat{f} : \mathbb{R} \to \mathbb{R}^2$ the control functions, which specify the movement of the predator and the prey, respectively. Then, the pursuit–evasion game is defined by the following equations (Hajek, 1975; Petrosjan, 1993):

$$\frac{d}{dt}\breve{\mathfrak{x}}(t) = \breve{f}(t), \quad \frac{d}{dt}\hat{\mathfrak{x}}(t) = \hat{f}(t) \tag{3.54}$$

where $|\breve{f}(t)| \leq \breve{v}$ and $|\hat{f}(t)| \leq \hat{v}$ for all $t \in [0,\infty)$. Then, for the distance $r(t) = \|\breve{\mathfrak{x}}(t) - \hat{\mathfrak{x}}(t)\|$, it holds true that

$$r(t) \leq r(0) - (\breve{v} - \hat{v})t = \breve{\mathfrak{x}}(0) - \hat{\mathfrak{x}}(0) - (\breve{v} - \hat{v})t. \tag{3.55}$$

Thus, the capture, that is, the locations of the predator and the prey such that $r(t) = 0$, occurs at some time τ, which is upper bounded as follows:

$$\tau \leq \|\breve{\mathfrak{x}}(0) - \hat{\mathfrak{x}}(0)\| / (\breve{v} - \hat{v}). \tag{3.56}$$

This inequality does not depend on the control functions $\breve{f}$ and $\hat{f}$, and, given the recognition distance $\text{r}(0)$ and maximal speed $\breve{\text{v}}$ of the forager, it provides an upper bound for the estimated time t_i of chasing the prey of type i that can move with maximal speed $\hat{\text{v}}_i$. Then, using the rates η_i of encountering the prey items of the types $i = 1, 2, \ldots, k$ and their expected net energies ξ_i, the forager is able to estimate the values $\xi_i\beta_{-i} - t_i\alpha_{-i}$. As a result, the forager obtains the required policy $\vec{\eth}$, which specifies whether the forager has to start chasing the prey of the type i or to leave it and maximizes the net energy intake rate $r_c(\vec{\eth})$.

The presented brief consideration demonstrates how the simple pursuit–evasion game provides the data, which can be incorporated into the foraging process. Detailed information regarding the pursuit–evasion games is presented in the books by Hajek (1975) and by Petrosjan (1993), and some additional considerations of the game theory methods in the framework of foraging theory are presented in a recent book of Stephens et al. (2007).

3.4 INFERENCES AND RESTRICTIONS

As indicated both in the Introduction (Section 1.4) and in the beginning of this chapter (Section 3.1), the theory of optimal foraging starts with the postulation that the animals forage optimally and then considers biologically feasible parameters, for which the optimization problem is formulated. Following this methodology, as indicated by Krebs and Kacelnic (Stephens et al. 2007, pp. xi–xii), Foraging theory has moved from simply testing the predictions of one kind of model to comparing the ability of a range of models to explain the data.

> At the same time, each model requires a priori specification of the implemented assumptions regarding the goals, available knowledge, and desired behavior of the animals.

However, as it is stressed by Pyke (1984), the assumptions such as "the currency of fitness, the information possessed by the animal, and any behavioral constraints" are often not justified. In addition, crucial to most of the models, "encounter rates with food types" should be determined "from the animal's point of view," which is certainly impossible. From this perspective, Pyke formulated eight criteria, which has to be satisfied by the acceptable models of the foraging behavior and then concludes (Pyke, 1984, p. 554):

> Unfortunately, there appear to be no studies that satisfy all of these criteria.

And the variety of models developed in the theory of optimal foraging during the last decades (Stephens et al., 2007) demonstrates that such a situation stays still.

In general, the models applied in the theory of optimal foraging and the algorithms considered earlier in particular are developed as follows: Given the certain living forager, at first, the biologically feasible parameters with corresponding assumptions are specified; then the appropriate optimization problem is formulated and solved; and, finally, the obtained solution is compared with the observable behavior of the considered forager. If the obtained solution is, in certain sense, equivalent to the observed behavior, then the model is accepted and is used for further predictions and explanations; otherwise, the parameters and assumptions are revised, and the consideration starts form the beginning using new parameters and assumptions.

Hence, the modeling in the foraging theory follows two general stages. The first stage is the same as it is used in the search and screening theory (see Section 3.2), where the appropriate parameters of the artificial agent are defined and the optimization problem is solved with respect to the desired behavior of the agent. The difference is that instead of the artificial and programmable agent, in the foraging theory, an abstract animal with certain assumed and estimated characteristics is used. The second stage, in contrast, is specific to the foraging theory and is based on the postulated optimality of the observed behavior of the living forager, with which the obtained solution is compared and is used for corrections of the model.

It is clear that in spite of the variety of fruitful results, such development of the models of foraging is constrained by the contemporary knowledge on the animals foraging behavior and consequent assumptions. That, on one hand, restricts the heuristic and predictive abilities of the models and, on the other hand, can result to erroneous reasoning led by the unjustified assumptions. In the next chapter, the models and algorithms of foraging are specified in opposite direction: the modelling starts with the observed behavior of the foraging animals and then an appropriate formal description of this behavior is developed.

3.5 SUMMARY

The foraging theory considers the problem of search from the other point of view. Namely, in this theory, it is assumed that the animals, which seek for food, act optimally, and then the research efforts are concentrated on the description of the animals movements in the given conditions.

General description of the animal's behavior with respect to the number of prey is provided by the models of spatial dynamics of populations (Section 3.2). There are several basic models of such dynamics, each of which defines certain Fokker–Planck equation that governs the probabilities of the animal's location in the points of its habitat (see, e.g., Equation 3.4 for Okubo model). Combination of the equations governing prey and predator results in the prey–predator system (see Equation 3.7), which forms a basis for general models of the search and foraging processes.

The main approach to the consideration of the animals' trajectories is foraging by patches, in which it is assumed that the prey is distributed in certain areas (patches) and the forager should decide when to stop searching in the current patch and to leave it for another one. The problem is usually considered in two settings: deterministic setting, in which the forager is informed about the patches and the quantity of prey in each patch, and stochastic setting, in which the forager is aware only about the probabilities of finding certain quantity of prey in each patch:

1. Deterministic foraging (Section 3.3.1) is based on the Charnov marginal value theorem, which determines conditions for maximization of the energy gain rate that is the ratio of the overall gained energy given by the energy required for movements in the patches and between them. The implementation of this theorem is given in the form of Algorithm 3.1.
2. Stochastic foraging (Section 3.3.2) deals with the same problem of finding the optimal leaving or, that is, the same, the optimal times of seeking in the patches, such that the gain rate is maximal. The solution of the problem is provided by the Oaten scheme, which is implemented in the form of Algorithm 3.2 and then extended by McNamara (see Example 3.3). The most general model of stochastic foraging by patches was suggested by Green (for the implementation of this model, see Algorithm 3.3 of foraging in discrete space and time).

The problem, which is closely related with the optimal foraging theory, is the problem of optimal preying (Section 3.3.3), which addresses the hunting process. Usually, this problem is considered in the framework of pursuit–evasion games, which define the conditions of successful hunting.

As it follows from the methods of optimal foraging, this theory assumes strong sensing abilities of the foragers and a wide knowledge about the environment that seriously restrict direct implementation of the theory for navigation and control of artificial mobile agents. Nevertheless, the algorithms and optimization criterions used in the models of optimal foraging provide useful insights to the development of bioinspired algorithms of search and foraging.

4 Models of Individual Search and Foraging

In the previous chapters, the trajectories of the searchers and foragers are obtained as a solution of certain optimization problem, and depend on the implemented constraints and assumptions. This chapter addresses the search and foraging in opposite direction: it starts with the class of trajectories, which are postulated as optimal, and then considers feasible models and algorithms, which result in such trajectories. The systematic considerations of such approach were initiated by Viswanathan et al. (1996), and during the last two decades, it was successfully applied in biological and ecological studies. The current chapter presents the ideas and methods that are used in this approach and considers their relation with the probabilistic algorithms of search and screening.

4.1 MOVEMENTS OF THE AGENTS AND THEIR TRAJECTORIES

In the above presented methods and algorithms, the trajectories of the searching and foraging agents were considered as direct results of the algorithms of search and path planning (Chapter 2) and as traces of the agent foraging in patchy environment (Chapter 3). According to the task, the obtained trajectories satisfy certain optimality criteria: maximal probability of detection of the target or minimal search time up to certain detection of the target in the case of search and screening problems or maximal expected intake energy rate in the case of foraging.

The other approach to the search and foraging problems follows the opposite direction. The consideration starts with the observed trajectories of the foraging animals, birds, or insects, which, with respect to their abilities and habitat, are postulated as evolutionary optimal foragers (Pyke, 1984), and then the formal as simple as possible model, which allows such trajectories, is developed and analyzed (Calengea, Draya, & Royer-Carenzia, 2009; Turchin, 1998).

Certainly, different living organisms follow different paths, which are described by different stochastic processes. In the simplest modeling case, the movement of the forager is considered as a Brownian random walk (Hoffmann, 1983), which in the framework of search and screening theory results in the Koopman random search formula (see Section 2.2.1; cf. the Oaten model in Section 3.3.2). The movements of the foraging ants usually are modeled by the correlated Brownian motion, which demonstrates a good correspondence with experimental results (Bovet & Benhamou, 1988). In contrast, Viswanathan et al. (1996) argued that the wander of the albatross has better correspondence with the Lévy flights, and the same is possibly true for the flying insects. The examples of the trajectories resulted by the indicated processes are shown in Figure 4.1.

For illustrative purposes, all shown trajectories both simulated and observed include 1000 points with the coordinates normalized for the square arena 100 × 100 units. In Brownian motion (Figure 4.1a), the step length is 3 units. In the Lévy flight (Figure 4.1b), parameter $\mu = 1.6$, minimal step length is 1 unit, and maximal step length is 140 units that is a rounded value of the arena diagonal. In the correlated Brownian motion (Figure 4.1c), the correlation coefficient between the consequent directions of the steps is 0.5. The trajectory of the ant (Figure 4.1d) was recorded in our laboratory experiments with the *Linepithema humile* (Argentinian) ants.

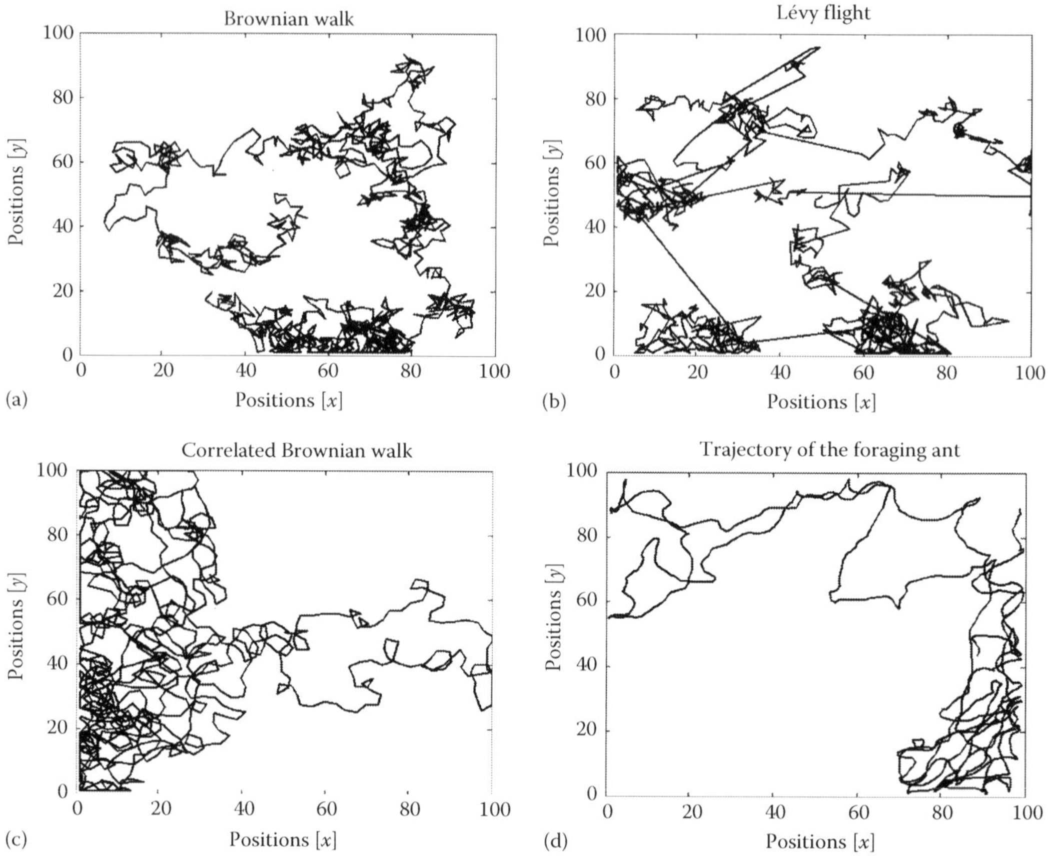

FIGURE 4.1 Trajectories of the agents defined (a) by the simple Brownian walk, (b) by the Lévy flight, (c) by correlated Brownian walk and (d) the observed trajectory of the *L. humile* worker ant.

Notice that "in large," the trajectory specified by the Lévy flight (Figure 4.1b) looks similar to the trajectories generated by the algorithms of foraging in the patchy environment considered in Section 3.3. However, in more close resolution, it is seen that the Lévy flight trajectory is scale invariant, while in the trajectory of foraging in the patchy environment is not, and its long distance jumps are defined by the location of the patches.

The Lévy flights demonstrate good correspondence with the long distance wanderings. In addition, in the models, which consider the short distance movements and, especially, the movements of the animals and insects, the dependence on the environment is defined, or certain mixed models, which combine different strategies of search and foraging, are used.

The indicated methods are developed at most in the framework of contemporary foraging theory (Viswanathan, da Luz, Raposo, & Stanley, 2011) and are aimed to provide the formal models for the observed movements of living organisms. At the same time, the similar trajectories of the search agents are specified by the methods of the real-time probabilistic search and path planning, which were developed in the framework of search and screening theory (Kagan & Ben-Gal, 2013). The goal of the discourse below is to consider the contemporary methods used in the foraging theory and search theory and to present the search and foraging algorithms, which implement these methods. For detailed overview of general methods, which are used for modeling spatial trajectories, see, for example, recent publication by Brillinger (2010).

4.2 BROWNIAN SEARCH AND FORAGING

As indicated earlier, the trajectories of the agents can be defined by different stochastic processes with respect to the considered problem. This section presents historically the first and the most general approach based on Brownian walks that provides a framework for consideration of foraging behavior of a wide range of species acting in their habitats and forms a basis for many search and screening methods. In addition, this approach is closely related with the considered methods of population dynamics in Section 3.2; the discourse below stresses this relation.

4.2.1 Simple Brownian Search

The simple Brownian search is a variant of the considered sequential search in Section 2.2.3 with single-point observed areas. The search strategy implies the equiprobable choice of the next location from the set of the neighboring points of the current location of the searcher. In the case of discrete time and space, such choice results in the random walk of the searcher, and in the case of continuous time and space, it defines simple Brownian motion. In the previous sections, such motion was formally applied for definition of the target's motion (Section 2.3.3) and was mentioned as the simplest policy of the searcher (Sections 3.2 and 3.3). The discourse below briefly recalls the main ideas of the Brownian motion on a plane; additional information on random walks and Brownian motion can be found in the books of Feller (1970–1971), Hughes (1995–1996), and Shiryaev (1995). Modern theory of Brownian motion is presented in recent books by Morters and Peres (2010) and by Lawler and Limic (2010), and the detailed consideration of the relation between the random walk model and diffusion equation was published by Lawler (2010).

Recall (see Section 3.2) that the function $v : \mathfrak{X} \times [0,T] \to \mathbb{R}^+$ stands for the probability density function (pdf) such that $v(\mathfrak{x},t)d\mathfrak{x}$ is a probability that the searcher is located in the neighborhood $[\mathfrak{x}, \mathfrak{x}+d\mathfrak{x})$ of the point $\mathfrak{x}$ at time $t \in [0,T]$, and $v(\mathfrak{x},t)dt$ is a probability that the searcher is located in the point $\mathfrak{x}$ during the period $[t, t+dt)$. In the case of discrete domain $\mathfrak{X} = \{\mathfrak{x}_1, \mathfrak{x}_2, \ldots, \mathfrak{x}_n\}$, the function v is a probability mass function, and $v(\mathfrak{x}_i,t)$ is a probability that the searcher is located in the point $\mathfrak{x}_i$ at time t, $t = 0, 1, 2,\ldots,T$. Since it is assumed that the searcher is certainly in the domain $\mathfrak{X}$, it holds true that $\int_{\mathfrak{X}} v(\mathfrak{x},t)\,d\mathfrak{x} = 1$ for each $t \in [0,T]$; for the discrete domain and discrete time, this condition is that $\sum_{i=1}^{n} v(\mathfrak{x}_i,t) = 1$ for each $t = 0, 1, 2,\ldots,T$.

4.2.1.1 Random Walk in Discrete Domain

Assume that the searcher acts in the gridded discrete domain $\mathfrak{X} = \{\mathfrak{x}_1, \mathfrak{x}_2, \ldots, \mathfrak{x}_n\}$, $n \in \mathbb{N}$, with the points $\mathfrak{x}_i = (i_x, i_y)$, $i_x = 1, 2,\ldots, n_x$, and $i_y = 1, 2,\ldots, n_y$ such that $n = n_x \times n_y$ and $i = ((i_x - 1)n_x + i_y)$. Being located at time t in the point $\breve{\mathfrak{x}}(t) = (i_x, i_y) \in \mathfrak{X}$ of the domain $\mathfrak{X}$, it observes this point and according to observation result $z(\{\breve{\mathfrak{x}}(t)\}) \in [0,1]$ decides whether to terminate or to continue searching. If the search continues, then the searcher chooses its next location $\breve{\mathfrak{x}}(t+1)$ from the set $\mathfrak{n}(\breve{\mathfrak{x}}(t)) \subset \mathfrak{X}$ of the neighboring points of $\breve{\mathfrak{x}}(t)$ such that $\mathfrak{n}(\breve{\mathfrak{x}}(t)) = \{\mathfrak{x} : |\mathfrak{x} - \breve{\mathfrak{x}}(t)| = 1\}$. In particular, if $\breve{\mathfrak{x}}(t) = (i_x, i_y)$ is an internal point of $\mathfrak{X}$, then (cf. Equation 2.82)

$$\mathfrak{n}\left(\breve{\mathfrak{x}}(t)\right) = \mathfrak{n}\left((i_x, i_y)\right) = \left\{(i_x+1, i_y), (i_x-1, i_y), (i_x, i_y+1), (i_x, i_y-1)\right\}. \tag{4.1}$$

If the point $\breve{\mathfrak{x}}(t)$ lies at the boundary of the domain, then the neighborhood $\mathfrak{n}(\breve{\mathfrak{x}}(t))$ includes three points, two boundary points and one internal point: if $\breve{\mathfrak{x}}(t)$ is the corner point, then $\mathfrak{n}(\breve{\mathfrak{x}}(t))$ includes only two boundary points. The possible steps of the searcher are illustrated in Figure 4.2.

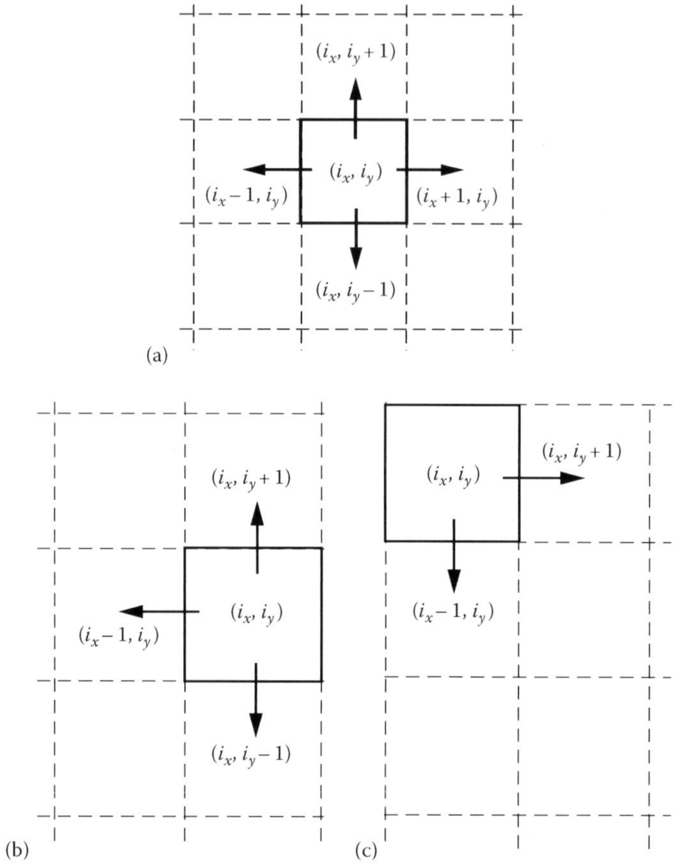

FIGURE 4.2 Possible steps of the randomly walking searcher: (a) current location is an internal point; (b) current location is at the boundary; (c) current location is in the corner.

Consider the internal point of the domain. Similar to the transition probabilities of the moving target (see Section 2.3.2), denote by $\breve{\rho} = \|\breve{\rho}_{ij}\|_{n\times n}$ the transition probabilities matrix of the searcher, where

$$\breve{\rho}_{ij} = Pr\{\breve{\mathfrak{x}}(t+1) = \mathfrak{x}_j \mid \breve{\mathfrak{x}}(t) = \mathfrak{x}_i\}, \quad i, j = 1, 2, \ldots, n, \quad t = 0, 1, 2, \ldots, \tag{4.2}$$

and $\breve{\mathfrak{x}}(t)$ and $\breve{\mathfrak{x}}(t+1)$ are locations of the searcher at the times t and t + 1, correspondingly. In the simple random walk, the points from the neighborhood $\mathfrak{n}(\breve{\mathfrak{x}}(t))$ are chosen with equal probabilities that are defined as follows:

$$\breve{\rho}_{ij} = Pr\{\breve{\mathfrak{x}}(t+1) = \mathfrak{x}_j \mid \breve{\mathfrak{x}}(t) = \mathfrak{x}_i\} = \begin{cases} 1/|\mathfrak{n}(\mathfrak{x}_i)| & \text{if } \mathfrak{x}_j \in \mathfrak{n}(\mathfrak{x}_i), \\ 0 & \text{otherwise,} \end{cases} \quad i, j = 1, 2, \ldots, n, \quad t = 0, 1, 2, \ldots, \tag{4.3}$$

where $|\mathfrak{n}(\mathfrak{x}_i)|$ stands for the number of points in the neighborhood $\mathfrak{n}(\mathfrak{x}_i)$ of $\mathfrak{x}_i$; as indicated earlier, for the internal point $\mathfrak{x}$ of 2D domain, $|\mathfrak{n}(\mathfrak{x})| = 4$.

Denote the neighboring points of $\mathfrak{x}$ by $\mathfrak{x} \pm 1_x = (i_x \pm 1, i_y)$ and $\mathfrak{x} \pm 1_y = (i_x, i_y \pm 1)$. Then, the probability $v(\mathfrak{x}, t)$ that at time t the searcher is located in the point $\mathfrak{x}$ is the following:

$$v(\mathfrak{x},t) = v(\mathfrak{x} \pm 1_x, t-1) Pr\{\breve{\mathfrak{x}}(t) = \mathfrak{x} \mid \breve{\mathfrak{x}}(t-1) = \mathfrak{x} \pm 1_x\}$$
$$+ v(\mathfrak{x} \pm 1_y, t-1) Pr\{\breve{\mathfrak{x}}(t) = \mathfrak{x} \mid \breve{\mathfrak{x}}(t-1) = \mathfrak{x} \pm 1_y\}. \quad (4.4)$$

If the neighboring points $\mathfrak{x} \pm 1_x = (i_x \pm 1, i_y)$ and $\mathfrak{x} \pm 1_y = (i_x, i_y \pm 1)$ of the internal point $\mathfrak{x}$ are also internal, then $|\mathfrak{n}(\mathfrak{x})| = |\mathfrak{n}(\mathfrak{x} \pm 1_x)| = |\mathfrak{n}(\mathfrak{x} \pm 1_y)| = 4$ and

$$v(\mathfrak{x},t) = \frac{1}{|\mathfrak{n}(\mathfrak{x})|}\left(v(\mathfrak{x} \pm 1_x, t-1) + v(\mathfrak{x} \pm 1_y, t-1)\right) = \frac{1}{4}\left(v(\mathfrak{x} \pm 1_x, t-1) + v(\mathfrak{x} \pm 1_y, t-1)\right). \quad (4.5)$$

In general, this equation usually is written in the form

$$v(\mathfrak{x},t) = \frac{1}{2d} \sum_{\mathfrak{x}' \in \mathfrak{X}, |\mathfrak{x}' - \mathfrak{x}| = 1} v(\mathfrak{x}', t-1), \quad (4.6)$$

where d is a dimension of the domain $\mathfrak{X}$; in the considered case of $\mathfrak{X} \subset \mathbb{R}^2$, it is $d = 2$.

Let us define the discrete Laplace operator V (Lawler, 2010):

$$V(v(\mathfrak{x},t)) = \frac{1}{2d} \sum_{\mathfrak{x}' \in \mathfrak{X}, |\mathfrak{x}' - \mathfrak{x}| = 1} \left(v(\mathfrak{x}',t) - v(\mathfrak{x},t)\right); \quad (4.7)$$

in the considered 2D case, where $d = 2$, this operator is

$$V(v(\mathfrak{x},t)) = \frac{1}{4}\left(v(\mathfrak{x} \pm 1_x, t) - 4v(\mathfrak{x},t) + v(\mathfrak{x} \pm 1_y, t)\right). \quad (4.8)$$

Finally, denote by

$$\partial_t v(\mathfrak{x},t) = v(\mathfrak{x},t) - v(\mathfrak{x},t-1) \quad (4.9)$$

the difference between the probabilities of the searcher's location in the point $\mathfrak{x}$ at time t and the previous time $t - 1$. Then, the discrete variant of the Fokker–Planck equation (3.6) with unit diffusion rate and zero drift is defined as follows (Lawler, 2010):

$$\partial_t v(\mathfrak{x},t) = V(v(\mathfrak{x},t)). \quad (4.10)$$

This equation, which is known as the discrete diffusion equation, specifies the movement of the searcher according to the simple random walk. In addition, notice that the operator defined by Equation 4.8 is a discrete representation of the operator $V\left(v; \left\|\breve{s}(\mathfrak{x},t,u,v)\right\|, \left\|\breve{m}(\mathfrak{x},t,u,v)\right\|\right)$, which specifies the searcher's motion in the system "prey–predator" with unit diffusion rate $\left\|\breve{s}(\mathfrak{x},t,u,v)\right\| = I$ and zero drift $\left\|\breve{m}(\mathfrak{x},t,u,v)\right\| = 0$ (see Section 3.2). In the terms of finite differences, Equation 4.8 defines a finite differences scheme at the node of 2D network with the step of two units in both directions.

Denote by S_t^d the d-dimensional random walk, which follows Equation 4.10. The searcher's policy according to the considered 2D random walk S_t^2 is summarized as follows:

Algorithm 4.1: (The Simple Random Walk Search)

Given a domain $\mathfrak{X} = \{\mathfrak{x}_1, \mathfrak{x}_2, \ldots, \mathfrak{x}_n\}$, do:

1. Observe the current location $\breve{\mathfrak{x}}(t)$ and decide either to terminate the search or to continue.
2. If the search continues, choose as the next location $\breve{\mathfrak{x}}(t+1)$ one of the neighboring points $\mathfrak{x} \in \mathfrak{n}(\breve{\mathfrak{x}}(t))$ arbitrary with the same probabilities.
3. Move to the chosen point $\breve{\mathfrak{x}}(t+1)$ and continue with its observation (line 1).

These actions define the trajectory $\vec{\mathfrak{a}} = \langle \breve{\mathfrak{x}}(0), \breve{\mathfrak{x}}(1), \breve{\mathfrak{x}}(2), \ldots \rangle$ of the searcher in the domain starting from the initial point $\breve{\mathfrak{x}}(0)$ at $t = 0$, and the probabilistic dynamics of the searcher is defined by the diffusion equation (4.10). Solution of this equation is obtained as follows.

Consider the 1D random walk S_t^1 with unit steps; usually for such random walk, the upper index is dropped and it is denoted by S_t. In such a case, Equation 4.10 becomes

$$\partial_t v(\mathfrak{x}, t) = \frac{1}{2}\left(v(\mathfrak{x} + 1_x, t) + v(\mathfrak{x} - 1_x, t)\right) = \frac{1}{2}\left(v(i+1, t) + v(i-1, t)\right). \tag{4.11}$$

Denote by ζ a Bernoulli random variable that obtains the values -1 and 1 with the probabilities

$$Pr\{\zeta = 1\} = p \quad \text{and} \quad Pr\{\zeta = -1\} = 1 - p = q; \tag{4.12}$$

and by $\zeta(t)$ the value of the variable ζ at time t. The steps of the searcher are illustrated in Figure 4.3.

Assume that the searcher starts at the internal point $\breve{\mathfrak{x}}(0)$ and the coordinates are defined in such a manner that $\breve{\mathfrak{x}}(0) = 0$. Then, the location of the searcher at the time t is defined as follows:

$$\breve{\mathfrak{x}}(t) = S_t = \zeta(0) + \zeta(1) + \zeta(2) + \cdots + \zeta(t). \tag{4.13}$$

The expected coordinate $\mathcal{E}(S_t)$ of the searcher at time t and its variation $\mathrm{Var}(S_t)$ at this time are

$$\mathcal{E}(S_t) = (p - q)t \quad \text{and} \quad \mathrm{Var}(S_t) = 4pqt. \tag{4.14}$$

Certainly, if $p = q = 1/2$, then $\mathcal{E}(S_t) = 0$ and $\mathrm{Var}(S_t) = 1$. An example of such random walk-up to the time t =10,000 is shown in Figure 4.4.

The probability $v(k, t) = Pr\{|i(t)| = k\} = Pr\{|\breve{\mathfrak{x}}(t) - \breve{\mathfrak{x}}(0)| = k\}$ that at time t the searcher is located in the point $i = \breve{\mathfrak{x}}(t)$ such that the distance between this point and the starting point $\breve{\mathfrak{x}}(0) = 0$ is equal to k units is governed by the binomial distribution (Feller, 1970–1971; Papoulis & Pillai, 2002)

$$v(k, t) = \binom{t}{k} p^k q^{t-k}. \tag{4.15}$$

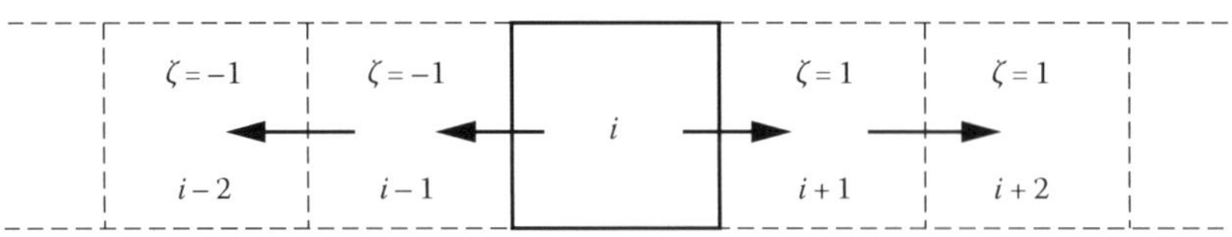

FIGURE 4.3 Steps of the searcher defined by 1D random walk.

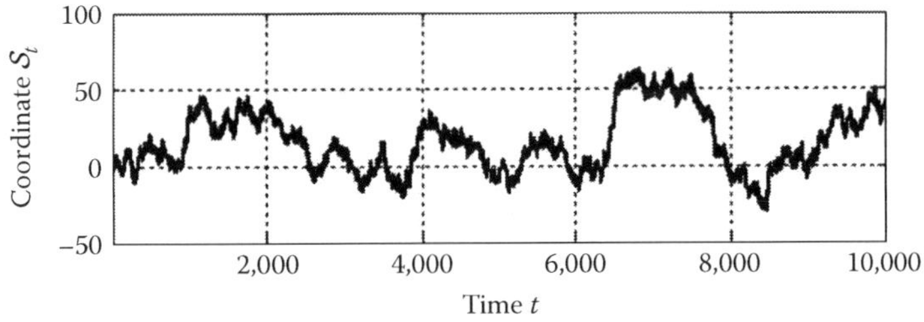

FIGURE 4.4 Example of simple 1D random walk.

Then, for large t and for k lying in the $(tpq)^{1/2}$ vicinity of tp, it holds true that

$$v(k,t) \simeq \frac{1}{\sqrt{2\pi tpq}} \exp\left(-\frac{(k-tp)^2}{2tpq}\right), \tag{4.16}$$

and, by denoting the coordinate $\mathfrak{x}$ in one dimension by $x = 2k-n$, for the probabilities $p = q = 1/2$, one obtains (Papoulis & Pillai, 2002)

$$v(\mathfrak{x},t) \simeq \frac{1}{\sqrt{\pi t/2}} \exp\left(-\frac{x^2}{2t}\right), \tag{4.17}$$

where the sign "$\simeq$" means that the ratio of values at the left and right sides tends to unity. The value specified by the Equation 4.17 is the probability of the searcher's location at the point $\breve{\mathfrak{x}}(t) = \mathfrak{x}$ at time t while it is governed by 1D random walk $\mathcal{S}_t$.

Finally, to obtain the 2D random walk $\mathcal{S}_t^2$ from two independent 1D random walks $\mathcal{S}_t$, let us define the steps of the searcher as

$$\breve{\mathfrak{x}}(t+1) = \breve{\mathfrak{x}}(t) + \zeta_{xy}(t) = \left(i_x + \zeta_x(t), i_y + \zeta_y(t)\right), \tag{4.18}$$

where ζ_x and ζ_y are Bernoulli random variables such that $Pr\{\zeta_x = -1\} = Pr\{\zeta_x = 1\} = 1/2$ and $Pr\{\zeta_y = -1\} = Pr\{\zeta_y = 1\} = 1/2$ and $\zeta_x(t)$ and $\zeta_y(t)$ are the values of these variables at time t.

Then, the searcher will step along the diagonals of the original grid, and the next cell is chosen with the probability $1/|\mathrm{n}(\mathfrak{x})|$ as earlier. Thus, the original random walk is obtained by clockwise rotation of the coordinates by $\pi/4$ radians and changing the length scale by $1/\sqrt{2}$, as it is illustrated in Figure 4.5. As in the preceding text, here it is assumed that $\breve{\mathfrak{x}}(t)$ is an internal point of the domain.

It is seen that the steps defined by two independent 1D random walks and the steps defined by the choice between the neighboring points in horizontal and vertical directions (Figure 4.2a) are equivalent with respect to the clockwise rotation of the domain by $\pi/4$ radians. Following this observation, the 2D random walk is usually defined as a pair $\mathcal{S}_t^2 = (\mathcal{S}_t, \mathcal{S}_t)$ of 1D random walks without stressing the need of rotation and scaling. In addition, notice that in the discrete domain, instead of rotation, appropriate substitutions of the indices can be applied.

4.2.1.2 Simple Brownian Motion

Let us continue the consideration of the searcher's movements in one dimension. However, in contrast to the random walk, assume that both the time and the space are continuous; to stress

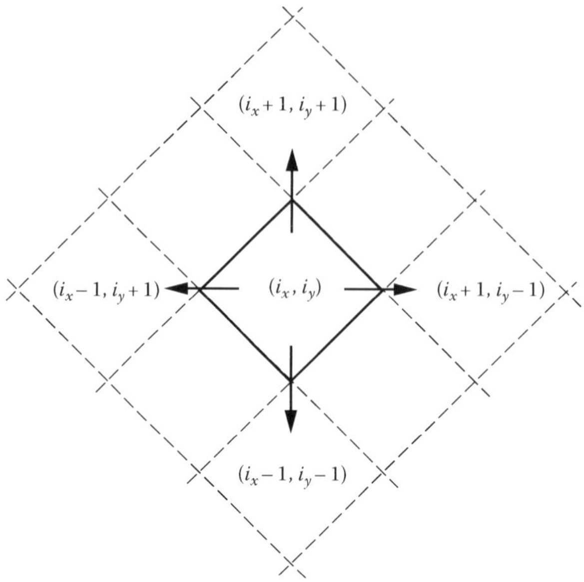

FIGURE 4.5 The steps of 2D random walk defined by two independent 1D random walks.

the difference between discrete and continuous time, the discrete time moments are denoted by τ. The direct formulation of Brownian motion based on the random walk is the following (Feller, 1970–1971).

Denote by δ the length of the step in the random walk $\mathcal{S}_t$ and by ν the number of steps per time unit. Certainly, in the simple random walk, both δ and ν are equal to one. Recall that $v(\mathfrak{x},\tau)$ is the probability that at the time moment τ, the searcher is located in the point $\mathfrak{x} = \breve{\mathfrak{x}}(\tau)$, and consider the probability $v(\mathfrak{x},t)$ that at time $t \simeq \tau/\nu$, the searcher is located in the neighborhood of the point $\mathfrak{x} \simeq \delta k$, where, as earlier, the sign "$\simeq$" means that the ratio of values at the left and right sides tends to unity. Similar to the Equation 4.4, in the 1D case, it is defined as follows (Feller, 1970–1971):

$$v(\mathfrak{x},t) = v\left(x+\delta, t-\frac{1}{\nu}\right)p + v\left(x-\delta, t-\frac{1}{\nu}\right)q, \tag{4.19}$$

where the probabilities p and q have the same meaning as in the preceding text (see Equation 4.12). By the second-order approximation over coordinate $\mathfrak{x}$ and the first-order approximation over time t and by avoiding the addendums that in limit tend to zero, one obtains the 1D diffusion equation

$$\frac{\partial}{\partial t}v(\mathfrak{x},t) = \frac{1}{2}\sigma^2\frac{\partial^2}{\partial x^2}v(\mathfrak{x},t) - \mu\frac{\partial}{\partial x}v(\mathfrak{x},t), \tag{4.20}$$

with the source in the point $\mathfrak{x} = 0$ that is the starting point of the searcher $\breve{\mathfrak{x}}(0)$; here, $\sigma^2 = \delta^2\nu$ is a diffusion rate and $\mu = (p-q)\delta\nu$ is a drift.

It is clear that the diffusion equation (4.20) is the Fokker–Planck equation with constant diffusion rate and drift (cf. Equations 2.103 and 2.104, which define the target's motion, and Equations 3.1 and 3.6). The solution of this equation is given by 1D Gaussian distribution (cf. Equation 2.105):

$$v(\mathfrak{x},t) = \frac{1}{\sqrt{2\pi t}\sigma}\exp\left(-\frac{(x-\mu t)^2}{2\sigma^2 t}\right), \tag{4.21}$$

where it is taken into account that $v(\mathfrak{x},0) = v(\breve{\mathfrak{x}}(0),0) = 1$. If the probabilities $p = q$, then the drift is $\mu = 0$, and Equation 4.20 is reduced to the equation

$$\frac{\partial}{\partial t}v(\mathfrak{x},t) = \frac{1}{2}\sigma^2\frac{\partial^2}{\partial x^2}v(\mathfrak{x},t), \tag{4.22}$$

with the solution

$$v(\mathfrak{x},t) = \frac{1}{\sqrt{2\pi t}\sigma}\exp\left(-\frac{x^2}{2\sigma^2 t}\right). \tag{4.23}$$

As it is expected, Equations 4.17 and 4.23 have the same form and differ only by the value of σ and the scale of time.

The stochastic process $\mathcal{W}_t$, which satisfies Equation 4.22 and defines the 1D Brownian motion in continuous space and time, is called the Wiener process. Since this process obtained a limiting process of the random walk $\mathcal{S}_\tau$, $\tau = 0, 1, 2,\ldots$, with zero mean and unit variance, it is often defined as $\mathcal{W}_t \simeq (1/\sqrt{\tau})\mathcal{S}_\tau\sqrt{\tau}$, while $\tau \to \infty$.

Formally, the Wiener process $\mathcal{W}_t$ is defined as follows (see, e.g., Bulinsky & Shiryaev, 2003; Lawler, 2010; Lawler & Limic, 2010):

1. With probability one $\mathcal{W}_0 = 0$.
2. For every t_1 and t_2 such that $0 \le t_1 < t_2 < \infty$, the random variable $\mathcal{W}_{t_2} - \mathcal{W}_{t_1}$ has a normal distribution with zero mean and variance $t_2 - t_1$ and is independent on all the variables $\mathcal{W}_{t_0}$, where $t_0 \le t_1$.
3. With probability one, the function $t \to \mathcal{W}_t$ is continuous.

Usually, the second property is called the independent increments property. From this property, it follows that for each $t > 0$, it holds true that the random variable $\mathcal{W}_t - \mathcal{W}_0$ has a normal distribution with zero mean and variance t, that is, $(\mathcal{W}_t - \mathcal{W}_0) \sim \mathcal{N}(0,t)$, and so $\mathcal{W}_t \sim \mathcal{N}(0,t)$. In addition, if t_1 and t_2 are such that $0 \le t_1 < t_2 < \infty$, then the covariance between $\mathcal{W}_{t_1}$ and $\mathcal{W}_{t_2}$ is (Bulinsky & Shiryaev, 2003; Shiryaev, 1995)

$$\sigma(\mathcal{W}_{t_1},\mathcal{W}_{t_2}) = \mathcal{E}\left[\left(\mathcal{W}_{t_1} - \mathcal{E}(\mathcal{W}_{t_1})\right)\left(\mathcal{W}_{t_2} - \mathcal{E}(\mathcal{W}_{t_2})\right)\right] = \sigma^2(\mathcal{W}_{t_1} - \mathcal{W}_0) = t_1. \tag{4.24}$$

In other words, for the Wiener process $\mathcal{W}_t$, it holds true that

$$\mathcal{E}(\mathcal{W}_t) = 0 \quad \text{for all } t \ge 0 \quad \text{and} \quad \sigma(\mathcal{W}_{t_1},\mathcal{W}_{t_2}) = \min\{t_1,t_2\} \quad \text{for all } t_1,t_2 \in [0,\infty). \tag{4.25}$$

These properties are often used as a definition of the Wiener process and corresponding Brownian motion. In the case of simple random walk $\mathcal{S}_\tau$ with $\tau = 0, 1, 2, \ldots$, the covariance is defined by the covariance matrix $\Sigma = \|\sigma_{ij}\|$, where covariance $\sigma_{ij} = \sigma(\mathcal{S}_i,\mathcal{S}_j)$ is $\sigma_{ij} = \sigma^2(\mathcal{S}_i)$ if $i = j$ and $\sigma_{ij} = 0$ otherwise.

Similar to the random walk considered in the preceding text, the 2D Wiener process $\mathcal{W}_t^2$, which governs a 2D Brownian motion, is defined as a pair $\mathcal{W}_t^2 = (\mathcal{W}_t,\mathcal{W}_t)$ of two 1D Wiener processes $\mathcal{W}_t$.

The diffusion equation, which corresponds to the simple 2D Brownian motion and defines the probabilities of the searcher's location in the points $\mathfrak{x} = (x, y)$ of 2D domain $\mathfrak{X} \subset \mathbb{R}^2$, is

$$\frac{\partial}{\partial t} v(\mathfrak{x}, t) = \frac{1}{2}\left(\breve{\sigma}_1^2 \frac{\partial^2}{\partial x^2} + \breve{\sigma}_2^2 \frac{\partial^2}{\partial y^2} \right) v(\mathfrak{x}, t), \tag{4.26}$$

where $\breve{\sigma}_1^2$ and $\breve{\sigma}_2^2$ stand for the diffusion rates. The solution of this equation is given by the 2D Gaussian distribution with the center in the point $\mathfrak{x} = (0,0)$ (cf. Equation 2.105):

$$v(\mathfrak{x}, t) = \frac{1}{2\pi\breve{\sigma}_1\breve{\sigma}_2 t} \exp\left(-\frac{x^2}{2\breve{\sigma}_1^2 t} - \frac{y^2}{2\breve{\sigma}_2^2 t} \right), \tag{4.27}$$

where it is taken into account that at the starting point the searcher's location probability is $v(\mathfrak{x}, 0) = 1$.

It is clear that the obtained diffusion equation is the simplest type of the Fokker–Planck equation, which was also applied for definition of target's location probabilities (see Equations 2.103 and 2.104) and of spatial dynamics of population (see Equations 3.1 and 3.6).

Notice that the similar, as earlier, reasoning regarding the Brownian motion of the searcher was applied in the pioneering Koopman works in search and screening theory (Koopman, 1946, 1956–1957) and formed a basis for the Koopman random search formula (see Equation 2.8). In the considered examples of the algorithms of foraging in patchy environment (see Section 3.3), the random walk was implemented for governing the predator in the patches; for detailed consideration of the relation between Lévy flights and foraging by patches, see Section 4.3.3.

In the terms of the searcher or predator behavior, the policy defined by the considered simple random walk and Brownian motion means the following (Hoffmann, 1983):

1. The searcher applies the same strategy in all the regions and does not use any information regarding the target's location.
2. The search strategy does not change in time, and the target is detected when the searcher reaches its location.
3. The searcher has no preferable direction (in long term), and its movements are independent.

In addition, as it follows from the properties of the random walk and Brownian motion, if there are no erroneous detections, then in long term, the searcher will find the static target with certainty.

Certainly, the considered search according to the simple random walk and Brownian motion specified by Algorithm 4.1 is not optimal and cannot be considered as a model of any realistic search behavior by the agents of reasonable geometric size and motion abilities. In fact, according to the simple Brownian search strategy, at each time moment, the agent chooses the direction of the next step with the same probabilities that implies that in any point, it is able to make a step in each possible direction. In particular, in the case of internal point of the gridded domain, such ability means that the agent can move forward, backward, left, and right without taking into account its current heading that even for the foraging bacteria (Passino, 2005) and, certainly, for animals (Calengea et al., 2009) and reasonable ground vehicles cannot be satisfied. However, the simple Brownian motion provides a general basis for definition of more specific search policies, which can be applied for description of the movement of different natural and artificial agents. The next sections address some of such policies.

4.2.2 Correlated Brownian Motion and the Model of the Agent

Consider the searcher acting in the continuous domain $\mathfrak{X} \subset \mathbb{R}^2$, and assume that its steps have the length δ, and the number of steps per time unit is ν. Then, the movements of the searcher are defined

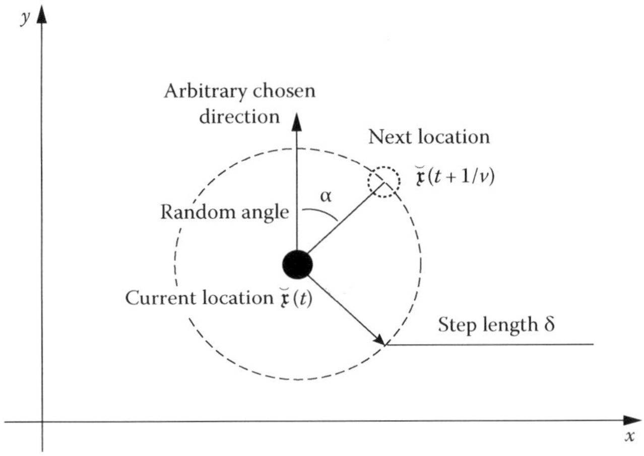

FIGURE 4.6 The choice of the next location for the randomly walking searcher in the continuous domain; the choice is based on the constant step length and arbitrary angles.

as follows. Assume that at time $t \in [0, T]$, the searcher is located in the internal point $\breve{x}(t) \in \mathfrak{X}$ such that the distance between this point and the bounds of the domain is greater than the length δ of the step. Then, the neighboring points of $\breve{x}(t)$ are located on the circle with the radius δ and the center in the point $\breve{x}(t)$, that is, $\mathfrak{n}(\breve{x}(t), \delta) = \{x : d(x, \breve{x}(t)) = \delta\}$, where $d(\cdot,\cdot)$ stands for the distance between the points of the domain. In such a case, the choice of the next location $\breve{x}(t+1/\nu)$, which is required by the simple Brownian search strategy, is equivalent to the random choice of the angle $\alpha \in [0, 2\pi)$ with respect to some definite direction and further specification of the intersecting points between the circle $\mathfrak{n}(\breve{x}(t), \delta)$ and its radius such that the angle between the radius and the indicated direction is α. The choice of the next searcher's location is illustrated in Figure 4.6.

Certainly, the presented angle-based model is equivalent to the considered step-based model of random walk $\mathcal{S}_t^2$ in Section 4.2.1, and in the limiting case with $\delta \to 0$ and $\nu \to \infty$, it results in the simple Brownian motion $\mathcal{W}_t^2$. The angle α is considered as a random variable governed by a uniform distribution on the interval $[0, 2\pi)$ or, that is the same, on the interval $[-\pi, \pi)$. The formal reasoning regarding the representation of the Brownian motion using the turn angles can be found, for example, in the books of Feller (1970–1971), Hughes (1995–1996), Lawler (2010), and Morters and Peres (2010).

Now, instead of the arbitrary defined direction, with respect to which the angle α is chosen, let us consider the heading of the agent and the angle, which were specified by its previous steps. As earlier, let δ be the length of the stepand ν be the number of steps per time unit, and denote the angle chosen at the time t by $\alpha(t)$. Assume that currently, the searcher is located in the internal point $\breve{x}(t)$ and is choosing the next location $\breve{x}(t+1/\nu)$ specified by the choice of the angle $\alpha(t)$. Consider its previous locations $\breve{x}(t-1/\nu)$, and $\breve{x}(t-2/\nu)$ and previously chosen angle $\alpha(t-1/\nu)$ and assume that the searcher chooses the angle $\alpha(t)$ with respect to its current heading. The scheme of such choice is shown in Figure 4.7.

Finally, assume that the angle $\alpha(t)$ is chosen with respect to the previous angle $\alpha(t-1/\nu)$ (Viswanathan et al., 2011). Then, the resulting motion is called correlated random walk, which in the limiting case with $\delta \to 0$ and $\nu \to \infty$ forms a correlated Brownian motion in two dimensions. The process, which governs such motion, is called correlated Wiener process and denoted by $\mathcal{W}_t^{(c)2}$, where $c \geq 0$ stands for the correlation coefficient; if $c = 0$, then the process $\mathcal{W}_t^{(c)2}$ reduced the Wiener process $\mathcal{W}_t^2$, which governs the simple Brownian motion. Similar to the Wiener process $\mathcal{W}_t^2$, the correlated Wiener process $\mathcal{W}_t^{(c)2}$ is represented by a pair of 1D Wiener processes that, is, $\mathcal{W}_t^{(c)2} = (\mathcal{W}_t^{(c_1)}, \mathcal{W}_t^{(c_2)})$, such that at least one of the correlation coefficients c_1 or c_2 is greater than zero.

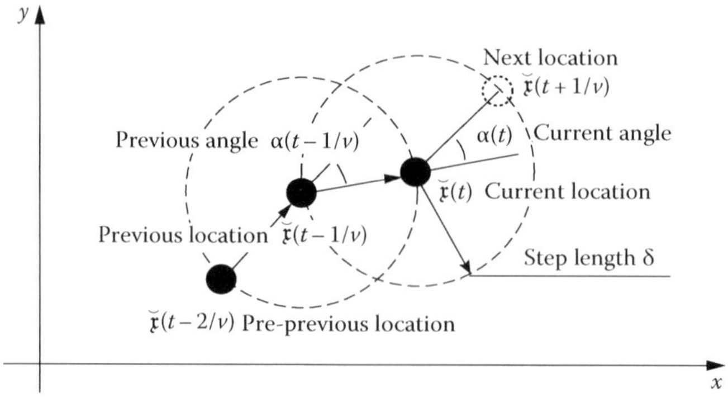

FIGURE 4.7 The choice of the next location in the continuous domain using the current heading of the searcher.

Formally, the 1D correlated Wiener process $\mathcal{W}_t^{(c)}$ is defined using the covariance function that for such process is

$$\sigma_c\left(\mathcal{W}_{t_1}^{(c)},\mathcal{W}_{t_2}^{(c)}\right)=\min\left\{t_1,t_2\right\}-ct_1t_2 \quad \text{for all } t_1,t_2\in\left[0,\infty\right). \tag{4.28}$$

Notice that since the Wiener process is scale independent that $\mathcal{V}_t=\left(1/\sqrt{r}\right)\mathcal{W}_t$ for any $r>0$ is again the Wiener process, in the definition of the correlated Wiener process, it is usually assumed that $c = 1$, and instead of Equation 4.28, the covariance function for all $t_1,t_2\in[0,\infty)$ is defined as (Shiryaev, 1995) and (Bulinsky & Shiryaev, 2003)

$$\sigma_1\left(\mathcal{W}_{t_1}^{(c)},\mathcal{W}_{t_2}^{(c)}\right)=\min\left\{t_1,t_2\right\}-t_1t_2.$$

Similar to Algorithm 4.1, let us outline the search according to the correlated random walk as follows (Viswanathan et al., 2011).

Algorithm 4.2: (Correlated Brownian Search)

Given a domain $\mathcal{X}=\{x_1,x_2,\ldots,x_n\}$, step length δ, steps rate ν and correlation coefficient $c\geq 0$, do:

1. Observe the current location $\breve{x}(t)$ and decide either to terminate the search or to continue.
2. If the search continues, choose as the next location $\breve{x}(t+1/\nu)$ as follows:
3. Calculate the angle $\alpha(t)=(1-c)r+c\alpha(t-1/\nu)$, where r is a random variable drawn according to the uniform distribution on the interval $[-\pi,\pi)$.
4. Choose the point $\breve{x}(t+1/\nu)$ such that $d(\breve{x}(t+1/\nu),\breve{x}(t))=\delta$ and the angle between interval $[\breve{x}(t+1/\nu),\breve{x}(t)]$ and $[\breve{x}(t),\breve{x}(t-1/\nu)]$ is $\alpha(t)$, where positive angle means the clockwise turn and negative angle means the counterclockwise turn with respect to $[\breve{x}(t),\breve{x}(t-1/\nu)]$ (see Figure 4.7).
5. Move to the chosen point $\breve{x}(t+1/\nu)$ and continue with its observation (line 1).

Certainly, if $c = 0$, then this algorithm results in the simple Brownian motion and for the simple grid with the unit step is reduced to Algorithm 4.1. In Section 4.1, Algorithm 4.2 was already used for creating the trajectories (a) and (c) shown in Figure 4.1.

In the terms of the 1D diffusion equation (see Equation 4.20), which governs the pdf v of the searcher's location in the points of the domain, the correlation is represented by the drift μ that is the difference between the probabilities of the movement's left or right (see Equation 4.12 and Figure 4.3).

Consequently, for the 2D motion $\mathcal{W}_t^{(c)2}$ in the domain $\mathfrak{X} \subset \mathbb{R}^2$ with the points $\mathfrak{x} = (x, y)$, the corresponding diffusion equation is as follows (cf. Equation 4.26 and already mentioned Fokker–Planck equations (2.103), (2.104), (3.1) and (3.6)):

$$\frac{\partial}{\partial t} v(\mathfrak{x}, t) = \frac{1}{2}\left(\breve{\sigma}_1^2 \frac{\partial^2}{\partial x^2} + \breve{\sigma}_2^2 \frac{\partial^2}{\partial y^2}\right) v(\mathfrak{x}, t) - \left(\breve{\mu}_1 \frac{\partial}{\partial x} + \breve{\mu}_2 \frac{\partial}{\partial y}\right) v(\mathfrak{x}, t), \tag{4.29}$$

and its solution is given by the time-dependent Gaussian distribution

$$v(\mathfrak{x}, t) = \frac{1}{2\pi \breve{\sigma}_1 \breve{\sigma}_2 t} \exp\left(-\frac{(x - \breve{\mu}_1 t)^2}{2\breve{\sigma}_1^2 t} - \frac{(y - \breve{\mu}_2 t)^2}{2\breve{\sigma}_2^2 t}\right), \tag{4.30}$$

where, similar to the Equation 4.27, it is taken into account that at the beginning of the search, the probability of the searcher's location in the initial point is $v(\mathfrak{x}, 0) = 1$.

The searcher's trajectories and the dynamics of its location probabilities are illustrated by the following example.

Example 4.1

Assume that the searcher acts in the square domain $\mathfrak{X}$ of the size $n = 100 \times 100$ with the impermeable bounds. In addition, assume that the location probabilities at the time $t = 0$ are defined by the binormal distribution with the center in the point $\breve{\mathfrak{x}}(0) = (50, 50)$ and equivalent variances $\sigma_1 = \sigma_2 = 10$ along both axes. This distribution is shown in Figure 4.8a.

Starting from the initial distribution, the dynamics of the searcher's location probabilities is defined by the Equation 4.29 with the diffusion rates $\breve{\sigma}_1 = \breve{\sigma}_2 = 0.3$ and drift parameters $\breve{\mu}_1 = \breve{\mu}_2 = -0.3$; the negative drifts mean that the center of distribution moves toward zero. The resulting distribution is at the time $t = 100$ as shown in Figure 4.8b. It is seen that the center of distribution moves toward the point $\mathfrak{x} = (0, 0)$ and is flattened according to the diffusion rates.

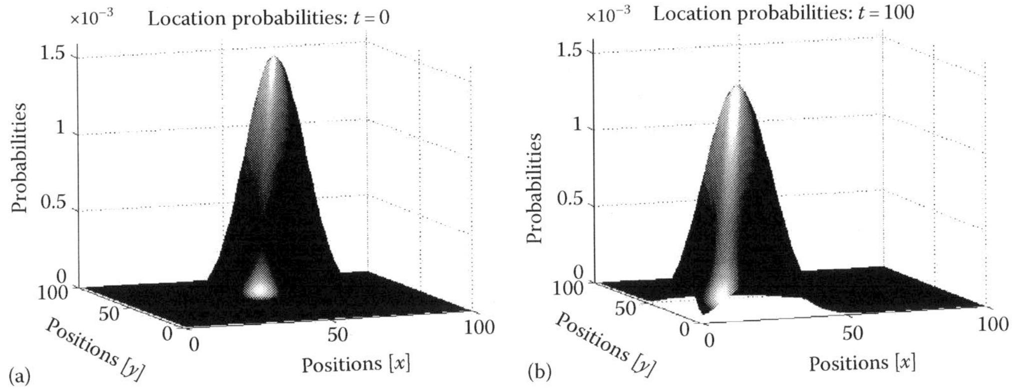

FIGURE 4.8 Location probabilities specified by the diffusion equation (a) at initial time $t = 0$ and (b) at the time $t = 100$.

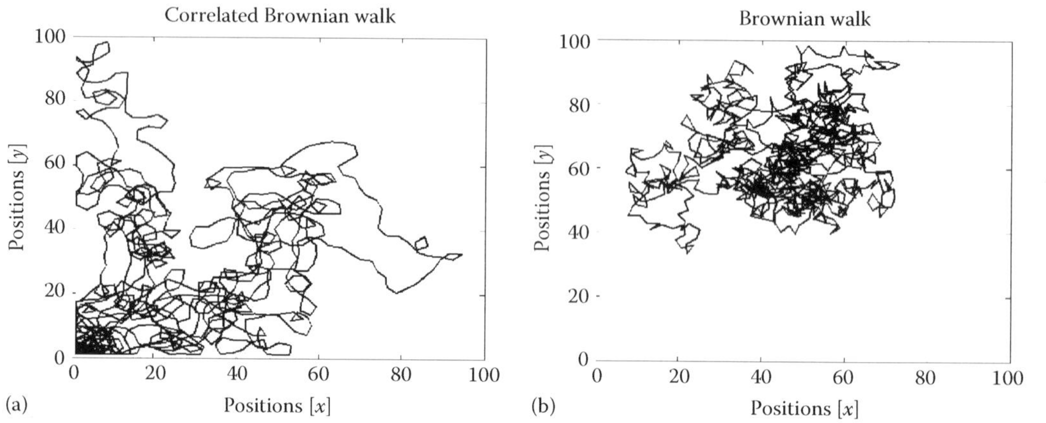

FIGURE 4.9 Trajectories of the searcher (a) according to the correlated Brownian motion and (b) according to the simple Brownian motion.

The obtained distribution of the location probabilities corresponds to the correlated Brownian motion searcher. Assume that the searcher starts in the point $\breve{x}(0) = (50,50)$. The trajectory of the searcher up to the time $t = 1000$ created by Algorithm 4.2 with the step $\delta = 3$ and the correlation coefficient $c = 1-\left(\mu_1^2+\mu_2^2\right)^{1/2} = 0.576$ is shown in Figure 4.9a. For comparison, the trajectory of simple Brownian motion (with $c = 0$) is shown in Figure 4.9b.

As it is expected, the correlated Brownian motion with the indicated correlation coefficient c tends to the corner point (see also Figure 4.1c), while the trajectory of the simple Brownian motion is distributed symmetrically around the starting point $\breve{x}(0) = (50,50)$ (see also Figure 4.1a). In the last case, Equation 4.29, which specifies the dynamics of location probabilities, is reduced to Equation 4.26, which defines the diffusion without drift. ■

In 1988, Bovet and Benhamou (1988) suggested to apply the correlated Brownian motion for modeling the animals' movement and demonstrated good correspondence between the trajectories resulted by such motion and the foraging trajectories of the *Serrastruma lujae* ants. The same results were reported later by several authors (see, e.g., Haefner & Crist [1994], Bartumeus et al. [2005]; for overview and discussion, see Bartumeus & Catalan [2009] and Section 11.1 in the book by Viswanathan et al. [2011]) and were observed in our experiments with *Messor ebeninus* and *L. humile* (Argentinian) ants.

As indicated earlier, the better correspondence between the animals' motion and correlated Brownian motion is based on the fact that such model takes into account the current heading of the animal, or, in biological terms, "the cephalocaudal polarization and the bilateral symmetry leading to a tendency to go forward" (Bovet & Benhamou, 1988, p. 420). To stress such tendency, let us consider a simple model of the ant (Kagan, Rybalov, Sela, Siegelmann, & Steshenko, 2014); this model is used for constructing the mobile agent and for simulations of swarm dynamics.

In the model, it is assumed that the ant has a certain geometric size and that its motion has resulted by consequent steps of the left and right side legs as it follows from widely accepted odometry and stepping scheme (Wohlgemuth, Ronacher, & Wehner, 2001; Wittlinger, Wehner, & Wolf, 2006, 2007). In addition, it is assumed that the ant's sensors are located at the same constant width as the tips of the legs and on constant positions relatively to the center of the ant such that the independent movements of the sensors are not considered. The correspondence between the real (*M. ebeninus*) ant and the model is illustrated in Figure 4.10.

Following the implemented stepping scheme, when the ant steps by the left side legs, it turns around the tip of the middle leg at the right side, and, similarly, when the ant steps by the right side

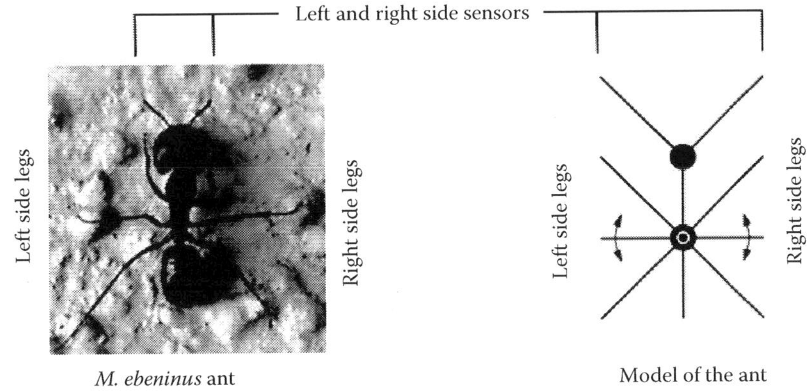

FIGURE 4.10 Correspondence between the real (*M. ebeninus*) ant (photo by S. Khodorov) and the model used in the mobility analysis.

legs, it turns around the tip of the middle leg at the left side. Then, at each time t, the single turn of the ant is restricted by maximal possible displacements of the legs, and the range of the relative angle $\alpha(t)$ (see Figure 4.7) is less than 2π. The turns of the ant and resulting trajectory are clarified in Figure 4.11.

It is clear that the step of the real ants is more complex and allows a wider variety of simple motions than it is shown in Figure 4.11. However, the model represents the main property of the

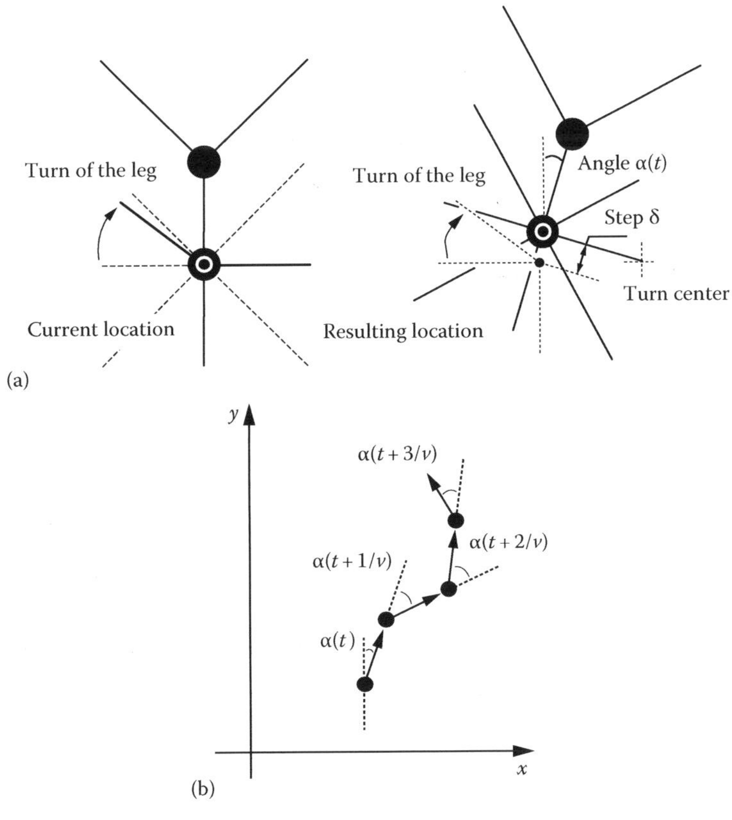

FIGURE 4.11 (a) Turns of the ant and (b) an example of steps of the resulting trajectory.

ant's motion, that is, its tendency to follow forward and inability to make direct steps left or right. As a result, while in the limiting case with the step length $\delta \to 0$ and, accordingly, the steps' rate $\nu \to \infty$, the trajectory of the ant tends to the trajectory of the correlated Brownian motion with the correlation coefficient specified by the ant's size. In addition, notice that such model can be applied for modeling the ant's motion "in small" and certainly is not applicable for modeling the movement in a large scale, which is at most governed by external factors like environmental states and the behavior of the neighboring ants.

The consideration presented earlier can be directly extended for modeling the individual movement of the other natural or artificial nonflying and nonjumping mobile agents; for overview and discussion, see, for example, Turchin (1998) and Viswanathan et al. (2011). The corresponding diffusion equation (4.29) and more general Fokker–Planck equations (2.103), (2.104), (3.1), and (3.6) can be applied for description of different biological multiagent systems (Romanovsky, Stepanova, & Chernavsky, 1984) including microorganisms' and plants, distribution (Turchin, 1998) and dynamics of the bacteria colonies (Ben-Jacob, Cohen, Golding, & Kozlovsky, 2001; Ben-Jacob & Levine, 2001). Notice that in the last case, instead of a single diffusion equation, a system of two equations called reaction–diffusion system is used. Such a system was already mentioned in Section 3.2 in consideration of the system "target–searcher" or "prey–predator" (see Equation 3.7); in Section 5.4, the reaction–diffusion and corresponding dynamics will be considered in details.

Certainly, the considered simple and correlated Brownian motion does not exhaust the variety of possible models of the movements of mobile agents, which can be defined using different covariance functions. For example, if in contrast to the Equations 4.25 and 4.28 the covariance function of the process $\mathcal{Y}_t$ is defined as

$$\sigma\left(\mathcal{Y}_{t_1},\mathcal{Y}_{t_2}\right)=\exp\left(-\left|t_2-t_1\right|\right) \quad \text{for all } t_1,t_2\in\left(-\infty,\infty\right),$$

then for each $t \in (-\infty, \infty)$, the variable $\mathcal{Y}_t$ has a normal distribution, and the future and the past states are independent with respect to $\mathcal{Y}_t$. Such process is known as Gauss–Markov process (Shiryaev, 1995).

The next section describes a process that forms a basis for fractional dynamics analysis (Metzler & Klafter, 2000, 2004). This process, known as fractional Brownian motion, is considered as one of the perspective models of optimal foraging, which extends the model based on the correlated Brownian motion and allows the description of the foraging behavior in a large scale.

4.2.3 Fractional Brownian Motion

Let us start with the 1D domain $\mathcal{X} \subset \mathbb{R}$ and consider the continuous-time process $\mathcal{W}_t^{(H)}$, $t \geq 0$, with zero mean $\mathcal{E}\left(\mathcal{W}_t^{(H)}\right) = 0$ and the covariance function (Bulinsky & Shiryaev, 2003):

$$\sigma_H\left(\mathcal{W}_{t_1}^{(H)},\mathcal{W}_{t_2}^{(H)}\right)=\frac{1}{2}\left(t_1^{2H}+t_2^{2H}-\left|t_1-t_2\right|^{2H}\right),\quad H\in\left(0,1\right),\quad t_1,t_2\in\left[0,\infty\right). \tag{4.31}$$

The value $H \in (0, 1)$ is called the Hurst parameter, and the process $\mathcal{W}_t^{(H)}$ is known as fractional Wiener process, which defines the fractional Brownian motion. If $H = 1/2$, then the function σ_H is reduced to the defined covariance function by Equation 4.25 of the simple Brownian motion (Bulinsky & Shiryaev, 2003; Hahn, Kobayashi, & Umarov, 2010).

The fractional Wiener process $\mathcal{W}_t^{(H)}$ was introduced in 1940 by Kolmogorov (1940). In 1968, Mandelbrot and van Ness (1968) obtained the representation of this process as integral with respect to the simple Wiener process $\mathcal{W}_t$ and, consequently, named it as fractional Brownian motion; this name will be used both for the process $\mathcal{W}_t^{(H)}$ and for the corresponding motion.

Let $\mathcal{W}_\tau$ be a 1D Wiener process. Then, for any Hurst parameter $H \in (0, 1)$ and $t \geq 0$, the fractional Brownian motion $\mathcal{W}_t^{(H)}$ in Lévy form is defined as follows (Mandelbrot & van Ness, 1968):

$$\mathcal{W}_t^{(H)} = \frac{1}{\Gamma\left(H + \frac{1}{2}\right)} \int_0^t \left((t-\tau)^{H-\frac{1}{2}}\right) d\mathcal{W}_\tau, \tag{4.32}$$

where $\Gamma(\cdot)$ is a gamma function, that is, $\Gamma(z) = \int_0^\infty y^{z-1} e^{-z} dy$ for any complex number z. Notice again that if $H = 1/2$, then $\mathcal{W}_t^{(H=1/2)} = \int_0^t d\mathcal{W}_\tau$, and the process $\mathcal{W}_t^{(H)}$ is reduced to the Wiener process $\mathcal{W}_t$. The integral appearing in Equation 4.32 is called Riemann–Liouville fractional integral of $\mathcal{W}_\tau$; for detailed consideration of fractional calculus, see, for example, the book by Samko, Kilbas, and Marichev (2002).

It is clear that similar to the Wiener process $\mathcal{W}_t$, the fractional Brownian motion $\mathcal{W}_t^{(H)}$ is continuous and $\mathcal{W}_0^{(H)} = 0$ with probability one. In addition, for the process $\mathcal{W}_t^{(H)}$, it holds true that (Bulinsky & Shiryaev, 2003; Mandelbrot & van Ness, 1968)

1. For any $a > 0$ and any time $t \geq 0$, two random variables $\mathcal{W}_{at}^{(H)}$ and $a^H \mathcal{W}_t^{(H)}$ have the same distribution
2. For every t_1 and t_2 such that $0 \leq t_1 < t_2 < \infty$, the random variable $\mathcal{W}_{t_2}^{(H)} - \mathcal{W}_{t_1}^{(H)}$ has the same distribution as the random variable $\mathcal{W}_{t_2-t_1}^{(H)}$
3. For every t_1 and t_2, $0 \leq t_1, t_2 < \infty$, it holds true that $\mathcal{E}\left(\mathcal{W}_{t_2}^{(H)} - \mathcal{W}_{t_1}^{(H)}\right)^2 = |t_2 - t_1|^{2H}$

The first property represents the self-similarity of the fractional Brownian motion, and the second property is usually called the stationary increments property. Notice that the stationary increments property does not imply the independency of the increments; thus, the current step of fractional Brownian motion is likely to follow the trend of the previous steps.

The 1D fractional Brownian motion with different Hurst parameters H is illustrated in Figure 4.12. Figure 4.12a shows the Brownian motion $\mathcal{W}_t$ that is the fractional motion with $H = 0.5$, with respect to which the fractional Brownian motions were calculated. Figure 4.12b shows the

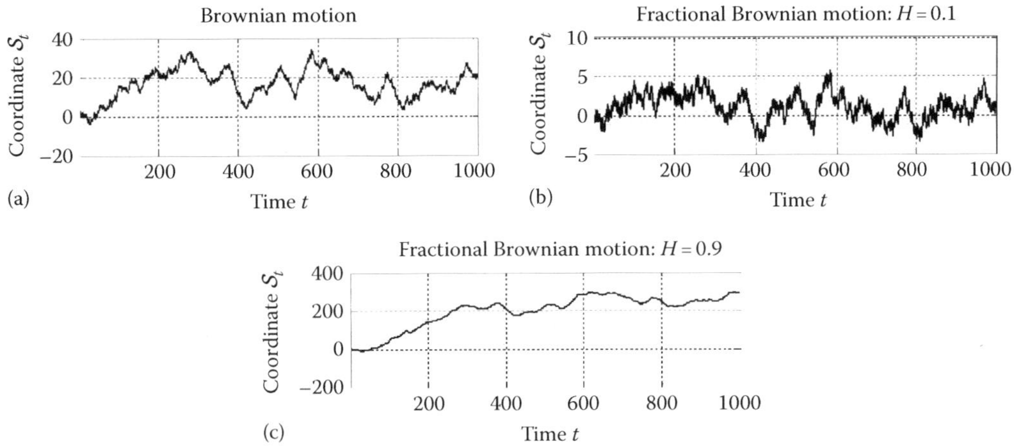

FIGURE 4.12 One-dimensional fractional Brownian motion with different Hurst parameters: (a) simple Brownian motion, $H = 0.5$; (b) fractional Brownian motion, $H = 0.1$; and (c) fractional Brownian motion, $H = 0.9$.

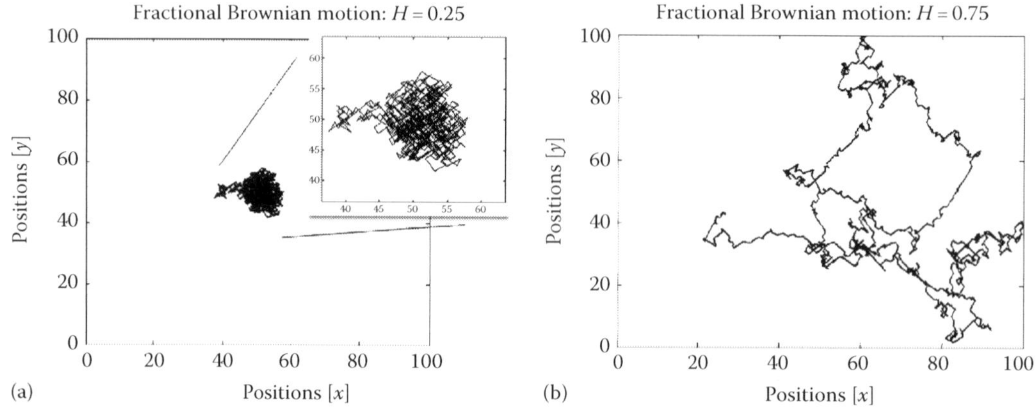

FIGURE 4.13 Trajectories of 2D fractal Brownian walk with the Hurst parameter (a) $H = 1/4$ and (b) $H = 3/4$. The starting point in both cases is $\mathfrak{x}_0 = (50,50)$, and in Figure (b), the trajectory is scaled following the size of the domain.

fractal Brownian motion with $H = 0.1$, and Figure 4.12c shows the fractal Brownian motion with $H = 0.9$. The time in all three figures is $t = 1 \ldots 1000$.

In the figure, it is seen that the smaller Hurst parameter leads to the motion with more frequent direction changes and less removal from the origin, while the motion with the larger Hurst parameter preserves the trend of the motion that leads to fast removal from the origin.

Now, let us consider the 2D fractal Brownian walk $\mathcal{W}_t^{(H)2}$ in the domain $\mathfrak{X} \subset \mathbb{R}^2$ with the points $\mathfrak{x} = (x, y)$. Similar to the considered simple and correlated Brownian walks, we assume that the 2D walk $\mathcal{W}_t^{(H)2}$ is a pair $\mathcal{W}_t^{(H)2} = \left(\mathcal{W}_t^{(H_1)}, \mathcal{W}_t^{(H_2)}\right)$ of 1D fractal Brownian walks such that at least one of the Hurst parameters H_1 or H_2 differs from 1/2. The 2D Brownian walk for the Hurst parameters $H_1 = H_2 = 1/4$ and $H_1 = H_2 = 3/4$ is illustrated in Figure 4.13. In both cases, the time is $t = 1, \ldots, 1000$.

As it is expected, the walk with smaller Hurst parameter is concentrated around the initial point, while the walk with greater Hurts parameter tends to move to the distant points.

Likewise, in simple and correlated Brownian motion, the location probabilities for the fractional Brownian motion $\mathcal{W}_t^{(H)2}$ are governed by the Fokker–Planck equation, which in general is defined as follows (Hahn et al., 2010) (for 1D case, see also Bassler, Gunaratne, & McCauley, 2006; McCauley, Gunaratne, & Bassler, 2007):

$$\frac{\partial}{\partial t} v(\mathfrak{x}, t) = Ht^{2H-1}\left(\frac{\partial^2}{\partial x^2} + \frac{\partial^2}{\partial y^2}\right) v(\mathfrak{x}, t). \tag{4.33}$$

In other words, the diffusion rates $\breve{\sigma}_1^2$ and $\breve{\sigma}_2^2$ for the fractional Brownian motion are proportional to the Hurst parameter and the Hurst power of time; in Equation 4.33, they are equal and $\breve{\sigma}_1^2 = \breve{\sigma}_2^2 = 2Ht^{2H-1}$ (cf. diffusion equation (4.26) for the simple 2D Brownian motion), that is, they stand for the diffusion rates.

The dynamics of the location probabilities governed by Equation 4.33 with different Hurst parameters is illustrated by the Example 4.2.

Example 4.2

Similar to Example 4.1, assume that the searcher acts in the square domain $\mathfrak{X}$ of the size $n = 100 \times 100$ with the impermeable bounds and that initial location probabilities are defined by the binormal distribution with the central point $\breve{\mathfrak{x}}(0) = (50,50)$ and equivalent variances $\sigma_1 = \sigma_2 = 10$. This distribution is shown in Figure 4.14a (see also Figure 4.8a).

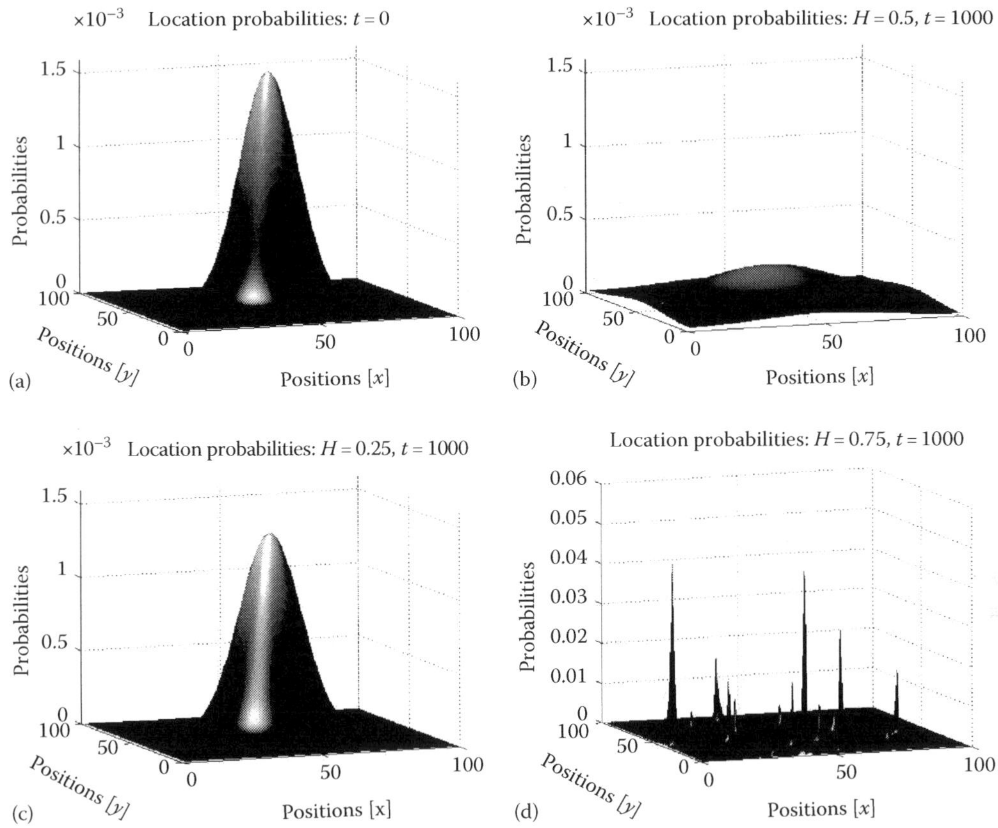

FIGURE 4.14 Location probabilities for the 2D fractal Brownian motion specified by the diffusion equation with different Hurst parameters. Figure (a) shows initial location probabilities, and the other figures show location probabilities at $t = 1000$ with the Hurst parameters (b) $H = 1/2$ (simple Brownian motion), (c) $H = 1/4$, and (d) $H = 3/4$ (anomalous diffusion).

Figure 4.14b–d demonstrates the resulting location probabilities at $t = 1000$ for different Hurst parameters. In Figure 4.14b, this parameter is $H = 1/2$ that corresponds to the simple Brownian motion (for the examples of the trajectories of such motion, see Figures 4.1a and 4.9b). Figure 4.14c shows the resulting location probabilities for the Hurst parameter $H = 1/4$ (an example of the corresponding trajectory is shown in Figure 4.13a), and Figure 4.14d shows these probabilities for the Hurst parameter $H = 3/4$ (an example of the corresponding trajectory is shown in Figure 4.13b).

It is seen that in the case of simple Brownian motion (Figure 4.14b), the location probabilities follow simple nondrifted diffusion. However, for the fractional Brownian motion, the smaller Hurst parameter leads to the concentration of the probabilities in the central point (Figure 4.14c), and the larger Hurst parameter leads to the sharp concentrations of the probabilities in different points of the domain (Figure 4.14d). The last case is known as anomalous diffusion that is used for modeling behavior of nonlinear dynamical systems (Metzler & Klafter, 2000, 2004). ■

Equation 4.33 specifies noncorrelated fractional Brownian motion. In the case of correlated fractional Brownian motion $\mathcal{W}_t^{(c,H)2}$, instead of Equation 4.33, the dynamics of location probabilities is defined by the diffusion equation with drift parameters $\breve{\mu}_1$ and $\breve{\mu}_2$ (Hahn et al., 2010):

$$\frac{\partial}{\partial t} v(\mathfrak{x},t) = Ht^{2H-1}\left(\frac{\partial^2}{\partial x^2} + \frac{\partial^2}{\partial y^2}\right) v(\mathfrak{x},t) - \left(\breve{\mu}_1 \frac{\partial}{\partial x} + \breve{\mu}_2 \frac{\partial}{\partial y}\right) v(\mathfrak{x},t), \tag{4.34}$$

where, as in the preceding text , it is assumed that the diffusion rates are equal and defined by the Hurst parameter as $\breve{\sigma}_1^2 = \breve{\sigma}_2^2 = 2Ht^{2H-1}$ (cf. diffusion equation (4.29) for the correlated 2D Brownian motion). It is clear that $\breve{\mu}_1 = \breve{\mu}_2 = 0$, and then this equation is reduced to the Equation 4.33 of noncorrelated fractional Brownian motion.

For the Hurst parameter $H \leq 1/2$, the dynamics of the location probabilities according to Equation 4.34 is specified as a combination of the non-drifted fractional Brownian motion and correlated Brownian motion, and for the Hurst parameter $H > 1/2$, the location probabilities are concentrated faster than in the case of noncorrelated fractional Brownian motion.

In 2005, Klafter and Sokolov (2005) suggested to apply the fractional Brownian motion for modeling the foraging behavior of birds and animals and argued that such motion together with the Lévy flights can provide complete description of the individual dynamics of complex systems. The next algorithm follows this direction and uses recently discovered fractional Brownian motion with variable Hurst parameters (Ryvkina, 2013).

Let $\mathcal{W}_t^{(H)}$ be a 1D fractional Wiener process, in which the Hurst parameter $H \in (0, 1)$ varies in time according to certain external conditions. Then, the correlation of the increments of $\mathcal{W}_t^{(H)}$ changes with respect to H that results in the changes of the regularity of the process either continuously or by jumps. The Fokker–Planck equation corresponding to such process is defined as follows (Ryvkina, 2013) (cf. Equation 4.33):

$$\frac{\partial}{\partial t} v(\mathfrak{x},t) = \left(\frac{dH}{dt} \ln t + \frac{H}{t} \right) t^{2H} \frac{\partial^2}{\partial x^2} v(\mathfrak{x},t), \tag{4.35}$$

where it is assumed that the derivative dH/dt exists.

In the case of 2D motion over $\mathfrak{X} \subset \mathbb{R}^2$, the process $\mathcal{W}_t^{(H)2} = \left(\mathcal{W}_t^{(H_1)}, \mathcal{W}_t^{(H_2)}\right)$ is defined by the pair of 1D processes with variable Hurst parameters, and the corresponding Fokker–Planck equation is defined, respectively.

Assume that in the process $\mathcal{W}_t^{(H)2}$, the Hurst parameters H_1 and H_2 depend on the location probabilities of the target. Then, the short and long jumps of the trajectory of $\mathcal{W}_t^{(H)2}$ are specified by these probabilities that allow application of the fractional Brownian motion for definition of the search process. In particular, in discrete domain and discrete time, the algorithm of search is outlined as follows.

Algorithm 4.3: (Fractional Brownian Search)

Given a domain $\mathfrak{X} = \{\mathfrak{x}_1, \mathfrak{x}_2, \ldots, \mathfrak{x}_n\}$ and the target's location probabilities $\hat{u}(\mathfrak{x},t)$ do:

1. Observe the current location $\breve{\mathfrak{x}}(t)$ and decide either to terminate the search or to continue.
2. If the search continues, define the Hurst parameters $H_1(\mathfrak{x},t)$ and $H_2(\mathfrak{x},t)$ with respect to $\hat{u}(\mathfrak{x},t)$.
3. Chose the next point $\breve{\mathfrak{x}}(t+1)$ according to the fractional Wiener process $\mathcal{W}_t^{(H)2}$.
4. Move to the chosen point $\breve{\mathfrak{x}}(t+1)$ and continue with its observation (line 1).

Notice that like Brownian search Algorithms 4.1 and 4.2 presented earlier, this algorithm does not define an optimal search and is considered only as a method of description of the motion of the natural foragers. The actions of the algorithm are illustrated by the following example.

Example 4.3

As mentioned earlier, assume that the searcher acts in the square domain $\mathfrak{X}$ of the size $n = 100 \times 100$ with the impermeable bounds. In addition, assume that the target's location probabilities $\hat{u}(\mathfrak{x},t)$ over all times $t = 0,\ldots,1000$ are defined by ten binormal distributions with random centers and the variances $\sigma_1 = \sigma_2 = 10$ and do not depend on the actions of the searcher.

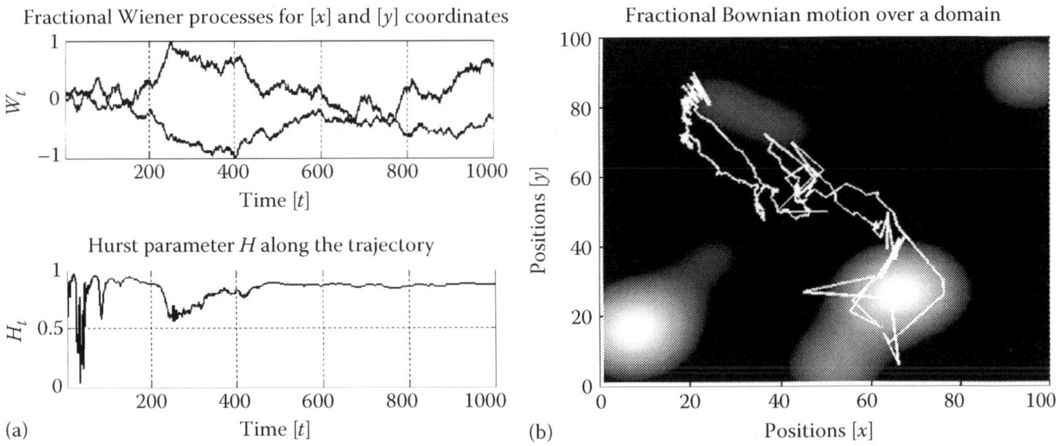

FIGURE 4.15 Fractional Brownian motion with the Hurst parameter varying according to the environmental states: (a) fractional Wiener processes in the horizontal and vertical directions and the Hurst parameter along the trajectory and (b) trajectory of fractional Brownian motion over a domain.

Assume that the Hurst parameter is specified as $H(\mathfrak{x},t) = 1 - \hat{u}(\mathfrak{x},t)$. The location of the searcher $\breve{\mathfrak{x}}(t) = (\breve{x}(t), \breve{y}(t))$ at each time t is defined by its coordinates $\breve{x}(t) = s_x t^{-H(\mathfrak{x},t)} \mathcal{W}_{xt}^{(H)}$ and $\breve{y}(t) = s_y t^{-H(\mathfrak{x},t)} \mathcal{W}_{yt}^{(H)}$, where $\mathcal{W}_{xt}^{(H)}$ and $\mathcal{W}_{yt}^{(H)}$ are the process over x and y axes and s_x and s_y are normalization coefficients depending on the size of the domain. The processes $\mathcal{W}_{xt}^{(H)}$ and $\mathcal{W}_{yt}^{(H)}$ and the Hurst parameter of the trajectory are shown in Figure 4.15a, and the trajectory of the searcher over a domain is shown in Figure 4.15b. In Figure 4.15b, white color stands for greater location probabilities and black color stands for smaller location probabilities; the searcher's trajectory is depicted in white.

The searcher starts at the initial point $\breve{x}(0) = (50,50)$, in which the target's location probability is close to 1/2. Thus, the first step of the searcher follows a simple Brownian motion. At the next location, the location probability is less than 1/2, so the Hurst parameter is close to one, and the searcher follows fractional Brownian motion with the positively correlated increments. At the time $t \sim 50$, the searcher enters the region with large location probabilities; thus, the Hurst parameter becomes close to zero, and the searcher follows the fractional Brownian motion with negatively correlated increments. The trajectory in this region consists of the long jumps (see the trajectory at the upper-left corner in Figure 4.15b). Then the searcher enters the region with small location probabilities and again follows the fractional Brownian motion with positively correlated increments. ■

The example demonstrates that the search based on fractional Brownian motion results in the trajectory of the searcher which includes both short and long jumps with respect to the target's location probabilities. This trajectory has the same form as a trajectory, obtained by Algorithms 3.1 through 3.3 of foraging by patches considered in Section 3.3 that allows a definition of the forager motion using appropriate fractional Brownian motion with variable Hurst parameter.

Additional degree of freedom in modeling foraging behavior using fractional Brownian motion is provided by the multifractional Brownian motion (Peltier & Lévy Vehel, 1995); for multifractional phenomena in general, see, for example, Stanley and Meakin (1988) and Havlin et al. (1993). In such motion, both diffusion rates and drifts are defined by the Hurst parameters that result in the varying steps of the forager, and usually, this property is modeled by the processes called Lévy walks and Lévy flights. In the next section, these processes and their application to the foraging behavior are considered in detail.

4.3 FORAGING BY LÉVY FLIGHTS

The models considered earlier of the searcher's motion, especially the suggested Bovet and Benhamou model based on correlated Brownian motion (Bovet & Benhamou, 1988), demonstrate good correspondence with the observed behavior of the nonflying and nonjumping animals and insects. However, as it was indicated by Bovet and Benhamou, such models less agree with the observed movement of birds and flying insects, in particular, of bumblebees (*Bombus flavifrons*). To describe the behavior of such foragers, in 1996, Viswanathan et al. in their seminal paper (Viswanathan et al., 1996) suggested to use the processes called Lévy walks and Lévy flights and demonstrated that such processes result in the trajectories that highly correspond with the observed trajectories of wandering albatrosses. The further success of Lévy-like processes for modeling animals' movements allowed Buchanan to declare them as a kind of "mathematical mirror to animals nature" (Buchanan, 2008). The discourse that follows addresses the Lévy walks and Lévy flights in details and considers search algorithms based on these processes. In addition, it includes the recent critics of this approach and presents the methods of search using combined strategies.

4.3.1 LÉVY FLIGHTS

In the simple and correlated Brownian walks considered earlier (see Sections 4.2.1 and 4.2.2), the direction of the agent's motion is chosen by random according to specified rule, while the step length in all directions is constant. In the case of fractional Brownian motion considered in Section 4.2.3, the step length is considered as random variable such that at each time, its value is specified with respect to the previous step length and the correlation between the step lengths is defined by the Hurst exponent. In contrast, in the Lévy walks also widely known as Lévy flights (Mandelbrot, 1983; Viswanathan et al., 2011), the step lengths are considered as independent or correlated random variables governed by a definite distribution that is usually the Pareto distribution. Often, for distinguishing between Lévy walks and Lévy flights, it is said that in Lévy walk, the agent has finite speed so it follows the chosen direction and the step's distance in finite time, while in the Lévy flight, it can conduct immediate jumps to any specified distance. All the processes in the following textare considered as Lévy flights, and the terms flight and walk are used interchangeably.

Let us start with the 1D domain $\mathfrak{X} \subset \mathbb{R}$. In the domain $\mathfrak{X}$, the discrete time Lévy flight $\mathcal{L}_t^{(\gamma)}$, $t = 0, 1, 2, \ldots$, with the parameter $\gamma \in (1,3]$ is defined as follows. Denote by δ the length of the step, and assume that the choice of the length δ is governed by the Pareto distribution. Namely, the probability $p(\delta)$ that the agent's step is of the length δ is proportional to the γ power of δ, that is (Mandelbrot, 1983; Viswanathan et al., 2011),

$$p(\delta) \sim \delta^{-\gamma}, \quad \delta \in [\delta_{\min}, \infty), \tag{4.36}$$

where $\delta_{\min} > 0$ is a minimal step length. Given minimal length $\delta_{\min}$ and the value of the parameter γ, this probability is defined directly as (James, Plank, & Edwards, 2011)

$$p(\delta) = \left((\gamma - 1)\delta_{\min}^{\gamma-1}\right)\delta^{-\gamma}, \quad \delta \in [\delta_{\min}, \infty),$$

The value $\gamma = a + 1$ is also known as fractal dimension of the Lévy flight (Mandelbrot, 1983) and $a \in (0,2]$ as a stability or Lévy index (Pavlyukevich, 2007; Viswanathan et al., 2011).
Formally, the Lévy flight $\mathcal{L}_t^{(\gamma)}$, $t = 0, 1, 2, \ldots$, has the following properties:

1. The differences $\mathcal{L}_{t+1}^{(\gamma)} - \mathcal{L}_t^{(\gamma)}$ are independent on t.
2. For every t_1 and t_2 such that $0 \le t_1 < t_2 < \infty$, the random variable $\mathcal{L}_{t_2}^{(\gamma)} - \mathcal{L}_{t_1}^{(\gamma)}$ is independent on all the variables $\mathcal{L}_{t_0}^{(\gamma)}$, where $t_0 \le t_1$.
3. With probability one, $\mathcal{L}_t^{(\gamma)}$ has infinite variance.

In other words, similar to the Wiener process $\mathcal{W}_t$, the process $\mathcal{L}_t^{(\gamma)}$ is stationary (property 1) and has independent increments (property 2).

Examples of the 1D Lévy flight $\mathcal{L}_t^{(\gamma)}$ with different values of the parameter γ are shown in Figure 4.16. In the figure, similar to the random walk, the direction of the step is chosen randomly with equal probabilities of moving left or right; thus, the coordinate $\breve{\mathfrak{x}}(t) = \mathcal{S}_t$ is defined by

$$\breve{\mathfrak{x}}(t) = \mathcal{S}_t = \zeta(0)\delta(0) + \zeta(1)\delta(1) + \zeta(2)\delta(2) + \cdots + \zeta(t)\delta(t),$$

where $Pr\{\zeta(t) = -1\} = Pr\{\zeta(t) = 1\} = 1/2$ and $\delta(t)$ is chosen according to the probability given by Equation 4.36, $t = 0, 1, 2, \ldots$

It is seen that the greater parameter γ results in greater increments of the flight $\mathcal{L}_t^{(\gamma)}$ and for all γ the variance of $\mathcal{L}_t^{(\gamma)}$ increases with time, as it is prescribed by property 3.

Now, let us consider 2D Lévy flight $\mathcal{L}_t^{(\gamma)2}$, $1 < \gamma \le 3$, $t = 0, 1, 2, \ldots$, in the domain $\mathfrak{X} \subset \mathbb{R}^2$. Similar to the correlated Brownian motion, denoting by α a random angle of the agent's turn with respect to its current direction (see Figure 4.7), as in the previous text, it is assumed that at each time t, the value $\alpha(t) \in [-\pi, \pi)$ is chosen randomly according to uniform distribution on the interval $[-\pi, \pi)$. The 2D Lévy flight with different values of the parameter γ is illustrated in Figure 4.17.

In the figure, since the domain is of the finite size 100 × 100 points, maximal value of the step is bounded by $\delta_{max} = 25$; minimal step length is $\delta_{max} = 1$. In all the cases, the time is $t = 1, \ldots, 1000$. In the left side of the figure, the pdf and the cumulative distribution function of the corresponding Pareto distributions with the shape parameters $\gamma - 1$ are depicted.

It is seen that the greater parameter γ results in more frequent long-term flights of the agent, while for smaller γ, the agent at most moves by short jumps. Such property forms a basis for consideration of the Lévy flights as a basic model of foraging behavior.

In the presented 2D Lévy flights $\mathcal{L}_t^{(\gamma)2}$, it is assumed that the time t is discrete. Now, let us consider the Lévy flight $\mathcal{L}_t^{(\gamma)2}$ in continuous time $t \in [0, \infty)$ and address the Fokker–Planck equation, which governs the location probabilities of the agent. In contrast to the equations, which correspond to the Brownian motion, the Fokker–Planck equation, which generates the Lévy flight, essentially uses fractional integrals and derivatives (Viswanathan et al., 2011). In the preceding text, fractional integral was already used in the definition of fractional Brownian motion given by Equation 4.32; here, these concepts are presented in detail.

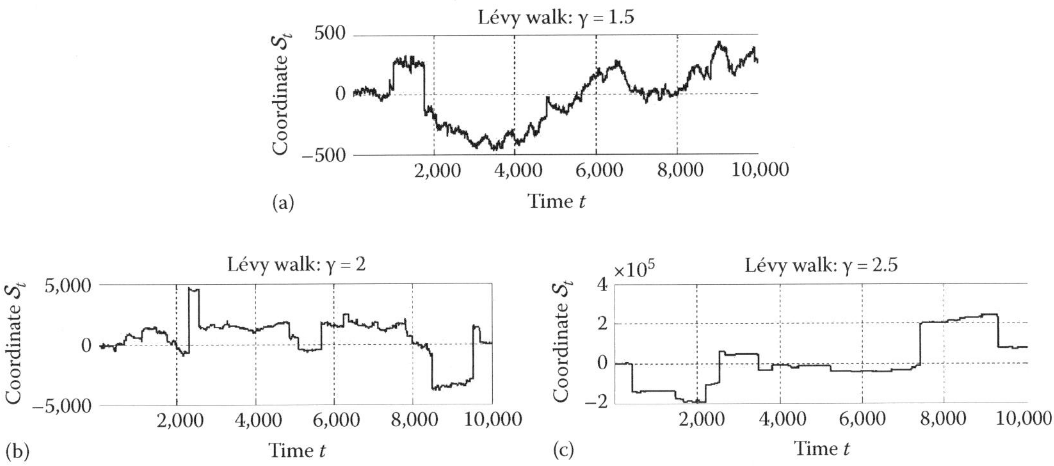

FIGURE 4.16 One-dimensional Lévy flights with different parameters γ: (a) $\gamma = 1.5$, (b) $\gamma = 2.0$, and (c) $\gamma = 2.5$. The direction is chosen randomly with equal probabilities of moving left or right.

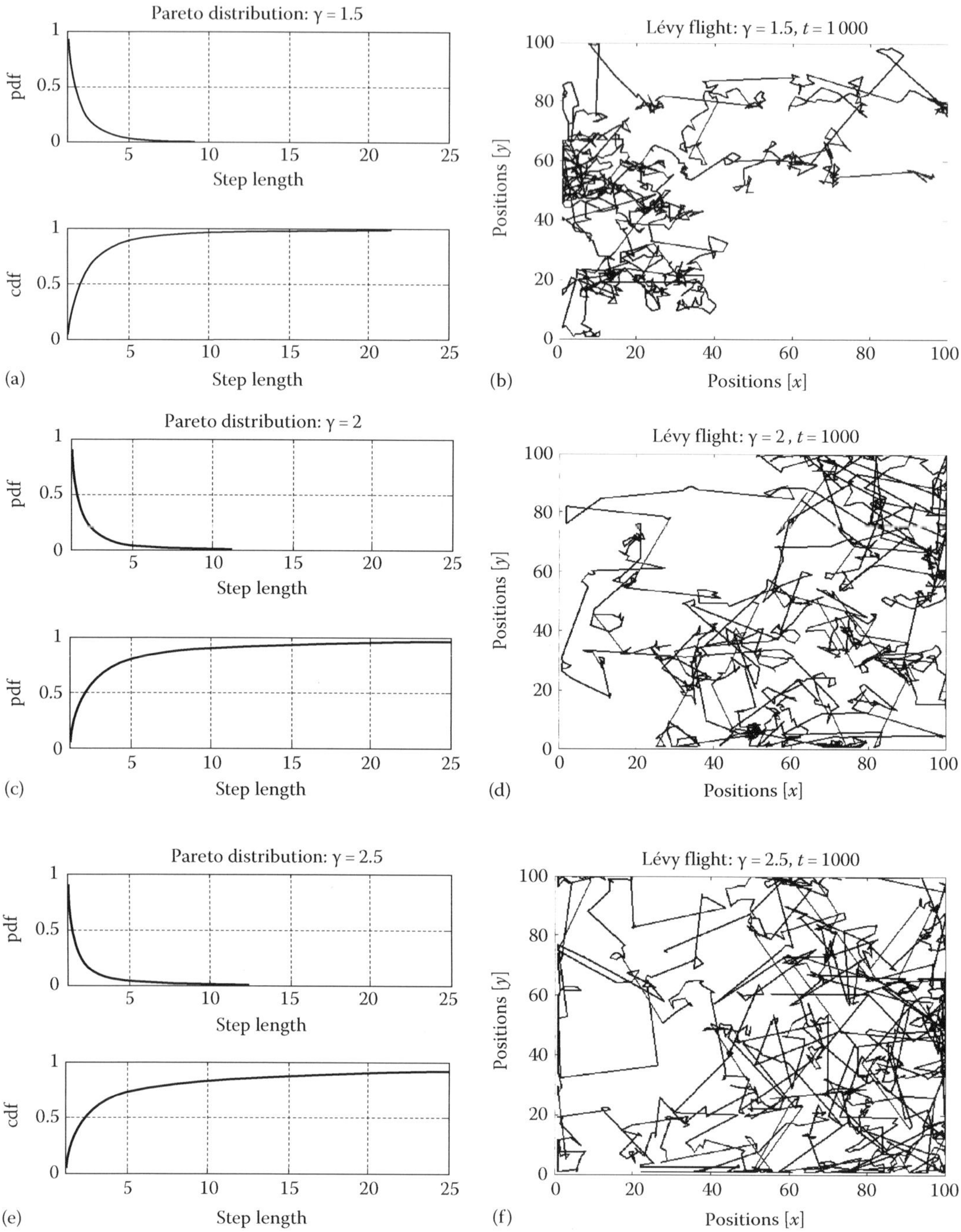

FIGURE 4.17 The probability density functions and cumulative distribution functions of the Pareto distributions of the step lengths (a), (c) and (e), and corresponding 2D Lévy flights (b), (d) and (f) with different values of the parameter γ. In all the cases, the size of the domain is 100×100, minimal step length is $\delta_{min} = 1$, and maximal step length is $\delta_{max} = 25$.

Assume that the agents move over a domain $\mathfrak{X} \subset \mathbb{R}^2$ according to the Lévy flight $\mathcal{L}_t^{(\gamma)2}$ in continuous time $t \in [0,\infty)$, where, as in the preceding text, $1 < \gamma \leq 3$ and the Lévy index is $a = \gamma - 1$. Then, the Fokker–Planck equation, which governs the probabilities of agent's location in the points $\mathfrak{x} \in \mathfrak{X}$, is defined as follows (Chechkin, Sokolov, & Klafter, 2012; Metzler & Klafter, 2000; Viswanathan et al., 2011):

$$\frac{\partial}{\partial t} v(\mathfrak{x},t) = \breve{\sigma}_a^2 |\nabla|^a (v(\mathfrak{x},t)) = \left(\breve{\sigma}_{1a}^2 \frac{\partial^a}{\partial |x|^a} + \breve{\sigma}_{2a}^2 \frac{\partial^a}{\partial |y|^a} \right) v(\mathfrak{x},t), \tag{4.37}$$

where $|\nabla|^a \equiv -(-\nabla^2)^{a/2}$ is a fractional Laplacian, $\partial^a/\partial|x|^a$ and $\partial^a/\partial|y|^a$ are Riesz fractional derivatives, and $\breve{\sigma}_a^2 = (\breve{\sigma}_{1a}^2, \breve{\sigma}_{2a}^2)$ are diffusion rates; it is assumed to have the same value for both x and y axes, that is, $\breve{\sigma}_{1a}^2 = \breve{\sigma}_{2a}^2$.

In general, for the function $f : \mathbb{R}^n \to \mathbb{R}^n$, the Riesz derivative $\mathcal{D}^a f = d^a f/d|x|^a$, $0<a<n$, is defined as follows (Kilbas, Srivastava, & Trujillo, 2006; Samko et al., 2002):

$$\mathcal{D}^a f = \frac{1}{\varkappa(a,n)} \int_{\mathbb{R}^n} \mathfrak{x}^{-1-a} (\mathfrak{I} - \mathfrak{P}_{\mathfrak{x}})^n f d\mathfrak{x}, \quad 0 < a < n, \tag{4.38}$$

where

$$\varkappa(a,n) = -\Gamma(-a) \sum_{k=0}^{n} (-1)^k \binom{n}{k} k^a = \int_0^\infty \frac{(1-e^{-\theta})^n}{\theta^{1+a}} d\theta = \frac{n}{a} \int_0^1 \frac{(1-\theta)^{n-1}}{(\ln(1/\theta))^a} d\theta, \tag{4.39}$$

$\mathfrak{I}$ is an identity operator, and $\mathfrak{P}_{\mathfrak{x}}$ is the Poison operator:

$$\mathfrak{P}_{\mathfrak{x}} f(\mathfrak{x}') = \int_{\mathbb{R}^n} \mathfrak{P}(\zeta,\mathfrak{x}) f(\mathfrak{x}' - \zeta) d\zeta, \quad \mathfrak{x} > 0, \tag{4.40}$$

with the kernel

$$\mathfrak{P}(\zeta,\mathfrak{x}) = \pi^{-(n+1)/2} \Gamma\left(\frac{n+1}{2}\right) \mathfrak{x} \left(|\zeta|^2 + \mathfrak{x}^2\right)^{-(n+1)/2}; \tag{4.41}$$

everywhere above $\Gamma(\cdot)$ stands for the gamma function, $\Gamma(\eta) = \int_0^\infty \beta^{\eta-1} e^{-\beta} d\beta$, $-\infty < \eta < \infty$.

It is clear that for $n = 1$ and the function $f : \mathbb{R} \to \mathbb{R}$, Equation 4.38 becomes

$$\mathcal{D}^a f = \frac{1}{\Gamma(-a)} \int_0^\infty x^{-1-a} \left(f(x) - \int_{-\infty}^{\infty} \frac{x}{|\zeta|^2 + x^2} f(x-\zeta) d\zeta \right) dx, \quad 0 < a < 1 \tag{4.42}$$

Another version of the Riesz derivative $\mathcal{D}^a f$, $f : \mathbb{R} \to \mathbb{R}$, also known as Riesz–Feller derivative, is defined as follows (Gorenflo & Mainardi, 1998; Gorenflo, Mainardi, Moretti, Pagnini, & Paradisi, 2002):

$$\mathcal{D}_\theta^a f = -\left(c_+(a,\theta) \mathcal{J}_+^{-a} f + c_-(a,\theta) \mathcal{J}_-^{-a} f\right), \quad 0 < a \leq 2, \tag{4.43}$$

Where

$$c_{+}(a,\theta)=\sin\left(\frac{\pi}{2}(a-\theta)\right)\Big/\sin(\pi a),\quad c_{-}(a,\theta)=\sin\left(\frac{\pi}{2}(a+\theta)\right)\Big/\sin(\pi a), \tag{4.44}$$

and inversed integral operators $\mathcal{J}_{\pm}^{-a}$ are defined using the Riemann–Liouville integrals as follows:

$$\mathcal{J}_{\pm}^{-a}f=\begin{cases}\pm\dfrac{d}{dx}\mathcal{J}_{\pm}^{1-a}f, & \text{if } 0<a\le 1,\\ \dfrac{d^2}{dx^2}\mathcal{J}_{\pm}^{2-a}f, & \text{if } 1<a\le 2,\end{cases} \tag{4.45}$$

where for any $b\in(0,2]$ (cf. Equation 4.32, which defines the fractional Brownian motion),

$$\mathcal{J}_{+}^{b}f(x)=\frac{1}{\Gamma(b)}\int_{-\infty}^{x}(x-\zeta)^{1-b}f(\zeta)d\zeta,\quad \mathcal{J}_{-}^{b}f(x)=\frac{1}{\Gamma(b)}\int_{x}^{\infty}(\zeta-x)^{b-1}f(\zeta)d\zeta. \tag{4.46}$$

The consideration below follows this definition with $\theta=0$ that, in particular, leads to the equalities $\mathcal{D}_0^1 f=df/dx$ and $\mathcal{D}_0^2 f=d^2f/dx^2$. For detailed consideration of the Riemann–Liouville integrals and the Riesz derivatives see the books by Samko et al. (2002) and by Kilbas et al. (2006). Different applications of these operators are presented by in the book by Tarasov (2010) and in the collective monograph edited by Klafter, Lim, and Metzler (2012).

The dynamics of the location probabilities governed by Equation 4.37 with the Riesz–Feller fractional derivatives given by Equation 4.43 is illustrated by the following example.

Example 4.4

As in the previous text, it is assumed that $\mathfrak{X}$ is a square domain of the size $n=n_x\times n_y$, $n_x=n_y=100$, with the impermeable bounds and that the initial location probabilities are defined by the binormal distribution with the central point $\mathfrak{x}(0)=(50,50)$ and equivalent variances $\sigma_1=\sigma_2=10$ (see Figures 4.14a and 4.8a).

Numerical simulations of Equation 4.37 are implemented using the finite differences scheme, which is built as follows (Gorenflo & Mainardi, 1998; Gorenflo et al., 2002):

Given the time step τ, Equation 4.37 is replaced by usual finite difference equation

$$\frac{1}{\tau}\big(v(\mathfrak{x},t+\tau)-v(\mathfrak{x},t)\big)=\left(\breve{\sigma}_{1a}^2\frac{\partial^a}{\partial|x|^a}+\breve{\sigma}_{2a}^2\frac{\partial^a}{\partial|y|^a}\right)v(\mathfrak{x},t),$$

where fractional derivatives $\partial^a/\partial|x|^a$ and $\partial^a/\partial|y|^a$ are defined using the Riesz–Feller derivatives $\mathcal{D}_\theta^a$ by x and y, respectively, and $\breve{\sigma}_{1a}^2$ and $\breve{\sigma}_{2a}^2$ are diffusion rates; it is assumed that $\breve{\sigma}_{1a}^2=\breve{\sigma}_{2a}^2=1$.

Given the space step h, which is the same for both x and y coordinates, the space derivative $\mathcal{D}_\theta^a$ is approximated as follows:

For $a=1$, 2 the derivative $\mathcal{D}_\theta^a$ is approximated by usual finite differences schemes, namely,

- If $a=1$, then

$$\mathcal{D}_\theta^a v(\mathfrak{x},t)\approx\frac{1}{2}\left(\frac{v((x+h,y),t)-v((x-h,y),t)}{2h}+\frac{v((x,y+h),t)-v((x,y-h),t)}{2h}\right).$$

- If $a = 2$, then

$$\mathcal{D}_\theta^a v(\mathfrak{x},t) \approx \frac{1}{2}\left(\frac{v((x+h,y),t)-2v((x,y),t)+v((x-h,y),t)}{h^2}+\frac{v((x,y+h),t)-2v((x,y),t)+v((x,y-h),t)}{h^2}\right).$$

Otherwise, the definition of the derivative $\mathcal{D}_\theta^a$ given by Equation 4.37 is applied. The Riemann–Liouville integrals $\mathcal{J}_\pm^{-a}$ appearing in the definition of the derivative $\mathcal{D}_\theta^a$ and defined by Equations 4.45 and 4.46 are approximated by the following series:

- If $0 < a < 1$, then

$$\mathcal{J}_\pm^{-a} v(\mathfrak{x},t) = \frac{1}{2h^a}\left(\sum_{k=0}^{n_x}(-1)^k\binom{a}{k}v((x\mp k,y),t)+\sum_{k=0}^{n_y}(-1)^k\binom{a}{k}v((x,y\mp k),t)\right).$$

- If $1 < a < 2$, then

$$\mathcal{J}_\pm^{-a} v(\mathfrak{x},t) = \frac{1}{2h^a}\left(\sum_{k=0}^{n_x}(-1)^k\binom{a}{k}v((x+1\mp k,y),t)+\sum_{k=0}^{n_y}(-1)^k\binom{a}{k}v((x,y+1\mp k),t)\right).$$

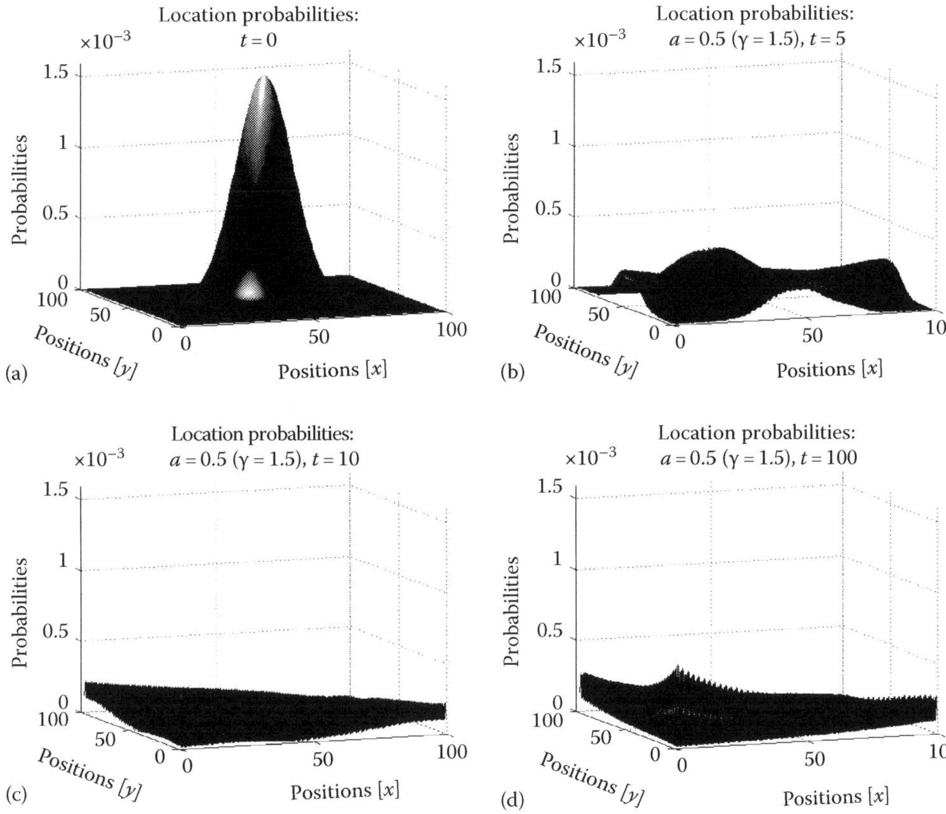

FIGURE 4.18 Location probabilities for the 2D Lévy flight with the Lévy index $a = 0.5$. Figure (a) shows initial location probabilities at $t = 0$, and Figures (b), (c), and (d) show location probabilities at times $t = 5$, $t = 10$, and $t = 100$, respectively.

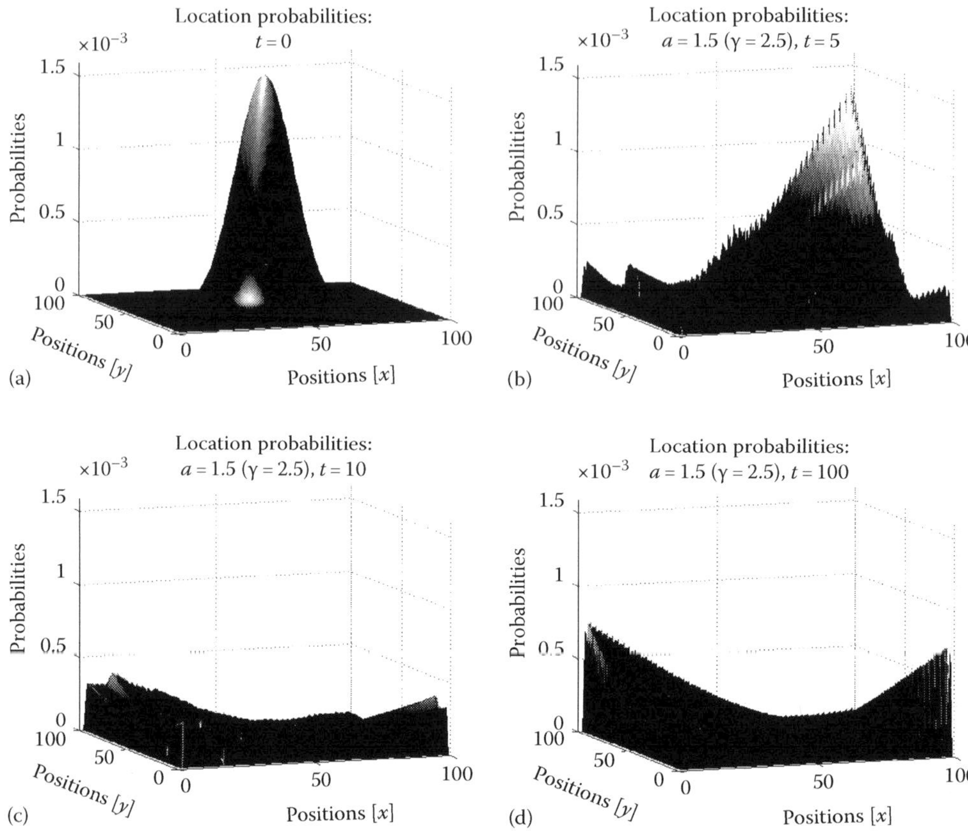

FIGURE 4.19 Location probabilities for the 2D Lévy flight with the Lévy index a = 1.5. Figure (a) shows initial location probabilities at t = 0, and the Figures (b), (c), and (d) show location probabilities at times t = 5, t = 10, and t = 100, respectively.

The agent's location probabilities for the Lévy flights $\mathcal{L}_t^{(\gamma)2}$ with the Lévy indices $a = 0.5$ and $a = 1.5$ (parameters $\gamma = 1.5$ and $\gamma = 2.5$, correspondingly), which were calculated using the presented scheme in the previous text, are shown in Figures 4.18 and 4.19.

It is seen that because of the nonzero probability of the long jumps in the Lévy flight $\mathcal{L}_t^{(\gamma)2}$, $\gamma = a + 1 = 1.5$ (see also Figure 4.17a), the location probabilities are quickly distributed over the domain, and already at the time, $t = 10$ their distribution is close to uniform. However, since the bounds are impermeable, the continuation of the process leads to the oscillations such that the probabilities accumulated at the bounds are slightly greater than in the internal points of the domain. This process is better seen in Figure 4.19.

Since the larger Lévy index ($a = 1.5$ in Figure 4.19 in contrast to $a = 0.5$ in Figure 4.18) leads to more probable long jumps in the Lévy flight $\mathcal{L}_t^{(\gamma)2}$, $\gamma = a + 1 = 2.5$ (see also Figure 4.17c), the location probabilities are quickly accumulated at one of the impermeable bounds (Figure 4.19b). Then they oscillate between the distribution, which is similar to the distribution shown in Figure 4.19c, and the distribution, in which the probabilities again accumulated at the bounds as it is illustrated in Figure 4.19d. ■

Notice that if function f is sufficiently good, and in the considered domain, its Fourier transform $\mathfrak{F}$ exists, and then the Riesz derivative $\mathcal{D}^a f$ is defined as follows (Kilbas, Srivastava, & Trujillo, 2006; Metzler & Klafter, 2000; Samko et al., 2002; Tarasov, 2010):

$$\mathcal{D}^a f = \mathfrak{F}^{-1}\left(|k|^a \left(\mathfrak{F}f\right)(k)\right), \quad 0 < a < 1, \tag{4.47}$$

where $k = 0, 1, 2\ldots$ is a wave mode. Such derivative, also called the Weil derivative (Samko et al., 2002), allows obtaining exact solutions of the Fokker–Planck equation (4.37), where diffusion coefficient $\breve{\sigma}_a^2$ is constant (Metzler & Klafter, 2000) or varies with respect to environmental states (Srokowski, 2009), as well as consideration of Lévy flights in external force fields (Jespersen, Metzler, & Fogedby, 1999).

For completeness, let us also notice that similar to the Lévy flight, the considered fractional Brownian motion $\mathcal{W}_t^{(H)2}$ can be defined using fractional derivative. However, in contrast to the Lévy flight, in the case of fractional Brownian motion, the Fokker–Planck equation includes fractional derivative in time and, instead of the previously used Equation 4.33, is defined as follows (Ibe, 2013):

$$\frac{\partial^a}{\partial |t|^a} v(\mathfrak{x},t) = \left(\breve{\sigma}_1^2 \frac{\partial^2}{\partial x^2} + \breve{\sigma}_2^2 \frac{\partial^2}{\partial y^2} \right) v(\mathfrak{x},t), \tag{4.48}$$

where parameter a is proportional by the Hurst exponent H, that is, $a \sim H \in (0, 1)$. Certainly, if $a = 1$, then Equation 4.48 is a standard diffusion equation with the diffusion rates $\breve{\sigma}_1^2$ and $\breve{\sigma}_2^2$ for x and y axes, respectively. Detailed consideration of such representation of the fractional Brownian motion is presented by Ibe in his recently published book (Ibe, 2013).

Certainly, in addition to the considered Lévy flights, in which the step lengths are governed by the Pareto distribution, the other flights can be defined. In particular, if the steps' lengths are governed by the Cauchy distribution, then the resulting motion forms the Cauchy flight, while their distribution according to the Gaussian law gives the so-called Rayleigh flights (Mandelbrot, 1983). However, in the tasks of search and foraging, the Lévy flights are considered as the most effective model of the agent's motion.

4.3.2 Search and Foraging by Lévy Flights

Probably, one of the first considerations of the Lévy flight as a process that governs the animals' movement appeared in 1988 in the paper by Othmer, Dunbar, and Alt (1988) in the framework of spatial dynamics of populations (see Section 3.2). In this paper, the process was called the kangaroo process, and the study treated it as a sequence of alternating paused and jumps such that either the time of pauses or the velocity of jumps is governed by certain probabilistic rules. The direct references to the Lévy flight as to the process that governs search and foraging behavior appeared in 1996 in the seminal paper by Viswanathan et al. (1996), who also reported that the wandering albatrosses follow the Lévy flight patterns.

In fact, since the Lévy flight includes long jumps, the probability of the Lévy flight's return to the previously visited area is smaller than such a probability of the Brownian walk (Ibe, 2013). Thus, during the same period, the Lévy flyer visits more areas than the Brownian walker, and following the assumption that the animals tend to maximize the probabilities of finding the prey makes Lévy flight more effective for the search and foraging tasks than the Brownian search models (Viswanathan et al., 1999).

Formally, the efficiency of the search is defined by the ratio of the number $n(\vec{\mathfrak{a}})$ of visited areas to the total length $l(\vec{\mathfrak{a}})$ of the searcher's path $\vec{\mathfrak{a}}$ (Viswanathan et al., 1999, 2011). Certainly, this value has the same meaning as the reward rate or net energy intake rate $r_e(\vec{\mathfrak{d}})$, which was applied for analysis of the foraging by patches in Section 3.3 and similarly is denoted by $r(\vec{\mathfrak{a}})$. Then,

$$r(\vec{\mathfrak{a}}) = \frac{n(\vec{\mathfrak{a}})}{l(\vec{\mathfrak{a}})}. \tag{4.49}$$

Using this ratio, the problem of search is defined in nonformal form as follows (cf. Problem 3.1).

Problem 4.1: (Random Search)

In a set of all possible models of random search, find a model that the results in the trajectory $\vec{a}$ such that the rate $r(\vec{a})$ reach its maximum.

Certainly, such definition of the problem is too general and cannot be applied directly for finding the appropriate search model. However, given two models of search, the rate defined by Equation 4.49 provides an effective criterion for comparison of these models and is used in the examples below.

Given that the searcher follows the Lévy flight, usually instead of the rate $r(\vec{a})$, the effectiveness of the search is defined by the expected reward rate $r(\vec{a})$ that is defined as follows (Viswanathan et al., 1999, 2011):

$$r(\vec{a}) = \frac{1}{(\langle\delta\rangle n(\vec{a}))}, \tag{4.50}$$

where

$\langle\delta\rangle$ is an average flight's length
$n(\vec{a})$ is a mean of the number of Lévy flights between two successive target areas

Then, following Problem 4.1, it is required to find the parameters of the Lévy flight such that the rate $r(\vec{a})$ reaches its maximum.

Denote by r_{vis} a vision distance of the searcher that is a maximal distance between the searcher's and the target's locations, at which the searcher detects the target. Given the searcher's location $\breve{x}(t)$ at time t, this distance defines the observed area $a(t)$ such that $a(t) = \{x : \|\breve{x}(t) - x\| \le r_{vis},\ x \in \mathcal{X}\}$, where $\mathcal{X}$ is a search domain and $\|x_1 - x_2\|$ stands for a distance between the points $x_1, x_2 \in \mathcal{X}$. In addition, denote by $\langle l\rangle$ an average distance between two successive locations of the target. In general, the search is conducted as follows (Viswanathan et al., 1999, 2011):

1. If the target is in the direct vision of the searcher, that is, $\|\breve{x}(t) - \hat{x}(t)\| \le r_{vis}$, then the searcher flights directly to the target's location $\hat{x}(t)$ or, in the case of multiple targets, to the nearest target's location.
2. If, in opposite, there is no target in direct vision of the searcher, that is, $\|\breve{x}(t) - \hat{x}(t)\| > r_{vis}$, then the searcher chooses both direction and step length at random according to the specified distribution and moves to the next point.

The required in the line 2 movement to the chosen point can be conducted in two different manners. Namely, if it is defined as a walk, then it is assumed that the searcher walks to the chosen point, and during the walk, it seeks for the target. Otherwise, if this motion is defined as flight, then the searcher flights directly to the chosen point. Following this section, it is assumed that the searcher acts according to the first type of movement; the second type will be considered in Section 4.3.3.

Assume that the search is conducted by the Lévy flight $\mathcal{L}_t^{(\gamma)2}$ with the parameter $\gamma \in (1,3]$, and the Lévy index is $a = \gamma - 1$. Let $n(\langle l\rangle)$ be a mean of the number of Lévy flights, which were conducted by the searcher for covering the average distance $\langle l\rangle$ between two successive locations of the target, and let $n(\langle l\rangle)$ be a true number of the Lévy flights between such target's locations. Consider two possible activities of the searcher (Viswanathan et al., 1999, 2011):

1. Destructive search, in which the searcher acts as a predator and "eats" the target or the prey when it is detected; in the case of probabilistic search, such a behavior means that the searcher zeroes the target's location probabilities in the observed areas.

2. Nondestructive search, in which the searcher counts the number of detected targets, but does not destroy them and does not zero the target's location probabilities; consequently, in such scenario, the searcher can revisit the previously visited areas.

In the case of destructive search, the target's location probabilities are governed by pdf u such that $u(x,t)$ is a conditional probability that the target or prey is located in the point x at time t given that it was not detected up to this time. In contrast, in the case of destructive search, these probabilities are governed by the function $\hat{u}$ such that $\hat{u}(x,t)$ is a joint probability that the target or prey is located in the point x at time t and was not detected up to this time. The discourse formally follows this convention and formally distinguishes between the functions u and $\hat{u}$. However, since the algorithms presented in this section consider the target's location probabilities as environmental states bounded by zero and unit and do not require any other probabilistic properties, the distinction between u and $\hat{u}$ does not have effects on the algorithms' actions.

In the destructive search, it holds true that the mean number $\mathrm{n}(\langle l\rangle)$ of the flights between two target's locations is

$$\mathrm{n}(\langle l\rangle) \sim \left(\frac{\langle l\rangle}{\mathrm{r}_{\mathrm{vis}}}\right)^{\gamma-1}, \quad 1<\gamma\leq 3. \tag{4.51}$$

Recall that in the nondestructive search, the searcher can revisit the previously visited areas. Thus, the true number $n(\langle l\rangle)$ of flights between two successive target's locations is not less than the mean number $\mathrm{n}(\langle l\rangle)$ of the flights, that is, $n(\langle l\rangle)\geq \mathrm{n}(\langle l\rangle)$. Assume that $n(\langle l\rangle)\sim\sqrt{\mathrm{n}(\langle l\rangle)}$. Then, from Equation 4.51, it follows that

$$n(\langle l\rangle) \sim \left(\frac{\langle l\rangle}{\mathrm{r}_{\mathrm{vis}}}\right)^{\frac{\gamma-1}{2}}. \tag{4.52}$$

Finally, since for the considered model of search it holds true that

$$\langle\delta\rangle = \left(\frac{\gamma-1}{2-\gamma}\right)\left(\frac{\langle l\rangle^{2-\gamma}-\mathrm{r}_{\mathrm{vis}}^{2-\gamma}}{\mathrm{r}_{\mathrm{vis}}^{1-\gamma}}\right)+\frac{\langle l\rangle^{2-\gamma}}{\mathrm{r}_{\mathrm{vis}}^{1-\gamma}},$$

and the value $\mathrm{n}(\vec{a})$ required by Equation 4.50 is exactly the mean number $\mathrm{n}(\langle l\rangle)$ defined by Equation 4.52, one obtains that the rate $r(\vec{a})$ reaches its maximum when the parameter γ is (Viswanathan et al., 1999, 2011)

$$\gamma \sim 2-\left(\ln\left(\frac{\langle l\rangle}{\mathrm{r}_{\mathrm{vis}}}\right)\right)^{-2}. \tag{4.53}$$

Notice again that in Equation 4.53, the average distance $\langle l\rangle$ between the target's locations, which is the characteristic of the domain, the vision distance $\mathrm{r}_{\mathrm{vis}}$, is the parameter, which characterizes the searcher's abilities, and resulting value γ is the parameter, which specifies the searcher's movement. It is clear that, in general, the value of the parameter γ is $\gamma\neq 3$, and so the motion of the searcher differs from the Brownian motion, and in the most cases when $\gamma<3$, the searcher moves according to the Lévy flight.

Additional argument in favor of the Lévy flights was provided by previous considerations of the groups of Lévy flyers, which demonstrated that the territory covered by a large number of Lévy flyers is greater than the territory covered by the same number of Brownian walkers

(Berkolaiko & Havlin, 1997; Larralde et al., 1992a, 1992b). The last property will be considered in Section 5.2 in the context of coalitional search and foraging.

As indicated in the previous text, after publication of the papers by Viswanathan et al. (1996, 1999), Lévy flights were applied for description of a wide class of search and foraging activities. For brief reviews, see, for example, the papers by Viswanathan et al. (2000) and by Burrow, Baxter, and Pitchford (2008), and for detailed consideration, see the recent book by Viswanathan et al. (2011).

The success of the Lévy flight model of search and foraging led to the formulation of the Lévy flight foraging hypothesis. This hypothesis was suggested in 2008 in a review (Viswanathan, Raposo, & da Luz, 2008, p. 135), and in its strong form it states that

> since Lévy flights (...) optimize random searches, biological organisms must have therefore evolved to exploit Lévy flights.

Later, the Lévy flight foraging hypothesis was formulated in a weaker form as follows (Viswanathan et al., 2011, pp. 85–86):

> Superdiffusive motion governed by fat-tailed propagators optimizes encounter rates under specific (but common) circumstances; hence some species must have evolved mechanisms that exploit these properties of Lévy walks.

Certainly, the foraging hypothesis assumes that Pyke's hypothesis (Pyke, 1984, p. 524) that "animals forage optimally" is true.

The algorithm of Lévy flight is obvious (cf. Algorithm 4.3 of fractional Brownian search).

Algorithm 4.4: (Lévy Flight Search)

Given a domain $\mathfrak{X} = \{\mathfrak{x}_1, \mathfrak{x}_2, \ldots, \mathfrak{x}_n\}$ and parameter $\gamma \in (1,3]$ do:

1. Observe the current location $\breve{\mathfrak{x}}(t)$ and decide either to terminate the search or to continue.
2. If the search continues, choose the next point $\breve{\mathfrak{x}}(t+1)$ according to the Lévy flight $\mathcal{L}_t^{(\gamma)2}$.
3. Move to the chosen point $\breve{\mathfrak{x}}(t+1)$ and continue with its observation (line 1).

Notice that in this algorithm, parameter γ is constant. The actions of Algorithm 4.4 with comparison with the Brownian search given by Algorithm 4.1 are illustrated by the following example.

Example 4.5

Likewise the previous examples, assume that the searcher acts in the square domain $\mathfrak{X}$ of the size $n = 100 \times 100$ with the impermeable bounds, and consider the motion of the searcher according to the simple Brownian walk $\mathcal{W}_t^2$ and the Lévy flight $\mathcal{L}_t^{(\gamma)2}$. Assume that the step δ of the searcher in the Brownian walk $\mathcal{W}_t^2$ is $\delta > 1$ and that it is also the mean of the step in the Lévy flight $\mathcal{L}_t^{(\gamma)2}$. Then, the Lévy index a of the flight $\mathcal{L}_t^{(\gamma)2}$ is $a = \gamma - 1 = \delta/(\delta - 1)$. In the example, it is specified that $\delta = 3$ and so $a = 1.5$. As in the previous text, it is assumed that both the Brownian search and the Lévy flight search start in the point $\breve{\mathfrak{x}}(0) = (50,50)$, and the search continues up to the time 1000, so $t = 0,\ldots,1000$. The trajectories of the searcher are shown in Figure 4.20; Figure 4.20a depicts the Brownian search trajectory, and Figure 4.20b depicts the Lévy flight search trajectory. The areas observed by the searcher are depicted by gray color.

Statistical comparison between the Brownian search and the Lévy flight search demonstrates the following. Over 1000 trials, the average number of the first-time visited areas by the Brownian searcher is 1531.78 and by Lévy flight searcher is 2195.224; thus, the Lévy flight search is 2195.224/1531.78 = 1.43 times more effective than the Brownian search. Distributions of the number of visited areas over 1000 trials are shown in Figure 4.21.

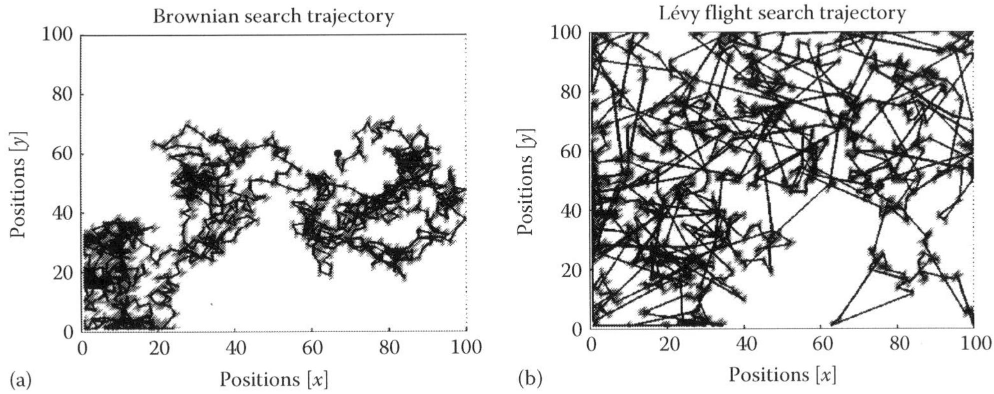

FIGURE 4.20 Trajectories of the searcher and observed areas: (a) Brownian search trajectory and (b) Lévy flight search trajectory. The observed areas along the trajectories are depicted by gray color.

Certainly, the obtained distributions are strongly different (as it is also confirmed by the Kolmogorov–Smirnov test).

For additional stressing, an advantage of the Lévy flight search in comparison with the Brownian search, let us consider the foraging in the domain, in which the prey is located in certain regions, while the rest of the domain is empty. In the regions with prey, it is assumed that each point $\mathfrak{x} \in \mathfrak{X}$ contains exactly one unit of prey. The trajectories of the searcher starting from the point $\breve{\mathfrak{x}}(0) = (50,50)$ are shown in Figure 4.22; as in the preceding text, the search continues up to the time 1000. Figure 4.22a depicts the Brownian search trajectory, and Figure 4.22b depicts the Lévy flight search trajectory. The regions, which include the prey, are depicted by gray color, and the empty regions are shown in black color.

The similar statistical comparison as earlier of the quantity eaten prey over 1000 trials demonstrates that the Brownian searcher eats 513.26 prey units at average and Lévy flight searcher eats 700.65 prey units at average; thus, the Lévy flight search in the considered domain is 700.65/513.26 = 1.37 times more effective than the Brownian search, which is similar to the previous result. Distributions of the eaten prey over 1000 trials are shown in Figure 4.23.

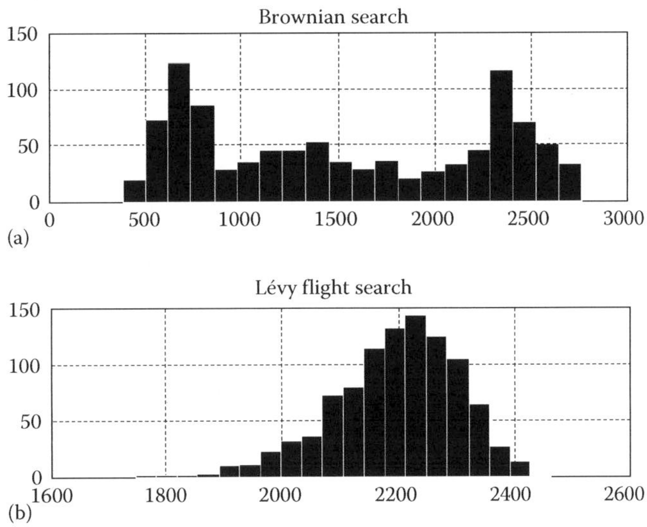

FIGURE 4.21 Distributions of the numbers of the first-time visited areas over 1000 trials: (a) Brownian search and (b) Lévy flight search.

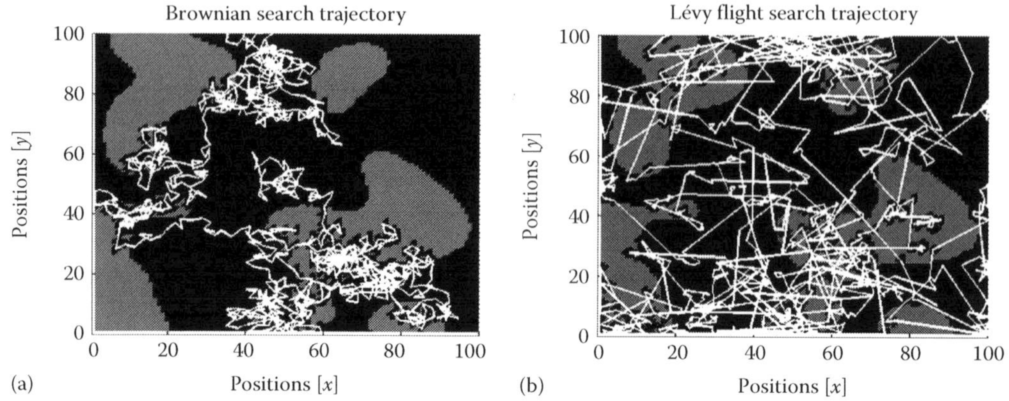

FIGURE 4.22 The domain with prey (gray regions) and the trajectories of the searcher and observed areas: (a) Brownian search trajectory and (b) Lévy flight search trajectory.

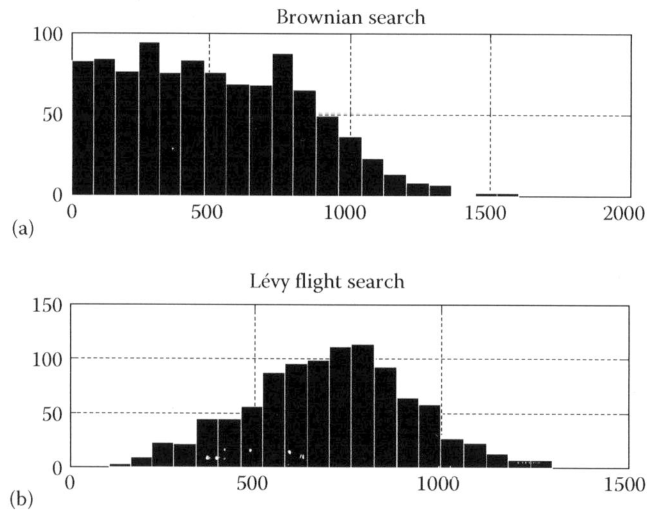

FIGURE 4.23 Distributions of the eaten prey over 1000 trials: (a) Brownian search and (b) Lévy flight search.

As in the preceding text, the obtained distributions are strongly different (as it is also confirmed by the Kolmogorov–Smirnov test). However, in contrast to the previous case, there are the trials of Brownian search with zero eaten prey, while the Lévy flight search does not demonstrate the trials with zero eaten prey. ■

The considered example demonstrates the effectiveness of the Lévy flight search with constant parameter γ in comparison with the simple Brownian search. In addition, it stresses that if the domain includes the empty regions, then Brownian search strategy does not guarantee finding at least one unit of the prey during long but finite time, while the Lévy flight search results in finding the prey during the same time. In addition, notice that the quantity of eaten prey by the Lévy flight searcher is proportional to the territory, which contains the prey. The last observation leads to the consideration of the search by the Lévy flight such that its characteristics depend on the states of the environment or the probabilities of finding the prey in the points of the domain that, in the terms of search and screening theory, are the target's location probabilities. Moreover, in the recent papers (Edwards et al., 2007; Humphries et al., 2010), the authors reported that the environmental states are the main factor that leads to the observed motion patterns of the predators.

There are several methods of specifying the characteristics of the Lévy flight search using the environmental states or target's location probabilities. In particular, in 2007, Pavlyukevich (2007) suggested to consider the search by Lévy flights with the varying parameter $\gamma \in (1,3]$. In particular, if the values of γ are specified with respect to the target's location probabilities $\hat{u}(\mathfrak{x},t)$ in points $\mathfrak{x} \in \mathfrak{X}$ of the domain, such flight leads to the following search algorithm (cf. Algorithm 4.3 of fractional Brownian search).

Algorithm 4.5: (Lévy Flight Search with Varying Index)

Given a domain $\mathfrak{X} = \{\mathfrak{x}_1, \mathfrak{x}_2, \ldots, \mathfrak{x}_n\}$ and the target's location probabilities $\hat{u}(\mathfrak{x},t)$ do:

1. Observe the current location $\breve{\mathfrak{x}}(t)$ and decide either to terminate the search or to continue.
2. If the search continues, define the parameter $\gamma(\mathfrak{x},t)$ with respect to $\hat{u}(\mathfrak{x},t)$.
3. Choose the next point $\breve{\mathfrak{x}}(t+1)$ according to the Lévy flight $\mathcal{L}_t^{(\gamma)2}$.
4. Move to the chosen point $\breve{\mathfrak{x}}(t+1)$ and continue with its observation (line 1).

Following the Lévy flight foraging hypothesis indicated earlier, this algorithm is optimal in the sense that it maximizes the encounter rate with the target with respect to the target's location probabilities. The actions of the algorithm are illustrated by the following example.

Example 4.6

Similar to the previous examples, assume that the searcher acts in the square domain $\mathfrak{X}$ of the size $n = 100 \times 100$ with the impermeable bounds. The states of the domain are defined according to the target's location probabilities $\hat{u}(\mathfrak{x},t)$. As in the preceding text, these probabilities are specified by ten binormal distributions with random centers and the variances $\sigma_1 = \sigma_2 = 10$. Then the Lévy indices $a(\mathfrak{x})$ for the points $\mathfrak{x} \in \mathfrak{X}$ are defined by $a(\mathfrak{x},t) = 2(1-u(\mathfrak{x},t))$ that are normalized such that $\min_{\mathfrak{x}} a(\mathfrak{x},0) = \varepsilon > 0$, $\varepsilon \ll 1$, and $\max_{\mathfrak{x}} a(\mathfrak{x},0) = 2$. Consequently, the parameters $\gamma(\mathfrak{x},t)$ are $\gamma(\mathfrak{x},t) = a(\mathfrak{x},t)+1$. The trajectory of the searcher defined by the Lévy flight $\mathcal{L}_t^{(\gamma)2}$, $t = 0,\ldots,1000$, starting from the point $\breve{\mathfrak{x}}(0) = (50,50)$ is shown in Figure 4.24b, and Figure 4.24a depicts the values of the parameter γ and the lengths of the steps along the trajectory.

According to the figure, during its motion, the searcher changes its behavior pattern according to the target's location probabilities. For the greater probabilities (areas in white color), the searcher tends to make short jumps, while for the smaller probabilities (regions in black color), it tends to make long flights. However, notice that in contrast to the fractional Brownian motion considered in Example 4.3, in the Lévy flight, the step lengths are independent random variables. ■

The motion presented by Example 4.6 mimics the foraging behavior of flying insects like honeybees and of birds. In certain sense, such behavior can also be used for modeling the movements of the mammals, which run by long-term tracks between the patches with prey and then search by short steps in the patches. Notice again that such technique is similar to the considered fractional Brownian search with varying Hurst parameter; however, in the case of the Lévy flight search, the target's location probabilities specify distribution of the step length, while in the case of fractional Brownian search, these probabilities define correlation between the step lengths.

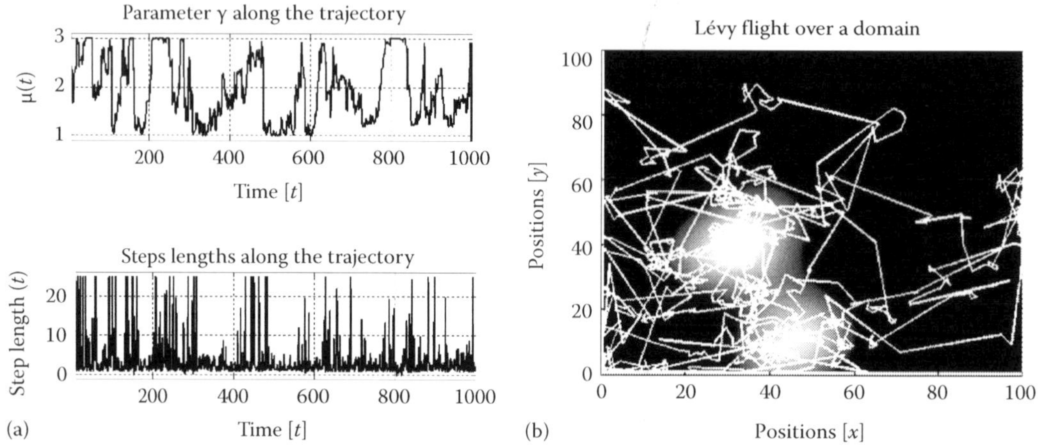

FIGURE 4.24 Lévy flight with the parameter γ varying according to the target's location probabilities: (a) the values of the parameter γ and the step lengths along the trajectory and (b) trajectory of the Lévy flight over a domain.

Additional possibility for utilizing the target's location probabilities is provided by specification of the Lévy flight with respect to environmental states (Srokowski, 2009) and to the external field (Jespersen et al., 1999). The Fokker–Planck equation, which defines the Lévy flight in nonhomogeneous environment, is specified as follows (Srokowski, 2009):

$$\frac{\partial}{\partial t} v(\mathfrak{x},t) = \breve{s}(\mathfrak{x})|\nabla|^{a}\left(v(\mathfrak{x},t)\right) - \breve{m}(\mathfrak{x})\nabla v(\mathfrak{x},t)$$

$$= \left(\breve{s}_1(\mathfrak{x})\frac{\partial^a}{\partial|x|^a} + \breve{s}_2(\mathfrak{x})\frac{\partial^a}{\partial|y|^a}\right)v(\mathfrak{x},t) - \left(\breve{m}_1(\mathfrak{x})\frac{\partial}{\partial x} + \breve{m}_2(\mathfrak{x})\frac{\partial}{\partial y}\right)v(\mathfrak{x},t), \qquad (4.54)$$

where, similar to the operator $V\left(v;\|\breve{s}\|,\|\breve{m}\|\right)$ defined in Section 3.2, function $\breve{s}$ specifies the diffusion rates in the points $\mathfrak{x}$ of the domain $\mathfrak{X}$ and function $\breve{m}$ specifies the drifts in these points. Notice that in this equation, the drift term, which corresponds to the influence of the external field, implements usual partial derivatives of the first order, while the diffusion term, which specifies the Lévy flight itself, is defined by the Riesz derivative. The next example illustrates the dynamics of the searcher's location probabilities governed by Equation 4.54 with constant diffusion rates $\breve{s}_1(\mathfrak{x}) = \breve{s}_2(\mathfrak{x}) = \text{const}$ and the drifts defined by the gradients of the target's location probabilities in the points $\mathfrak{x}$.

Example 4.7

As in the preceding text, it is assumed that $\mathfrak{X}$ is a square domain of the size $n = 100 \times 100$ with the impermeable bounds. Similar to the examples presented in Sections 2.2.1 and 2.3.2 (see Figures 2.2 and 2.7), assume that the target's location probabilities are defined by two binormal distributions with the centers in the points $\mathfrak{x}_1 = (30,70)$ and $\mathfrak{x}_2 = (70,30)$ and deviations $\sigma = (\sigma_x, \sigma_y)$ for the axes x and y are $\sigma_1 = (30, 30)$ and $\sigma_2 = (20, 20)$, respectively. Distribution $\hat{u}$ of the target's location probabilities is shown in Figure 4.25a. It is assumed that the target's location probabilities are constant in time and do not depend on the searcher's location probabilities.

Similar to the previous examples, assume that initial location probabilities are defined by the binormal distribution with the central point $\breve{\mathfrak{x}}(0) = (50,50)$ and equivalent variances $\sigma_1 = \sigma_2 = 10$; these probabilities are shown in Figure 4.25b.

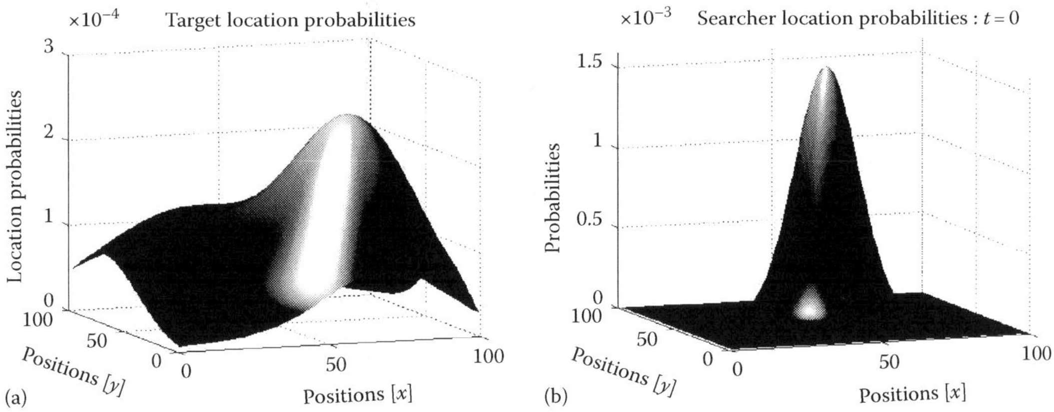

FIGURE 4.25 (a) Target's location probabilities and (b) initial searcher's location probabilities at $t = 0$.

Consider the dynamics of the searcher's location probabilities governed by Equation 4.54 with constant diffusion rates $\breve{s}_1(\mathbf{x}) = \breve{s}_2(\mathbf{x}) = \breve{\sigma}_1^2 = \breve{\sigma}_2^2 = \frac{1}{2}$ and drifts $\breve{m}_1(\mathbf{x}) = \frac{\partial}{\partial x}\hat{u}(\mathbf{x},t)$ and $\breve{m}_2(\mathbf{x}) = \frac{\partial}{\partial y}\hat{u}(\mathbf{x},t)$. In other words, the equation, which governs the searcher's location probabilities, is written as follows:

$$\frac{\partial}{\partial t}v(\mathbf{x},t) = \frac{1}{2}\left(\frac{\partial^a}{\partial|x|^a} + \frac{\partial^a}{\partial|y|^a}\right)v(\mathbf{x},t) - \left(\frac{\partial}{\partial x}\hat{u}(\mathbf{x},t)\frac{\partial}{\partial x} + \frac{\partial}{\partial y}\hat{u}(\mathbf{x},t)\frac{\partial}{\partial y}\right)v(\mathbf{x},t),$$

where $\partial^a/\partial|x|^a$ and $\partial^a/\partial|y|^a$ are Riesz fractional derivatives of the order a; if $a = 2$, then the searcher's location probabilities follow usual diffusion with the indicated drifts. The results of numerical simulations of the dynamics of the searcher's location probabilities are shown in Figure 4.26. Figure 4.26a shows the dynamics of the probabilities with $a = 2$ that corresponds to the diffusion with drift (or correlated Brownian motion), and Figure 4.26b shows the dynamics with $a = 1.5$ that corresponds to the correlated Lévy flight. In both cases, the searcher starts with initial distribution shown in Figure 4.25b, and the drift is defined by the target's location probabilities shown in Figure 4.25a. ■

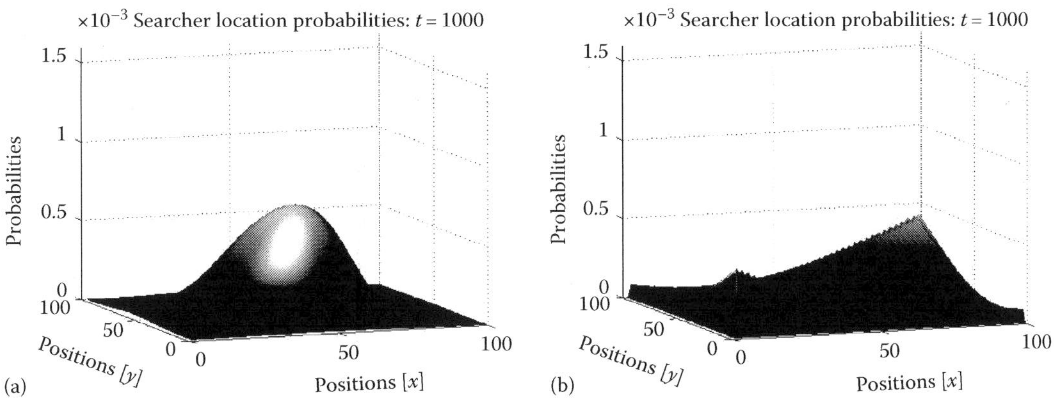

FIGURE 4.26 The searcher's location probabilities: (a) probabilities defined by the diffusion with drift at $t = 1000$ and (b) probabilities that correspond to the drifted Lévy flight with the Lévy index $a = 1.5$ at $t = 100$. In both cases, the drift is defined by the gradient of the same target's location probabilities.

As it is expected, the probabilities defined by the drifted diffusion and shown in Figure 4.26a follow the target's location probabilities. In contrast, the probabilities, which correspond to the Lévy flight (see Figure 4.26b), do not follow the target's location probabilities. However, notice because of the drifts, the resulting distribution shown in Figure 4.26b differs from the distribution shown in Figure 4.19d, which corresponds to the dynamics without drifts.

From the presented consideration earlier, it follows that the Lévy flight search highly outperforms the Brownian search in the sense of the number of the first-time visited areas (Example 4.5). In addition, the presented examples demonstrate that using the varying Lévy index, the motion can be controlled with respect to the target's location probabilities (Example 4.6) and that the distribution of the searcher's location probabilities is sensible to the drifts (Example 4.7). These observations can be summarized in the following unified algorithm.

Similar to the correlated Wiener process, formally denote the Lévy flight with drift (i.e., the correlated Lévy flight) by $\mathcal{L}_t^{(c,\gamma)2}$. Then the algorithm of the Lévy flight search with drifts and varying index is outlined as follows.

Algorithm 4.6: (Lévy Flight Search with Drifts and Varying Index)

Given a domain $\mathfrak{X} = \{\mathfrak{x}_1, \mathfrak{x}_2, \ldots, \mathfrak{x}_n\}$ and the target's location probabilities $u(\mathfrak{x}, t)$ do:

1. Observe the current location $\breve{\mathfrak{x}}(t)$ and decide either to terminate the search or to continue.
2. If the search continues, define the parameter $\gamma(\mathfrak{x}, t)$ and the drifts $\bar{m}_1(\mathfrak{x}, t)$ and $\bar{m}_2(\mathfrak{x}, t)$ with respect to $u(\mathfrak{x}, t)$.
3. Choose the next point $\breve{\mathfrak{x}}(t+1)$ according to the drifted Lévy flight $\mathcal{L}_t^{(c,\gamma)2}$.
4. Move to the chosen point $\breve{\mathfrak{x}}(t+1)$ and continue with its observation (line 1).

Notice that this algorithm differs from Algorithm 4.5 only in the definition of the drifts $\bar{m}_1(\mathfrak{x}, t)$ and $\bar{m}_2(\mathfrak{x}, t)$; in Algorithm 4.5 the drifts are zero, and in this algorithm, they depend on the target's location probabilities (also cf. Algorithm 4.3 of fractional Brownian search). The actions of Algorithm 4.6 are illustrated by the following example.

Example 4.8

The setting of the example is equal to the setting of Example 4.6. The searcher acts in the square domain $\mathfrak{X}$ of the size $n = 100 \times 100$ with the impermeable bounds. The states of the domain are defined according to the target's location probabilities $\hat{u}(\mathfrak{x}, t)$ that are specified by ten binormal distributions with random centers and the variances $\sigma_1 = \sigma_2 = 10$. Then the Lévy indices $a(\mathfrak{x},\ t)$ for the points $\mathfrak{x} \in \mathfrak{X}$ are defined by $a(\mathfrak{x}, t) = 2(1 - u(\mathfrak{x}, t))$ that are normalized such that $\min_{\mathfrak{x}} a(\mathfrak{x}, 0) = \varepsilon > 0$, $\varepsilon \ll 1$, and $\max_{\mathfrak{x}} a(\mathfrak{x}, 0) = 2$. Consequently, the parameters $\gamma(\mathfrak{x}, t)$ are $\gamma(\mathfrak{x}, t) = a(\mathfrak{x}, t) + 1$. Similar to Example 4.7, the drifts are defined by the gradient of u in the point $\mathfrak{x}$ at time t that defines the direction of the next step of the searcher, and it is assumed that the searcher follows the direction with maximal gradient (ties are broken randomly). In addition, similar to Example 4.5, the values $a(\mathfrak{x}, t)$ of the index in the observed points are decreased to small values $\varepsilon > 0$, $\varepsilon \ll 1$.

The trajectories of the searcher starting from the point $\breve{\mathfrak{x}}(0) = (50, 50)$ and up to the time $t = 1000$ are shown in Figure 4.27. For comparison, Figure 4.27b presents the trajectory of the drifted Brownian search, Figure 4.27c depicts the trajectory of the drifted Lévy flight search with constant Lévy index $a = 1.5$, and Figure 4.27d shows the Lévy flight search with the drifted Lévy index defined by the target's location probabilities. Figure 4.27a shows the domain and the starting point of the searcher.

Statistical comparison between the Brownian search and Lévy flight searches with constant and varying Lévy indices was conducted in the same manner as in Example 4.5. However, the number

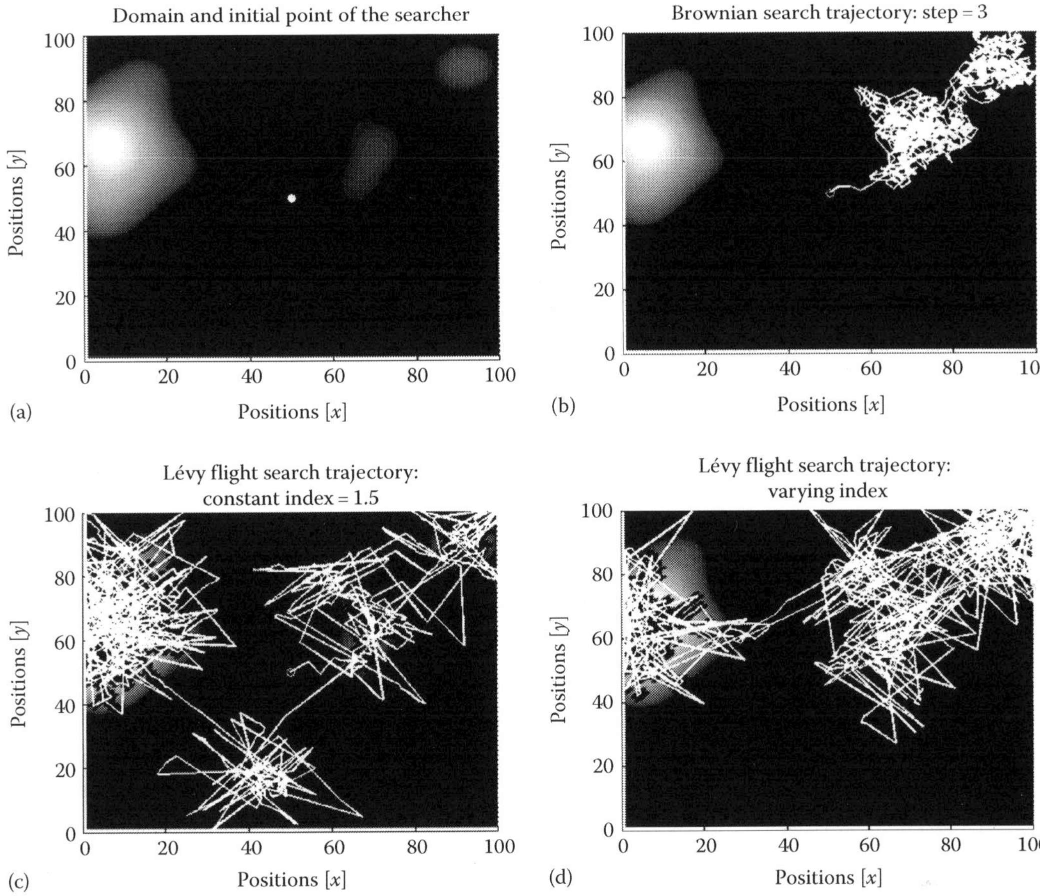

FIGURE 4.27 Domain and trajectories of the searcher governed by drifted Brownian walk and drifted Lévy flights with constant and varying Lévy indices. (a) Domain with target's location probabilities and initial point of the searcher $\breve{x}(0) = (50, 50)$, (b) drifted Brownian search with constant step length $\delta = 3$, (c) drifted Lévy flight search with constant Lévy index $a = 1.5$, and (d) drifted Lévy flight search with the varying Lévy index. In all the cases, the drifts are defined by the gradients of the target's location probabilities.

of the first-time visited areas was weighted with respect to the normalized Lévy indices $a(x, t)$ such that the weights are in the interval between 0 for observed areas and 1 for the areas with $a(x, t) = 2$. This number is referred as the weighted number of the first-time visited areas. Notice that if the index a is constant and so the weights are unit, then the indicated measure gives a number of the first-time visited areas, that is, the measure, which was used in Example 4.5. Over 1000 trials, the average weighted number of the first-time visited areas is the following. For the Brownian search with the step $\delta = 3$ (see Figure 4.27b). it is 795.588: for the Lévy flight search with constant index $a = 1.5$ (see Figure 4.27c). it is 1061.526: and for the Lévy flight search with the index $a(x,t) \in (0,2]$ varying with respect to the probabilities $1 - u(x,t)$ (see Figure 4.27d), it is 1078.237. Thus, both Lévy flight searches with drifts are nearly 1.35 times more effective than the considered Brownian search with drifts. Distributions of the weighted number of visited areas over 1000 trials are shown in Figure 4.28.

The Kolmogorov–Smirnov test (with significance value 0.05) shows that the distribution that corresponds to the Brownian search (see Figure 4.28a) is statistically different from both the distributions that correspond to both Lévy flight searches (see Figure 4.28b and c), while the distributions that correspond to the Lévy flight searches are statistically equivalent. ■

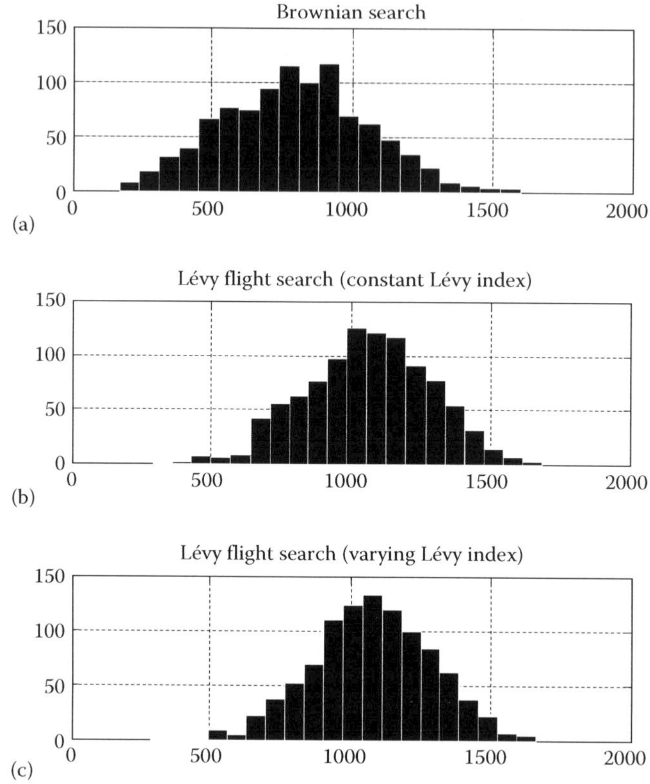

FIGURE 4.28 Distributions of the weighted numbers of the first-time visited areas over 1000 trials: (a) Brownian search, (b) Lévy flight search with constant Lévy index $a = 1.5$, and (c) Lévy flight search with varying Lévy index $a(\mathbf{x},t) \in (0, 2]$.

The considered example demonstrates that also in the case of drifted motion, the Lévy flight search is more effective that the Brownian search with the same rate as for the nondrifted motion. In addition, it shows that the linear dependence of the Lévy index on the environmental states does not influence the effectiveness of the search. However, as it follows from Examples 4.7 and 4.8, the dependence of the Lévy index on the environmental states results in the distribution of the searcher's location probabilities and, consequently, in the searcher's trajectory that differ from the distribution and trajectory, which correspond to the constant Lévy index.

The presented methods and algorithms provide a description of the searcher's movement that can be considered global in such a sense that the process, which governs the searcher's movement, is defined before starting the search (in the considered cases, there are different kinds of the Brownian walk or the Lévy flight). Then, the searcher moves according to the specified process with or without changing its characteristics with respect to the environmental states or the target's location probabilities. The only change of the movement's pattern is based on the properties of the Lévy flights that for certain values of the Lévy index are reduced to the Brownian motion, and any other changes of the movement's pattern are impossible.

Recently, such approach, especially the application of the processes with constant parameters, was strongly criticized in the series of papers by Plank and colleagues (Codling, Plank, & Benhamou, 2008; James, Plank, & Edwards, 2011; Plank & James, 2008), who indicated the crucial role of the environment. In particular, they argued that the observed Lévy flight patterns in the animal motion resulted not only by the process, that governs the animals' movement, but by these processes given the environmental states, while the processes themselves in homogeneous environment may differ from the Lévy flights. As indicated in the preceding text, the same conclusions were done by Edwards et al. (2007) and Humphries et al. (2010) based on field observations and analysis of the predators' motion. However, notice that later, the same authors also reported on the observed success of the Lévy flights in biological foraging (Humphries et al., 2012).

The search process, which includes several motion patterns, is often called composite search model (Plank & James, 2008) or the search with intermittent strategies. The next section briefly addresses some of such models, especially the model suggested in 2005–2007 by Benichou and collegues (Benichou et al., 2005, 2006, 2007) and then intensively studied (Benichou et al., 2011).

4.3.3 Intermittent Strategies and Relation with Foraging by Patches

The considered models of search provide the "global" description of the searcher's motion such that the process, which governs its activity, is specified beforehand and does not change during the search up to its termination. The dependence of the process on the environment was defined by certain relation of the parameters of the predefined process, in the considered cases the Lévy flight, while the type of the process was preserved. Certainly, such activity requires minimal sensing abilities of the searcher and does not require any decision making.

In this section, in opposite, it is assumed that the searcher is able both to perceive the states of the environment, in particular, the target's location probabilities, and to change the motion pattern with respect to the perceived information (Bartumeus, 2009; Bartumeus & Catalan, 2009). Such search model is known as composite search model (Plank & James, 2008) or the search with intermittent strategies (Benichou et al., 2005, 2006, 2007, 2011).

Assume that the searcher conducts a destructive search (see Section 4.3.2), that is, it destructs the detected targets or, in the case of probabilistic search, it zeroes the probabilities of the target's location in the observed areas. Consider the searcher, which interchanges between slow and fast motions and acts as follows (Benichou et al., 2011):

Phase 1: Slow motion with screening, during which the searcher is able to detect the target if it is in the direct vision of the searcher that, as in the preceding text, is $\left\|\breve{x}(t)-\hat{x}(t)\right\| \leq r_{\text{vis}}$, where $\breve{x}(t)$ is the searcher's location at time t, $\hat{x}(t)$ is the target's location at time t, and $r_{\text{vis}} < \infty$ is a vision distance.

In other words, the searcher detects the target if it is in the current observed area $a(t)$. Such motion can be conducted in three possible manners:

1. Static mode, in which the searcher is immobile and stays in its location $\breve{x}(t)$ for a certain period. If the target is in the observed area, $\hat{x}(t) \in a(t)$, the searcher detects it with some probability depending on the period, during which the searcher observes the area $a(t)$. This case is similar to the basic constrained search (see Section 2.2.1).
2. Diffusive mode, in which the searcher moves according to Brownian motion with certain diffusion rate and immediately detects the target if it is in the current observed area $a(t)$. This mode is the same as a simple Brownian search (see Section 4.2.1) and certainly can be extended to correlated fractional Brownian searches (see Sections 4.2.2 and 4.2.3).
3. Ballistic mode, in which the searcher during certain period moves by a straight line in an arbitrary direction and immediately detects the target if it is in the current observed area $a(t)$. Such behavior corresponds to the Lévy walk search (see Section 4.3). Notice that in contrast to the Lévy flight, the Lévy walker moves with the constant speed and screens the domain along its trajectory.

Phase 2: Fast relocation, during which the search moves fast by a straight line in an arbitrary direction and does not detect the target even if it is in the vision distance. This motion corresponds to the flights between the observed areas, as they are conducted by the Lévy flying searcher.

The alternations of slow screening and fast relocation are called intermittent strategies, and the goal of the searcher is to find the intermittence times such that they maximize the probability of finding the target in minimal expected time of search. Below, it is assumed that the searcher interchanges between static or slow diffusion motion with screening (phase 1) and fast relocation without screening (phase 2) and that its goal is to minimize the expected search time up to certain detection of the target.

In 2005, Benichou et al. considered such search in 1D domain (Benichou et al., 2005), and in 2006–2007, they extended their results to the search in 2D domain Bénichou et al., 2006, 2007). General procedures and additional studies of intermittent search in three dimensions were published in 2011 (Benichou et al., 2011). The consideration in the following text addresses the intermittent search in two dimensions.

Assume that the searcher moves in 2D domain $\mathfrak{X} \subset \mathbb{R}^2$, where, as earlier, each point $\mathfrak{x} = (x, y) \in \mathfrak{X}$ is defined by its Cartesian coordinates x and y. Similar to the search for a moving target (see Section 2.3.3), denote by $p_i(\breve{\mathfrak{x}}, t \mid \breve{\mathfrak{x}}_0, t_0)$ the probability of detecting the target during the period $[t_0, t)$ given that the searcher starts at t_0 in the point $\breve{\mathfrak{x}}_0 = \breve{\mathfrak{x}}(t_0)$ and its initial motion phase is i, where $i = 1$ stands for the phase of slow motion with screening and $i = 2$ stands for the phase of fast relocation without screening. Consequently, $q_i(\breve{\mathfrak{x}}, t \mid \breve{\mathfrak{x}}_0, t_0) = 1 - p_i(\breve{\mathfrak{x}}, t \mid \breve{\mathfrak{x}}_0, t_0)$ are the probabilities of nondetecting the target during the period $[t_0, t)$ starting from the point $\breve{\mathfrak{x}}$ with the motion phase $i = 1, 2$. Certainly, if the searcher starts at time $t_0 = 0$ in the point $\breve{\mathfrak{x}}(0)$ and does not change its motion pattern, then the probabilities $p_i(\breve{\mathfrak{x}}, t \mid \breve{\mathfrak{x}}_0, t_0)$ are reduced to the probability $p(\breve{\mathfrak{x}}, t) = p(\breve{\mathfrak{x}}, t \mid \breve{\mathfrak{x}}(0), 0)$ of detecting the target considered in Section 2.3.3.

By the same manner, denote by t the expected time of search up to certain detection of the target such that for some motion phase $i = 1, 2$, it holds true that $p_i(\breve{\mathfrak{x}}, t \mid \breve{\mathfrak{x}}_0, t_0) = 1$, and consequently, the survive probability is $q_i(\breve{\mathfrak{x}}, t \mid \breve{\mathfrak{x}}_0, t_0) = 1 - p_i(\breve{\mathfrak{x}}, t \mid \breve{\mathfrak{x}}_0, t_0) = 0$. Notice that i indicates the initial motion phase and that the target can be detected only at the phase $i = 1$ of slow motion with screening. Then, the expected search time t depends on the expected times t_i of slow ($i = 1$) and fast ($i = 2$) motions alternating during the expected search period $[t_0, t)$, and the problem is to define a proportion of the times t_1 and t_2 such that the time $t = t(t_1, t_2)$ reaches its minimum (Benichou et al., 2011). In other words, the problem of intermittent search is formulated as follows.

Problem 4.2: (Intermittent Search)

Given a domain $\mathfrak{X}$ and target's location pdf $u : \mathfrak{X} \times [0, \infty) \to \mathbb{R}^+$, find the expected times t_i of slow and fast motions of the searcher, $i = 1, 2$, such that

$$t^* = t\left(t_1^*, t_2^*\right) = \min_{0 \le t_1, t_2 < \infty} \left\{ t = t(t_1, t_2) < \infty \mid p_i(\breve{\mathfrak{x}}, t \mid \breve{\mathfrak{x}}_0, t_0) = 1 \right\}. \tag{4.55}$$

Notice that Equation 4.55 has the same meaning as Equation 2.111, but, in opposite to the last equation, which requires minimization of the expected time with respect to the trajectory of the searcher, Equation 4.55 requires minimization of the expected time with respect to the times of different motion types. Therefore, in spite of different formulation, such problem is similar to the problems considered by the theory of optimal foraging by patches presented in Section 3.3.

To obtain the solution of this problem, let us consider the dynamics of the survive probabilities $q_i(\breve{\mathfrak{x}}, t \mid \breve{\mathfrak{x}}_0, t_0)$, $i = 1, 2$ (Benichou et al., 2011).

In the case of slow diffusive motion of the searcher (phase $i = 1$), the survive probabilities $q_1(\breve{\mathfrak{x}}, t \mid \breve{\mathfrak{x}}_0, t_0)$ are governed by the following backward Kolmogorov equation:

$$\frac{\partial}{\partial t} q_1(\breve{\mathfrak{x}}, t \mid \breve{\mathfrak{x}}_0, t_0) = \left(\breve{s}_1(\mathfrak{x}) \frac{\partial^2}{\partial x_0^2} + \breve{s}_2(\mathfrak{x}) \frac{\partial^2}{\partial y_0^2} \right) q_1(\breve{\mathfrak{x}}, t \mid \breve{\mathfrak{x}}_0, t_0)$$
$$+ \frac{1}{t_1}\left(q_2(\breve{\mathfrak{x}}, t \mid \breve{\mathfrak{x}}_0, t_0) - q_1(\breve{\mathfrak{x}}, t \mid \breve{\mathfrak{x}}_0, t_0) \right) - \upsilon(\mathfrak{x}, t) q_1(\breve{\mathfrak{x}}, t \mid \breve{\mathfrak{x}}_0, t_0) \mathbf{1}\left(\breve{\mathfrak{x}}(t), \mathfrak{x}(t)\right), \tag{4.56}$$

where, as in the preceding text, $\breve{s}_1(\mathfrak{x})$ and $\breve{s}_2(\mathfrak{x})$ are the diffusion rates in the points $\mathfrak{x}$ of the domain $\mathfrak{X}$; $\upsilon(\mathfrak{x},t)$ is a detection rate that, similar to the basic two-cell model of search (see Section 2.3.1), specifies the velocity of changing the probability of detecting the target in the point $\mathfrak{x}$; and **1** is an indicator function such that $\mathbf{1}(\breve{\mathfrak{x}}(t),\mathfrak{x}(t))=1$ if $\|\breve{\mathfrak{x}}(t)-\hat{\mathfrak{x}}(t)\|\le r_{\text{vis}}$ and $\mathbf{1}(\breve{\mathfrak{x}}(t),\mathfrak{x}(t))=0$ otherwise. In this equation, the first term in the right side corresponds to the diffusion of the probabilities, the second term governs the intermittence between slow and fast motion, and the third term defines the changes of the probabilities with respect to detection.

In opposite, in the case of fast motion of the searcher (phase i = 2), the survive probabilities are $q_2(\breve{\mathfrak{x}},t\mid\breve{\mathfrak{x}}_0,t_0)$ governed by the Langevin equation:

$$\frac{\partial}{\partial t}q_2(\breve{\mathfrak{x}},t\mid\breve{\mathfrak{x}}_0,t_0)=\left(\breve{v}_1(\mathfrak{x})\frac{\partial}{\partial x_0}+\breve{v}_2(\mathfrak{x})\frac{\partial}{\partial y_0}\right)q_2(\breve{\mathfrak{x}},t\mid\breve{\mathfrak{x}}_0,t_0)$$
$$+\frac{1}{t_2}\left(q_1(\breve{\mathfrak{x}},t\mid\breve{\mathfrak{x}}_0,t_0)-q_2(\breve{\mathfrak{x}},t\mid\breve{\mathfrak{x}}_0,t_0)\right), \tag{4.57}$$

where $\breve{v}_1(\mathfrak{x})$ and $\breve{v}_2(\mathfrak{x})$ are the components of the velocity vector in the points $\mathfrak{x}$ of the domain $\mathfrak{X}$ (cf. the pursuit-evasion model considered in Section 3.3.3). The first term in the right side specifies the ballistic motion of the searcher, and the second term, as earlier, governs the intermittence between slow and fast motion. Notice that since during the fast motion the searcher cannot detect the target, the survive probabilities $q_2(\breve{\mathfrak{x}},t\mid\breve{\mathfrak{x}}_0,t_0)$ do not depend on the detection.

Denote by $\tau_i(\breve{\mathfrak{x}}_0)$, i = 1, 2, an expected first passage time to the target given that the searcher starts in the point $\breve{\mathfrak{x}}_0=\breve{\mathfrak{x}}(t_0)$ and its initial motion phase is i. Using the probabilities $q_i(\breve{\mathfrak{x}},t\mid\breve{\mathfrak{x}}_0,t_0)$, the first passage times $\tau_i(\breve{\mathfrak{x}}_0)$ are defined as follows:

$$\tau_i(\breve{\mathfrak{x}}_0)=\int_0^{\infty}q_i(\breve{\mathfrak{x}},t\mid\breve{\mathfrak{x}}_0,t_0)\,dt_0,\quad i=1,2. \tag{4.58}$$

Then, the corresponding dynamical equations for the times $\tau_i(\breve{\mathfrak{x}}_0)$ are (Benichou et al., 2011)

$$\left(\breve{s}_1(\mathfrak{x})\frac{\partial^2}{\partial x_0^2}+\breve{s}_2(\mathfrak{x})\frac{\partial^2}{\partial y_0^2}\right)\tau_1(\breve{\mathfrak{x}}_0)+\frac{1}{t_1}\left(\tau_2(\breve{\mathfrak{x}}_0)-\tau_1(\breve{\mathfrak{x}}_0)\right)-\upsilon(\mathfrak{x},t)\tau_1(\breve{\mathfrak{x}}_0)\mathbf{1}(\breve{\mathfrak{x}}(t),\mathfrak{x}(t))+1=0, \tag{4.59}$$

$$\left(\breve{v}_1(\mathfrak{x})\frac{\partial}{\partial x_0}+\breve{v}_2(\mathfrak{x})\frac{\partial}{\partial y_0}\right)\tau_2(\breve{\mathfrak{x}}_0)+\frac{1}{t_2}\left(\tau_1(\breve{\mathfrak{x}}_0)-\tau_2(\breve{\mathfrak{x}}_0)\right)+1=0. \tag{4.60}$$

Assume that the searcher starts the search by slow motion with screening (motion phase i = 1), and recall that the target can be detected also during the slow motion, that is, the search terminates in the phase i = 1. Then, the expected search time t is defined as an average of the time $\tau_1(\breve{\mathfrak{x}}_0)$ over all possible starting points $\breve{\mathfrak{x}}_0\in\mathfrak{X}$ of the searcher (Benichou et al., 2007, 2011):

$$t=\frac{1}{|\mathfrak{X}|}\int_{\mathfrak{X}}\tau_1(\breve{\mathfrak{x}}_0)\,d\breve{\mathfrak{x}}_0, \tag{4.61}$$

where $|\mathfrak{X}|$ is a volume of the domain $\mathfrak{X}$, and this time, $t=t(t_1,t_2)$ should be minimized with respect to the times t_1 and t_2.

Certainly, finding the general solution of the optimization problem defined by Equations 4.59 through 4.61 or, that is the same, by Equations 4.56 and 4.57, and the next Equation 4.62:

$$t = \frac{1}{|\mathfrak{X}|}\int_{\mathfrak{X}}\int_{0}^{\infty} q_1\left(\breve{x}, t \mid \breve{x}_0, t_0\right) dt_0\, d\breve{x}_0, \tag{4.62}$$

in the explicit form, is rather problematic. However, by the use of certain assumptions, Benichou and collegues succeeded to find an optimal proportion of the times t_1 and t_2 for different cases of the searcher's motion.

Assume that the search is conducted in the circle domain $\mathfrak{X}$ of the radius r_{dom} and that the diffusion rates and velocities appearing in Equations 4.59 and 4.60 (and Equations 4.56 and 4.57) are equal and constant for all points $x \in \mathfrak{X}$ and detection rate is also constant for all points $x \in \mathfrak{X}$ and times t, that is, $\breve{s}_1(x) = \breve{s}_2(x) = \breve{\sigma}^2 = \text{const}$, $\breve{v}_1(x) = \breve{v}_2(x) = \breve{v} = \text{const}$, and $\upsilon(x,t) = \upsilon = \text{const}$. Recall that the vision radius is denoted by r_{vis}. With respect to these parameters, the times t_1^* and t_2^*, which result in minimum search time $t^* = t(t_1^*, t_2^*)$ for the modes of slow motion (phase 1), are defined as follows (Benichou et al., 2006, 2007, 2011):

1. Static mode ($\breve{\sigma}^2 = 0$): The searcher starts with screening its location, then moves fast to the other location, screens this location, again moves fast to the next location, and so on up to the termination of the search. There are two possible relations between the parameters:
   1.1 If $r_{\text{vis}} \geq r_{\text{dom}}$, then the searcher detects the target from any initial point $\breve{x}(t_0)$ of the domain. Thus, the intermittence is not required, and the time of fast relocation is $t_2^* = 0$.
   1.2 If $r_{\text{vis}} < r_{\text{dom}}$, then the intermittence between static mode and fast relocation is necessary for finding the target, and in the limit $r_{\text{vis}} \ll r_{\text{dom}}$, the times t_1^* and t_2^* are

$$t_1^* = \sqrt{\frac{r_{\text{vis}}}{\upsilon\breve{v}}}\sqrt[4]{\frac{1}{8}\left(2\ln\left(\frac{r_{\text{dom}}}{r_{\text{vis}}}\right) - 1\right)}, \quad t_2^* = \frac{r_{\text{vis}}}{\breve{v}}\sqrt{\ln\left(\frac{r_{\text{dom}}}{r_{\text{vis}}}\right) - \frac{1}{2}}. \tag{4.63}$$

2. Diffusive mode ($\breve{\sigma}^2 > 0$): The searcher starts with slow diffusive motion with screening, then terminates slow motion, and moves fast to the other location; starting from this location, the searcher moves slowly with screening, then again terminates slow motion and moves fast to the next location, and so on up to the termination of the search. In this case, there are three possible relations between the parameters:
   2.1 If $r_{\text{vis}} < r_{\text{dom}} < \left(\breve{\sigma}^2/\breve{v}\right)$, then the time $t_1 \sim \left(r_{\text{dom}}^2/\breve{\sigma}^2\right)$, which is required for Brownian search around all domain, is less than the time $t_2 \sim \left(r_{\text{dom}}/\breve{v}\right)$, which is required for the search by ballistic motion. Thus, the intermittence is not required, the time of fast relocation is $t_2^* = 0$, and the searchers' optimal strategy is to conduct Brownian search.
   2.2 If $r_{\text{vis}} < \left(\breve{\sigma}^2/\breve{v}\right) < r_{\text{dom}}$, then the intermittence between slow motion and fast relocation leads to the shorter search time, and in the limit $r_{\text{vis}} \ll r_{\text{dom}}$, the times t_1^* and t_2^* are

$$t_1^* = \frac{r_{\text{dom}}^2}{\breve{\sigma}^2}\frac{4\ln a - 5 + c}{a^2\left(4\ln a - 7 + c\right)}, \quad t_2^* = \frac{r_{\text{dom}}}{\breve{v}}\frac{\sqrt{4\ln a - 5 + c}}{a}, \tag{4.64}$$

   where a is a solution of the equation

$$a = \frac{2\breve{v}r_{\text{dom}}}{\breve{\sigma}^2}f(a),\; f(a) = \frac{\sqrt{4\ln a - 5 + c}}{-8\left(\ln a\right)^2 + \left[6 + 8\ln\left(\frac{r_{\text{dom}}}{r_{\text{vis}}}\right)\right]\ln a - 10\ln\left(\frac{r_{\text{dom}}}{r_{\text{vis}}}\right) + 11 - c\left(\frac{c}{2} + 2\ln\left(\frac{r_{\text{vis}}}{r_{\text{dom}}}\right) - \frac{3}{2}\right)}$$

and $c = 4(\gamma - \ln 2)$ with $\gamma = -\int_0^{\infty} e^{-y} \ln y dy \approx 0.577$ standing for the Euler constant. An approximate value of a is

$$a \approx \frac{2\breve{v} r_{dom}}{\breve{\sigma}^2} f\left(\frac{\breve{v} r_{dom}}{2\breve{\sigma}^2 \ln\left(r_{dom} / r_{vis}\right)}\right).$$

2.3 Finally, if $\left(\breve{\sigma}^2/\breve{v}\right) < r_{vis} < r_{dom}$, then the intermittence strongly shortens the search time, and in the limit $r_{vis} \ll r_{dom}$, the times t_1^* and t_2^* of slow motion and fast relocation are

$$t_1^* = \frac{\breve{\sigma}^2}{2\breve{v}^2} \frac{\left(\ln\left(\frac{r_{dom}}{r_{vis}}\right)\right)^2}{2\ln\left(\frac{r_{dom}}{r_{vis}}\right) - 1}, \quad t_2^* = \frac{r_{vis}}{\breve{v}} \sqrt{\ln\left(\frac{r_{dom}}{r_{vis}}\right) - \frac{1}{2}}. \tag{4.65}$$

3. Ballistic mode: In general, the searcher starts with the motion with screening along a straight line, then moves fast without screening to some location, from which moves along a straight line with screening, then again moves fast to the next location, and so on up to the termination of the search. Such search can be conducted in two regimes: the first is without intermittence, that is, the searcher never changes its motion with screening to the fast relocation, and the second is with intermittence, in which the searcher periodically changes motion with screening to the fast relocation without screening.
   3.1 Search without intermittence ($t_2 \to 0$): The searcher moves by straight lines with periodical changing of its direction. As indicated in the preceding text, such search is equivalent to the search by Lévy walks and is effective if the velocity of the walk is sufficiently large and at least greater than $\left(\pi\breve{v} / 4\right)\sqrt{\ln\left(r_{dom}/r_{vis}\right)}$.
   3.2 Search with intermittence: In this case, the searcher interchanges between ballistic motion with screening and fast relocation without screening and the times t_1^* and t_2^* are

$$t_1^* \to 0, \quad t_2^* = \frac{r_{vis}}{\breve{v}} \sqrt{\ln\left(\frac{r_{dom}}{r_{vis}}\right) - \frac{1}{2}}, \tag{4.66}$$

and this case exhausts the list of possible modes of the searcher's motion.

To illustrate the presented results, let us consider an example of survive probability distribution with respect to the parameters corresponding to the diffusive mode 2.3.

Example 4.9

Assume that the searcher acts in the circle domain $\mathfrak{X}$ of the radius $r_{dom} = 50$ with the impermeable bounds and that the vision distance $r_{vis} = 1$. Assume also that, similar to the previous examples, the diffusion rates and velocities appearing in Equations 4.56 and 4.57 are $\breve{s}_1(\mathfrak{x}) = \breve{s}_2(\mathfrak{x}) = \breve{\sigma}^2 = 0.3$ and $\breve{v}_1(\mathfrak{x}) = \breve{v}_2(\mathfrak{x}) = \breve{v} = 0.5$. Then $\breve{\sigma}^2/\breve{v} = 0.3/0.5 = 0.6$, and it holds true that $\left(\breve{\sigma}^2/\breve{v}\right) < r_{vis} < r_{dom}$, which corresponds to the diffusive mode 2.3 of the searcher's motion. In this case, the times t_1^* and t_2^* of slow motion and fast relocation are specified by Equation 4.65 that gives $t_1^* = 0.404$ and $t_2^* = 3.694$. Finally, assume that the detection rate is $\upsilon(\mathfrak{x}, t) = \upsilon = 1$.

An initial distribution of the survive probabilities $q_1(\breve{\mathfrak{x}}, 0 \mid 0, 0) = q_2(\breve{\mathfrak{x}}, 0 \mid 0, 0)$ over the points $\mathfrak{x} \in \mathfrak{X}$ is defined by the binormally distributed searcher location probabilities concentrated around the center $\breve{\mathfrak{x}} = (50, 50)$ of the domain with the equal variances $\sigma_1 = \sigma_2 = 5$ for the axes x and y. The initial distribution of the survive probabilities and their distribution at time $t = 10$ are shown in Figure 4.29.

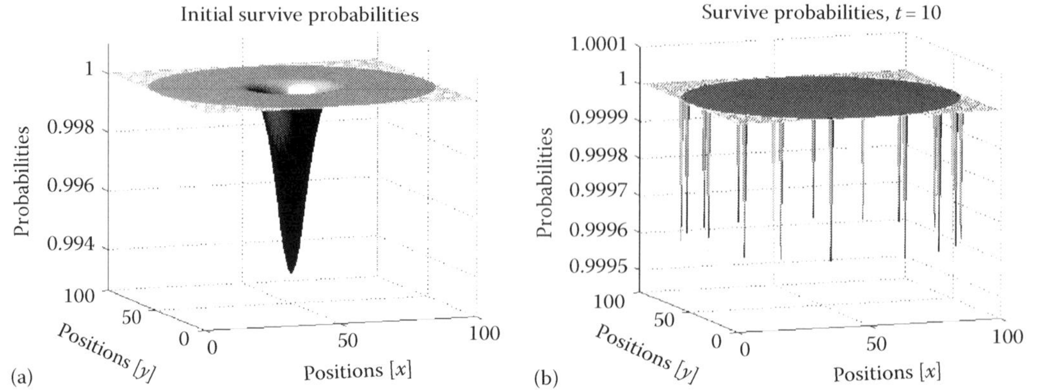

FIGURE 4.29 Distribution of the survive probabilities in intermittent search according to the diffusive mode with diffusion rates $\breve{s}_1(\mathfrak{x}) = \breve{s}_2(\mathfrak{x}) = \breve{\sigma}^2 = 0.3$ and velocities $\breve{v}_1(\mathfrak{x}) = \breve{v}_2(\mathfrak{x}) = \breve{v} = 0.5$. (a) Initial distribution of the survive probabilities and (b) distribution of the survive probabilities at the time $t = 10$.

At the beginning, the survive probabilities are distributed around the center with the minimum at the center of the domain, where the probability of the searcher's location is maximum. Then survive probabilities quickly disseminated over the domain such that their minimal values are concentrated in several points at the borders of the domain. In other words, the preferable behavior for the searcher is to seek for the target in the areas close to the boundaries of the domain, where the survive probabilities are minimal and corresponding probabilities of finding the target are maximal. ■

Certainly, the values presented earlier of optimal times t_1^* and t_2^* of slow motion and fast relocation, which provide the minimum search time $t^* = t(t_1^*, t_2^*)$, cannot be considered as strict solutions of Problem 4.2 of intermittent search but give certain phenomenological description of the searcher's activity. For additional information and reasoning on the intermittent strategies, see the original papers by Benichou et al. (2006, 2007, 2011).

Finally, let us consider the search procedure (Siegelmann, Kagan, & Ben-Gal, 2014) that combines the search by Lévy flights with varying indices (Algorithm 4.6) and the intermittent search and implements an approach of varying vision distance (Marthaler, Bertozzi, & Schwartz, 2004). In such procedure, the indices $a(\mathfrak{x},t) = \gamma(\mathfrak{x},t) - 1 = 2(1 - u(\mathfrak{x},t))$ are defined by the target's location probabilities $\hat{u}(\mathfrak{x},t)$ in the points $\mathfrak{x} \in \mathcal{X}$ of the domain at time t, and the direction of the steps are specified with respect to the target' location probabilities at the step distance from the current location of the search. This model does not require decision making regarding the next phase and mode of the search, but implies that the forager is able to perceive the state of the environment both in its current location $\breve{\mathfrak{x}}(t)$ and in the points $\mathfrak{x} \in \mathcal{X}$ such that $\left\|\breve{\mathfrak{x}}(t) - \mathfrak{x}\right\| = \delta(t)$, where $\delta(t)$ is a current step of the searcher. The algorithm of such search is outlined as follows (Siegelmann, Kagan, & Ben-Gal, 2014).

Algorithm 4.7: (Lévy Flight Search with Varying Parameters and Vision Distance)

Given a domain $\mathcal{X} = \{\mathfrak{x}_1, \mathfrak{x}_2, \ldots, \mathfrak{x}_n\}$ and the target's location probabilities $u(\mathfrak{x},t)$ or $\hat{u}(\mathfrak{x},t)$ do:

1. Observe the current location $\breve{\mathfrak{x}}(t)$ and decide either to terminate the search or to continue. In the case of destructive search, zero the current target's location probability, so $u(\breve{\mathfrak{x}}(t), t+1) = 0$. In the case of nondestructive search, the target's location probability remained without changes, so $u(\breve{\mathfrak{x}}(t), t+1) = \hat{u}(\breve{\mathfrak{x}}(t), t)$.

2. If the search continues, define parameter $\gamma(\mathfrak{x},t)$ with respect to $u(\breve{\mathfrak{x}}(t),t+1)$.
3. Using the value $\gamma(\mathfrak{x},t)$, draw the Lévy flight step $\delta(t)$ that is $p(\delta(t)) \sim \delta(t)^{-\gamma(\mathfrak{x},t)}$.
4. Choose the next point such that $\breve{\mathfrak{x}}(t+1) = \text{agrmax}_{\|\breve{\mathfrak{x}}(t)-\mathfrak{x}\|=\delta(t)}\{\hat{u}(\mathfrak{x},t+1)\}$.
5. Move to the chosen point $\breve{\mathfrak{x}}(t+1)$ and continue with its observation (line 1).

Notice that this algorithm does not assume long-term internal memory of the searcher and, for certain parameters, is reduced to the considered Lévy flight or Brownian search procedures. The trajectory of the searcher acting according to Algorithm 4.7 is illustrated by the following example.

Example 4.10

Assume that, similar to Example 4.8, the searcher acts in the square domain $\mathfrak{X}$ of the size $n = 100 \times 100$ with the impermeable bounds and that the states of the domain are defined according to the target's location probabilities $u(\mathfrak{x},t)$ that are specified by ten binormal distributions with random centers and the variances $\sigma_1 = \sigma_2 = 10$. The Lévy indices $a(\mathfrak{x}, t)$ for the points $\mathfrak{x} \in \mathfrak{X}$ are defined by $a(\mathfrak{x},t) = 2(1-u(\mathfrak{x},t))$ that are normalized such that $\min_{\mathfrak{x}} a(\mathfrak{x},0) = \varepsilon > 0$, $\varepsilon \ll 1$, and $\max_{\mathfrak{x}} a(\mathfrak{x},0) = 2$, and the parameters $\gamma(\mathfrak{x},t)$ are $\gamma(\mathfrak{x},t) = a(\mathfrak{x},t)+1$. Similar to the previous examples, it is assumed that the maximal value of the Lévy flight step is bounded by $\delta_{max} = 25$ and that the minimal step is $\delta_{max} = 1$. An example of the trajectory of the searcher acting according to the destructive search scenario is shown in Figure 4.30a. The distribution of the zeroed target's location probabilities that is the same as a quantity of eaten prey (see Example 4.5) and as weighted number of the first-time visited areas (see Example 4.8) over 1000 trials is shown in Figure 4.30b. The average weighted number of the first-time visited areas over 1000 trials is 1205.227, and the average path of the searcher is 4360.316.

It is seen that most of the time, the searcher moves in the regions with higher target's location probabilities with flights between such regions as it is prescribed by the intermittent search strategies. However, in contrast to the intermittence search, which requires changes of the motion phases, in Algorithm 4.7, the searcher follows the same Lévy flight process with varying index and vision distance.

A comparison of the results provided by Algorithm 4.7 with the results considered in Example 4.8 (Algorithm 4.6) shows that Algorithm 4.7 outperforms previous search procedures, and, according to Kolmogorov–Smirnov test, the difference between the average numbers of visited regions is statistically significant.

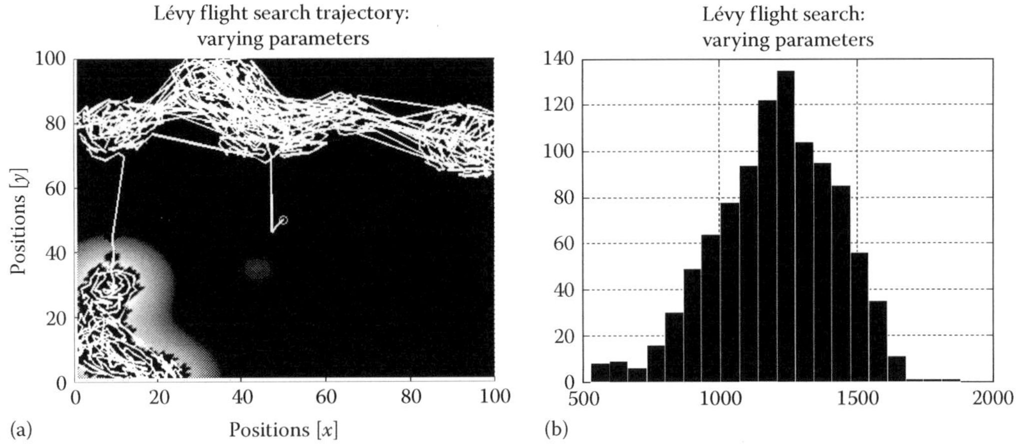

FIGURE 4.30 Destructive search according to the Lévy flight with varying index and vision distance: (a) trajectory of the searcher acting according to Algorithm 4.7 with destructive search scenario and (b) distribution of the zeroed target's location probabilities over 1000 trials.

In addition, notice that the trajectory resulted by Algorithm 4.7 has the same form as trajectories of foraging by patches (see Section 3.3) based on the marginal value theorem, which provides a solution of the problem of foraging under general assumption that the forager acts as economical as possible. ■

The formulated algorithm finalizes the consideration of the motion models based on the observed trajectories of natural foragers. Certainly, the presented procedures do not exhaust the list of such models, as well as the indicated motion patterns do not represent all the variety of the animals' foraging behavior. Nevertheless, they demonstrate the basic ideas and methods used in such tasks and provide a link between optimal foraging theory and the algorithms of search applied to navigation of mobile robots. The next section presents several algorithms of probabilistic search, which were developed in the framework of mobile robots' control and result in the same motion patterns as the algorithms based on Lévy flights considered earlier.

4.4 ALGORITHMS OF PROBABILISTIC SEARCH AND FORAGING

The methods and algorithms in the preceding text implement biomimetic approach to the search and foraging processes. These methods follow the assumption that the searcher is not informed about the target's location probabilities over a complete domain, and its motion is based on local information about such probabilities and predefined stochastic process, which governs long-distance relocations. In contrast, in the case of navigation of artificial mobile agents, like unmanned vehicles and mobile robots, the agents can be supported by the complete map target's location probabilities at least at the beginning of search, and the agents can consider their long-distance relocations with respect to such map. This section presents probabilistic algorithms of search by mobile agents, which are based on the knowledge of the complete map of the target's location probabilities.

The algorithms of search in discrete time and space presented in Section 4.4.1 follow the informational search approach (Kagan & Ben-Gal, 2006, 2013) and implement both destructive and nondestructive search scenarios. The first scenario corresponds to the errorless sensing (Goren, Kagan, & Ben-Gal, 2010; Kagan, Goren, & Ben-Gal, 2010), and the second represents the search with possible errors in detecting the targets (Chernikhovsky et al., 2012; Kagan, Goren, & Ben-Gal, 2012). Another approach is applied for the search over a topographic terrain with shadowing and restricted visibility (Israel, Khmelnitsky, & Kagan, 2012). Formally, this approach was mentioned in Section 2.3.3 in the consideration of search for a mobile target; the algorithmic implementation of such method is considered in Section 4.4.5. Notice that the indicated algorithms are applicable for the search of both static and moving targets, so the discourse below does not distinguish between such searches and considers both the cases in the same framework.

The probabilistic and informational approach to the search in discrete time and space was independently reported in recent works by Indelman et al. (Indelman, Carlone, & Dellaert, 2013; Indelman, Carlone, & Dellaert, 2013) and by Cetnarowicz et al. (Cetnarowicz et al., 2013; Sosnicki et al., 2013; Zabinska et al., 2013). In addition to the search tasks, Cetnarowicz et al. implemented this approach for navigation of mobile robots in the presence of obstacles; such problems are noticed in Section 4.4.6.

4.4.1 SEARCH FOR STATIC AND MOVING TARGETS

The presented algorithms of search for static and moving targets in discrete time and space follow the direction of search and screening theory considered in Section 2 and apply the same notation.

Assume that the searcher and the target act in the discrete domain $\mathfrak{X} = \{\mathfrak{x}_1, \mathfrak{x}_2, \ldots, \mathfrak{x}_n\}$; as in the preceding text, it is assumed that $\mathfrak{X}$ is a squire grid such that each point $\mathfrak{x}_i$, $i = 1, 2, \ldots, n$, is defined by a pair (i_x, i_y) of indices $i_x = 1, 2, \ldots, n_x$ and $i_y = 1, 2, \ldots, n_y$ such that $n = n_x \times n_y$ and $i = \left(\left(i_x - 1\right)n_x + i_y\right)$.

The target's location probabilities $u(\mathfrak{x}_i, 0) = u(\mathfrak{x}_i)$, $i = 1, 2, \ldots, n$, at the beginning of search at $t = 0$ are known to the searcher and depend on the searcher's actions. It is assumed that the target is certainly in the domain $\mathfrak{X}$, so for each time $t = 0, 1, 2, \ldots$ up to finding the target, it holds true that $\sum_{i=1}^{n} u(\mathfrak{x}_i, t) = 1$. Notice again that $u(\mathfrak{x}, t)$ are conditional probabilities that the target is in the point $\mathfrak{x}$ at time t given that it was not detected up to this time, while $\hat{u}(\mathfrak{x}, t)$ stands for the joint probability that the target is in the point $\mathfrak{x}$ at time t and was not detected up to this time. In contrast to the previously presented procedures, the algorithms given in this section distinguish these probabilities and apply the probabilistic rules in their actions.

Similar to Section 2.3.2 (see Equation 2.78), it is assumed that the movement of the target over the domain $\mathfrak{X}$ is governed by a Markov process with transition probabilities matrix $\left\|\hat{\rho} = \hat{\rho}_{ij}\right\|_{n \times n}$, where $\hat{\rho}_{ij} = Pr\left\{\hat{\mathfrak{x}}(t+1) = \mathfrak{x}_j \mid \hat{\mathfrak{x}}(t) = \mathfrak{x}_i\right\}$, $i,j = 1, 2, \ldots, n$, $t = 0, 1, 2, \ldots$, and $\sum_{j=1}^{n} \hat{\rho}_{ij} = 1$ for each $i = 1, 2, \ldots, n$. In the case of static target, transition probabilities matrix $\hat{\rho}$ is a unit matrix, that is, $\hat{\rho}_{ij} = 1$ for $i = j$ and $\hat{\rho}_{ij} = 0$ for $i \neq j$, and in the case of the target moving according to the Brownian walk, the diagonal elements of the matrix $\hat{\rho}$ are zero, that is, $\hat{\rho}_{ij} = 0$ for $i = j$ and $\hat{\rho}_{ij} \geq 0$ for $i \neq j$.

The searcher moves over the domain and seeks for a target by observing the observed areas $a \subset \mathfrak{X}$ (see Figure 2.8) and obtaining observation results $z(a) \in \{0, 1\}$ such that the value $z(a) = 1$ indicates that the target is in the area a and the value $z(a) = 0$ indicates that the target is not in this area a (see Sections 2.2.2 through 2.2.4). The observations are conducted by application of the detection function φ, which defines the probability $\varphi(a,\kappa)$ of detecting the target in the area a given that the target is located in this area, and the search effort applied to a is $\kappa(a,t) \in [0,\infty)$ (see Section 2.2.4). It is assumed that the observed areas include the neighboring points $\mathfrak{x} \in \mathfrak{X}$ of the current searcher's locations $\breve{\mathfrak{x}}(t)$ such that $a(t) = \left\{\mathfrak{x} : \left\|\breve{\mathfrak{x}}(t) - \mathfrak{x}\right\| \leq r_{\text{vis}}\right\}$, where $r_{\text{vis}} > 0$ is a vision distance, and that the detection probability does not depend on the search effort, that is, $\varphi(a, \kappa) = \varphi(a(t))$ for any κ. If the detection is errorless, then $\varphi(a(t)) \in \{0,1\}$, and if detection errors are allowed, then $\varphi(a(t)) \in [0,1]$.

The goal of the searcher is the same as in the search for a moving target in Section 2.3 and, according to Problems 2.4, is defined as follows (see also Problem 2.3 of sequential search).

Problem 4.3: (Static and Moving Target Search by Areas)

Find the strategy of choosing the areas $a(t) \subset \mathfrak{X}$, $t = 0, 1, 2, \ldots$, such that the search terminates in a minimal average (over possible target moves) number of steps, and the centers $\breve{\mathfrak{x}}(t)$ of the areas $a(t)$ form a continuous trajectory $\vec{a} = \left\langle \breve{\mathfrak{x}}(0), \breve{\mathfrak{x}}(1), \breve{\mathfrak{x}}(2), \ldots \right\rangle$ with respect to the metric in $\mathfrak{X}$.

According to the termination rule, this problem can be formulated either as a problem of finding minimal search time up to certain detection of the target or as a problem of finding maximal probability of detecting the target in a certain period. The algorithms considered in Sections 4.4.2–4.4.5 deal with the first formulation of the problem.

4.4.2 Search for Static and Moving Targets: Errorless Detection

Let us start with the algorithm of search with errorless detection, which was developed in 2010 in collaboration with Gal Goren (Goren, Kagan, & Ben-Gal, 2010). The discourse at most follows the paper by Kagan et al. (Kagan, Goren, & Ben-Gal, 2010); for additional information and corresponding framework of informational search techniques, see the book by Kagan and Ben-Gal (Kagan & Ben-Gal, 2013).

Assume that similar to the intermittent search, the detection probability $\varphi(a(t))$ is defined by the indicator function such that $\mathbf{1}(\breve{\mathfrak{x}}(t), \hat{\mathfrak{x}}(t)) = 1$ if $\|\breve{\mathfrak{x}}(t) - \hat{\mathfrak{x}}(t)\| \le r_{\text{vis}}$ and $\mathbf{1}(\breve{\mathfrak{x}}(t), \mathfrak{x}(t)) = 0$ otherwise, $t = 0, 1, 2,\ldots$ In terms of observed areas $a(t)$, it means that

$$\varphi(a(t)) = 1 \quad \text{if } \hat{\mathfrak{x}}(t) \in a(t), \quad \text{and} \quad \varphi(a(t)) = 0 \quad \text{otherwise}. \tag{4.67}$$

The search is conducted as follows (Kagan & Ben-Gal, 2006; Kagan, Goren, & Ben-Gal, 2010) (cf. Section 2.3.2). The searcher deals with the target's location probabilities $u(\mathfrak{x}_i, t)$ and $\hat{u}(\mathfrak{x}_i, t)$ in the points $\mathfrak{x}_i \in \mathfrak{X}$, $i = 1, 2,\ldots, n$, such that at the initial time $t = 0$, it holds true that $\hat{u}(\mathfrak{x}_i, 0) = u(\mathfrak{x}_i, 0) = u(\mathfrak{x}_i)$ and probabilities $u(\mathfrak{x}_i)$ are known by the searcher. Following the approach of sequential search and search by areas (see Sections 2.2.3 and 2.2.4), at each time t, the searcher chooses an observed area $a(t) \subset \mathfrak{X}$ and checks whether the target is in $a(t)$ or not. If an observation result is $z(a) = 1$ with the detection probability $\varphi(a(t)) = 1$, then the search terminates; otherwise, the agent zeroes the probabilities in the points $\mathfrak{x} \in a(t)$ that results in the set of probabilities $\hat{u}(\mathfrak{x}_i, t), i = 1, 2,\ldots, n$. By applying the Bayes rule to the probabilities $\hat{u}(\mathfrak{x}_i, t)$, the searcher obtains the updated target's location probabilities over the domain $\mathfrak{X}$. The updated probabilities are called observed probabilities and are denoted by $\bar{u}(\mathfrak{x}_i, t)$. To the observed probabilities $\bar{u}(\mathfrak{x}_i, t)$, the searcher applies some known or estimated rule of the target movement and obtains the target's location probabilities $u(\mathfrak{x}_i, t+1)$, $i = 1, 2,\ldots, n$, at the next time $t + 1$. In particular, if the target moves according to the indicated Markov process with transition probabilities matrix $\hat{\rho} = \|\rho_{ij}\|_{n \times n}$, then the probabilities $u(\mathfrak{x}_i, t+1)$ are defined as $u(\mathfrak{x}_i, t+1) = \sum_{j=1}^{n} \rho_{ij} \bar{u}(\mathfrak{x}_i, t)$, $i = 1, 2,\ldots, n$. Given the target's location probabilities $u(\mathfrak{x}_i, t+1)$, $i = 1, 2,\ldots, n$, the searcher selects the next observed area $a(t+1) \subset \mathfrak{X}$ and the process continues.

Following Problem 4.3, the goal is to find the strategy for choosing the areas $a(t) \subset \mathfrak{X}$, $t = 0, 1, 2,\ldots$, which results in minimal average number of search steps. An existence of such a strategy and convergence of the search procedure are provided by the A*-type online algorithms (Ishida & Korf, 1995; Korf, 1990) and their informational successors (Kagan & Ben-Gal, 2006; Kagan & Ben-Gal, 2013; Kagan & Ben-Gal, 2014); the detailed consideration of these algorithms is given in the book by Kagan and Ben-Gal (2013). The next algorithm implements a probabilistic version of local search and uses the estimated global distances, while decision making utilizes distance estimations and the characteristics of target's movement.

Denote by D the set of possible directions, to which the searcher and the target can move one step at each time t, $t = 0, 1, 2,\ldots$ Since $\mathfrak{X}$ is discrete domain, the set D is certainly finite. In the simplest case, the set D consists of five directions, that is, $D = \{d_1, d_2, d_3, d_4, d_5\}$, where d_1 stands for "move forward," d_2 "move backward," d_3 "move right," d_4 "move left" and d_5 "stay in the current point" (cf. Brownian motion considered in Section 4.2.1). The target chooses its direction according to the transition probabilities matrix $\hat{\rho}$, and the searcher's choice follows the decision obtained according to the probability $\varphi(a(t))$ of detecting the target in the observed area $a(t)$ around the agent. The algorithm is outlined as follows.

Algorithm 4.8: (Search with Errorless Detection) (Kagan, Goren, & Ben-Gal 2010; Kagan & Ben-Gal 2013)

Given a discrete domain $\mathfrak{X} = \{\mathfrak{x}_1, \mathfrak{x}_2,\ldots, \mathfrak{x}_n\}$, initial target's location probabilities $u(\mathfrak{x}_i)$, $i = 1, 2, \ldots, n$, and target's transition probabilities matrix $\hat{\rho} = \|\rho_{ij}\|_{n \times n}$ do:

1. Set $t = 0$ and start with the probabilities $\hat{u}(\mathfrak{x}_i, t) = u(\mathfrak{x}_i, t) = u(\mathfrak{x}_i)$, $i = 1, 2, \ldots, n$, and initial observed area $a(t)$.
2. While $\varphi(a(t)) \neq 1$ do:
3. For all $\mathfrak{x}_i$ do: Set $\hat{u}(\mathfrak{x}_i, t) = u(\mathfrak{x}_i, t)$.

Calculate observed probabilities $\bar{u}(\mathfrak{x}_i,t)$, $i = 1, 2, \ldots, n$:

4. For all $\mathfrak{x}_i$ do: If $\mathfrak{x}_i \in a(t)$, then set $\hat{u}(\mathfrak{x}_i,t) = 0$.
5. For all $\mathfrak{x}_i$ do: Set $\bar{u}(\mathfrak{x}_i,t) = \hat{u}(\mathfrak{x}_i,t) \Big/ \sum_{j=1}^{n} \hat{u}(\mathfrak{x}_j,t)$.

Calculate location probabilities $u(\mathfrak{x}_i,t+1)$, $i = 1, 2, \ldots, n$:

6. For all $\mathfrak{x}_i$ do: Set $u(\mathfrak{x}_i,t+1) = \sum_{j=1}^{n} \hat{\rho}_{ij}\bar{u}(\mathfrak{x}_i,t)$.

Calculate weights β_k *for the directions* $d_k \in D$:

7. For all d_k do: Set $\beta_k = \sum_{\mathfrak{x} \in a_k} u(\mathfrak{x},t+1)$, where a_k is an area that will be observed if the agent will move by the direction d_k.

Estimate global probabilities $\hat{u}(\mathfrak{x}_i,t+1)$ *if needed*, $i = 1, 2, \ldots, n$:

8. If $\beta_k = 0$ for each k, then
9. For all $\mathfrak{x}_i$ do: Set $\hat{u}(\mathfrak{x}_i,t+1) =$ estimate probabilities$\left(\breve{\mathfrak{x}}(t),\ u(\mathfrak{x}_i,t+1),\ \hat{\rho}\right)$.
10. Else
11. For all $\mathfrak{x}_i$ do: Set $\hat{u}(\mathfrak{x}_i,t+1) = u(\mathfrak{x}_i,t+1)$.
12. End if.

Select the next observed area $a(t+1)$ *using estimated probabilities* $\hat{u}(\mathfrak{x}_i,t+1)$, $i = 1, 2, \ldots, n$:

13. Set $a(t+1) =$ next observed area$\left(\hat{u}(\mathfrak{x}_i,t+1)\right)$.

Move to the area $a(t+1)$ *and observe it*:

14. Set $t = t + 1$ and get $\varphi(a(t))$.
15. End while.
16. Return $a(t)$ for which $\varphi(a(t)) = 1$.

As indicated earlier, at each time t, the target moves to its next location according to the transition probabilities matrix $\hat{\rho}$, and since the target is not aware of the agent activities up to the successful termination of the search, the target's movements do not depend on the searcher actions.

The algorithm includes two key functions: the function estimate probabilities( ) provides global probabilities estimation and resolves an uncertainty for local decision making, and the function next observed area( ) conducts local decision making regarding the next observed area. Let us start with the function next observed area( ).

Similar to the Algorithm 4.7 of the Lévy flight search, Algorithm 4.8 implements local decision making based on maximum gradient of the expected target's location probabilities $u(\mathfrak{x}_i,t)$, $i = 1, 2, \ldots, n$, in the current searcher's location $\breve{\mathfrak{x}}(t)$. In other words, the agent chooses the next observed area such that its observation maximally affects the probabilities $u(\mathfrak{x}_i,t)$. In general, the distances between the probabilities $u(\mathfrak{x}_i,t)$ and $u_k(\mathfrak{x}_i,t)$, $i = 1, 2, \ldots, n$, where k corresponds to the indices of the direction $d_k \in D$, are calculated according to the Ornstein metric (Ornstein, 1971), which for pointwise comparison is

$$d\left(u_k(\mathfrak{x}_i,t),u(\mathfrak{x}_i,t)\right) = \left|u_k(\mathfrak{x}_i,t) - u(\mathfrak{x}_i,t)\right|. \tag{4.68}$$

Given the set D of possible directions, function next observed area( ) is outlined as follows:

next observed area$\left(\hat{u}(\mathfrak{x}_i,t)\right)$:

1. For all movements $d_k \in$ D do:
2. Determine the candidate observed area $a_k(t+1)$ according to d_k.

3. Set $\beta_k = \sum_{\mathfrak{x} \in a_k(t+1)} \hat{u}(\mathfrak{x},t)$.
4. For all $\mathfrak{x}_i$ do: if $\mathfrak{x}_i \in a_k(t+1)$ then set $\hat{u}_{k1}(\mathfrak{x}_i,t) = 1$.
5. For all $\mathfrak{x}_i$ do: set $\bar{u}_{k1}(\mathfrak{x}_i,t) = \hat{u}_{k1}(\mathfrak{x}_i,t) \Big/ \sum_{j=1}^{n} \hat{u}_{k1}(\mathfrak{x}_j,t)$.
6. For all $\mathfrak{x}_i$ do: if $\mathfrak{x}_i \notin a_k(t+1)$ then set $\hat{u}_{k0}(\mathfrak{x}_i,t) = 0$.
7. For all $\mathfrak{x}_i$ do: set $\bar{u}_{k0}(\mathfrak{x}_i,t) = \hat{u}_{k0}(\mathfrak{x}_i,t) \Big/ \sum_{j=1}^{n} \hat{u}_{k0}(\mathfrak{x}_j,t)$.
8. Set $\beta(a_k(t+1)) = \sum_{i=1}^{n} \beta_k d(\bar{u}_{k1}(\mathfrak{x}_i,t), \hat{u}(\mathfrak{x}_i,t)) + (1-\beta_k) d(\bar{u}_{k0}(\mathfrak{x}_i,t), \hat{u}(\mathfrak{x}_i,t))$.
9. End for.
10. Set $a(t+1) = \text{argmax}\{\beta(a_k(t+1))$ over all $a_k(t+1)\}$; ties are broken randomly.
11. Return $a(t+1)$.

The operations conducted by the function next observed area( ) have the same meaning as the corresponding operation of Algorithm 4.8. In particular, line 3 defines the weights for the movements in the directions $d_k \in D$; in lines 4–7, the observed probabilities for successful and unsuccessful observations are calculated; the first probabilities are denoted by $\bar{u}_{k1}(\mathfrak{x}_i,t)$ and the second by $\bar{u}_{k0}(\mathfrak{x}_i,t)$, and line 8 specifies the weighted Ornstein distance over successful and unsuccessful observations. Line 10 prescribes the choice of the next observed area for which the weighted Ornstein distance is maximal.

Local decision making regarding the next observed area, which is specified by the presented function *next observed area*(), is based on the probabilities $\hat{u}(\mathfrak{x}_i,t+1)$, $i = 1, 2,\ldots, n$. As it is defined by lines 8–11 of Algorithm 4.8, if there is no local uncertainty in the choice of the observed area, then these probabilities are set equal to the target's location probabilities $u(\mathfrak{x}_i,t+1)$ at the time $t + 1$, $i = 1, 2,\ldots, n$. Otherwise, if for all possible movements $d_k \in D$, the weights β_k are equal to zero, then this uncertainty is resolved by global estimation of the probabilities over the domain.

As indicated in line 9 of Algorithm 4.8, global estimation conducted using the function estimate probabilities() is applied to the target's location probabilities $u(\mathfrak{x}_i,t+1)$ at time $t + 1$ and results in the probabilities $\hat{u}(\mathfrak{x}_i,t+1)$, $i = 1, 2,\ldots, n$. The probabilities $\hat{u}(\mathfrak{x}_i,t+1)$ are calculated using the target's transition probabilities matrix $\hat{\rho}$ (either given or estimated) as the target's location probabilities after a certain number of steps m. In other words, the estimation implements m steps of the diffusion-like process governed by the transitions matrix $\hat{\rho}$ that allows to define the nonzero probabilities in the close neighborhood of the searcher with respect to the actual target's location probabilities $u(\mathfrak{x}_i,t+1)$, $i = 1, 2,\ldots, n$.

The number m of the steps of diffusion process is defined using the distance to the estimated target's location $\tilde{\mathfrak{x}}$ and the entropy measure (Cover & Thomas, 1991), which specifies the lack of available information, with respect to the size of the observed areas. An amount of the required information is defined as follows:

$$I(u) = \log_s(n) - \sum_{i=1}^{n} u(\mathfrak{x}_i,t+1) \log_s\left(1/u(\mathfrak{x}_i,t+1)\right), \tag{4.69}$$

where

s is a size of the observed areas
$\log_s$ stands for the logarithm base s, and, conventionally, it is assumed that $0 \cdot \log_s(0) = 0$

The estimated target's location $\tilde{\mathfrak{x}}(t)$ at time t is defined as point that corresponds to the center of distribution of the target's location probabilities, which is calculated as a "center of gravity" for the probabilities $u(\mathfrak{x}_i,t)$, $i = 1, 2,\ldots, n$. Over the gridded domain $\mathfrak{X}$ of the size $n = n_x \times n_y$, the coordinates of the estimated target's location $\tilde{x}(t) = (\tilde{\imath}_x(t), \tilde{\imath}_y(t))$ are calculated by the function estimated location(), which is outlined as follows:

$$\text{estimated location}\left(u(x_i, t)\right):$$

1. Set $\tilde{\imath}_x(t) = 0$ and $\tilde{\imath}_y(t) = 0$.
2. For $i_x = 1,\ldots,n_x$ do:
3. For $i_y = 1,\ldots,n_y$ do:
4. Set $\tilde{\imath}_x(t) = \tilde{\imath}_x(t) + i_x u(x_i, t)$, where $i = ((i_x - 1)n_x + i_y)$.
5. End for.
6. End for.
7. For $i_y = 1,\ldots,n_y$ do:
8. For $i_x = 1,\ldots,n_x$ do:
9. Set $\tilde{\imath}_y(t) = \tilde{\imath}_y(t) + i_y u(x_i, t)$, where $i = ((i_x - 1)n_x + i_y)$.
10. End for.
11. End for.
12. Set $\tilde{\imath}_x(t) = [\tilde{\imath}_x(t)]$ and $\tilde{\imath}_y(t) = [\tilde{\imath}_y(t)]$, where $[a]$ stands for the integer number closest to the real number a.
13. Return $\tilde{x}(t) = (\tilde{\imath}_x(t), \tilde{\imath}_y(t))$.

Finally, the function estimate probabilities( ) that applies the information lack $I(u)$ and the estimated target's location $\tilde{x}$ and provides global estimation of the probabilities given the current location $\breve{x}(t)$ of the searcher is outlined as follows:

$$\text{estimate probabilities}\left(\breve{x}(t), u(x_i, t+1), \hat{\rho}\right):$$

1. Set $\tilde{x}(t+1) = \text{estimated location}\left(u(x_i, t+1)\right)$.

Calculate distance between $\breve{x}(t)$ and $\tilde{x}(t+1)$ in the metric in $\mathfrak{X}$:

2. Set $d = \left\| \breve{x}(t) - \tilde{x}(t+1) \right\|$.

Calculate the estimated number m of the diffusion steps:

3. $m = \lceil I(u) \rceil \cdot [d]$, where $\lceil a \rceil$ stands for a closest integer greater than or equal to the real number a and $[a]$ stands for the integer number closest to the real number a.
4. For all x_i do: Set $\hat{u}(x_i, t+1) = u(x_i, t+1)$.

Calculate m steps of diffusion-like process governed by the transition matrix $\hat{\rho}$:

5. Do m times:
6. For all x_i do: Set $\tilde{u}(x_i) = \hat{u}(x_i, t+1)$.
7. For all x_i do: Set $\hat{u}(x_i, t+1) = \sum_{j=1}^{n} \hat{\rho}_{ij} \tilde{u}(x_i)$.
8. Done.
9. Return $\hat{u}(x_i, t+1)$, $i = 1, 2, \ldots, n$.

The actions of Algorithm 4.8 are illustrated by the following example.

Example 4.11

As earlier, assume that the search is conducted in the square domain $\mathfrak{X}$ of the size $n = 100 \times 100$ with the impermeable bounds and that the nonzero target's location probabilities are distributed sparsely such that in most of the points $x \in \mathfrak{X}$, these probabilities $u(x, t)$ are zero. Assume also that the distances are defined by the Manhattan metric that for any two points $\mathfrak{x}_1 = (x_1, y_1)$ and $\mathfrak{x}_2 = (x_2, y_2)$ of the domain $\mathfrak{X}$, is

$$d(\mathfrak{x}_1, \mathfrak{x}_2) = |x_1 - x_2| + |y_1 - y_2|,$$

and that the movements are conducted according to the indicated set $D = \{d_1, d_2, d_3, d_4, d_5\}$ of possible directions such that d_1 stands for "move forward," d_2 "move backward," d_3 "move right," d_4 "move left," and d_5 "stay in the current point."

The searcher starts in the boundary point $\breve{\mathfrak{x}}(0) = (15,1)$. The target's initial location is drawn randomly with respect to its location probabilities; in the considered example, it is $\hat{\mathfrak{x}}(0) = (16,60;$ thus initial distance between the searcher and the target is $d(\breve{\mathfrak{x}}(0), \mathfrak{x}(0)) = 60$. In the case of static target, the target stays in its location $\hat{\mathfrak{x}}(0)$ during all the search period, and in the case of Brownian target, it walks randomly starting from the point $\hat{\mathfrak{x}}(0)$ and up to the termination of search.

An example of the searcher's trajectories for the search for static and Brownian target is shown in Figure 4.31, where it is assumed that the searcher observes squire areas around its location $\breve{\mathfrak{x}}(t)$ and the size of the observed areas is $|a(t)| = 9$, $t = 0, 1, 2, \ldots$.

It is seen that the trajectory of the searcher seeking for a static target is similar to the trajectory of the foraging by patches (see Section 3.3, especially Figures 3.5c and 3.7c), and moreover, it has the same form as trajectories governed by the Lévy walks. In opposite, the trajectory obtained for the moving target search consists of long ballistic movements, which change the direction with respect to the estimated target's location probabilities. Notice that in the considered example, the number of search steps in the static target search, that is 825, and it is greater than the number of search steps in the moving target search, that is, 673, which indicated the movements of the search in the regions with nonzero target's location probabilities. ■

The number of steps in the search according to Algorithm 4.8 depends both on the size $|a(t)|$, $t = 0, 1, 2, \ldots$, of the observed areas (it is as smaller as larger the observed area), and on the type of the target's motion (Kagan, Goren, & Ben-Gal, 2010). For the static target, the number of search steps is, certainly, smaller than the number of search steps in the search for a moving target, and the number of steps in the search for a Markovian target, which is allowed to stay in its location, is smaller than the number of steps in the search of a Brownian target. In particular, in 1000 trials

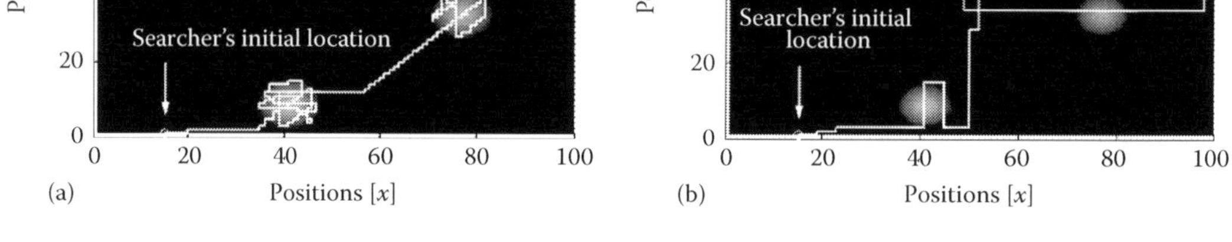

FIGURE 4.31 Trajectories of the searcher obtained by Algorithm 4.8 in (a) the search for a static target and (b) the search for a target moving according to the Brownian walk. The regions with zero target's location probabilities are depicted by black color, and nonzero probabilities are depicted by gray color such that the brighter tone corresponds to the higher location probability.

of the algorithm over the domain of the size $n = 50 \times 50 = 2500$ and with the same setting as in Example 4.11, it resulted in the following statistics (Kagan, Goren, & Ben-Gal, 2010):

	Average Number of Search Steps		
Type of the Target	$\lvert a(t)\rvert = 3\times 3 = 9$	$\lvert a(t)\rvert = 5\times 5 = 25$	$\lvert a(t)\rvert = 7\times 7 = 49$
Static target	125.4	85.5	78.2
Markovian target	510.8	254.6	155.4
Brownian target	563.0	268.5	158.8

Additional analysis of the search in the indicated settings demonstrates that in the case of search for a static target, the average number 125.4 of search steps required by Algorithm 4.8 is very close to the average number 124.04 of search steps required by the optimal Huffman algorithm acting in the same setting. In the case of search for a moving target, Algorithm 4.8 is comparable with Algorithms 2.3 and 2.4 considered in Section 2.3.2, but since they act in different settings, in particular, errorless detection in Algorithm 4.8 and erroneous detection in Algorithms 2.3 and 2.4, the direct comparison of the algorithms is meaningful.

The presented algorithm follows the methodology of the destructive search and zeroes the target's location probabilities in the observed areas up to detecting the target. However, it does not restrict the searcher to follow via such areas for shortening the search trajectory. In contrast, the next two algorithms with detection errors follow the methodology of the nondestructive search, which allows revisiting the previously observed areas.

4.4.3 Search for Static and Moving Targets: False-Negative Detection Errors

In the algorithm considered earlier, it is assumed that the detection is errorless, and direct generalization of the algorithm for the search with erroneous detections fails. However, this algorithm forms a basis for the algorithm of search with false-negative detection errors that applies two instances of the target's location probabilities and uses both detection probability function φ and observation function z, which is used as a detection probability for errorless detections (Kagan, Goren, & Ben-Gal, 2012).

In contrast to Algorithm 4.8, assume that the detection probabilities $\varphi(a(t))$, $t = 0, 1, 2, \ldots$, are specified as follows:

$$\varphi(a(t)) \le 1 \quad \text{if } \hat{\mathfrak{x}}(t) \in a(t) \quad \text{and} \quad \varphi(a(t)) = 0 \quad \text{otherwise.} \tag{4.70}$$

In addition, similar to the procedures of search and screening considered in Sections 2.2.2 through 2.2.4, let $z(a(t))$, $t = 0, 1, 2, \ldots$, be an observation result of the area $a(t)$ such that

$$z(a(t)) = 1 \quad \text{if } \hat{\mathfrak{x}}(t) \in a(t), \quad \text{and} \quad z(a(t)) = 0 \quad \text{otherwise.} \tag{4.71}$$

Certainly, in the case of errorless detection, the values $\varphi(a(t))$ and $z(a(t))$ are formally equivalent that was utilized in the Algorithm 4.8; in the next algorithm, these values are distinguished, and $\varphi(a(t))$ is considered as a probability of obtaining the observation result $z(a(t)) = 1$.

The key idea of the algorithm of search with false-negative detection errors is the following. The searcher starts with two instances, $u_1(\mathfrak{x}_i, t)$ and $u_2(\mathfrak{x}_i, t)$, $i = 1, 2, \ldots, n$, $t = 0, 1, 2, \ldots$, of the target's location probabilities. Corresponding joint probabilities are denoted by $\hat{u}_1(\mathfrak{x}_i, t)$ and $\hat{u}_2(\mathfrak{x}_i, t)$, respectively. The searcher starts with the initial target's location probabilities $u(\mathfrak{x}_i, 0) = u(\mathfrak{x}_i)$, $i = 1, 2, \ldots, n$, and follows Algorithm 4.8 with errorless detection using function z

over the first instance $u_1(\mathfrak{x}_i,t)$ of the probabilities. In parallel, it updates the second instance of the probabilities $u_2(\mathfrak{x}_i,t)$ with respect to the erroneous detections using function φ. When the searcher is uncertain regarding its motion according to the probabilities $u_1(\mathfrak{x}_i,t)$, it updates them by the probabilities $u_2(\mathfrak{x}_i,t)$ and continues the search according to Algorithm 4.8. The updating of the probabilities $u_1(\mathfrak{x}_i,t)$ with the probabilities $u_2(\mathfrak{x}_i,t)$ is conducted when local estimations regarding the target's location become weaker than corresponding global estimation. The actions of the algorithm are outlined as follows (in the outline, the functions are the same as in the Algorithm 4.8).

Algorithm 4.9: (Search with False-Negative Detection Errors)
(Kagan, Goren, & Ben-Gal, 2012)

Given a discrete domain $\mathfrak{X} = \{\mathfrak{x}_1, \mathfrak{x}_2, \ldots, \mathfrak{x}_n\}$, initial target's location probabilities $u(\mathfrak{x}_i)$, $i = 1, 2, \ldots, n$, and target's transition probabilities matrix $\|\hat{\rho} = \rho_{ij}\|_{n\times n}$ do:

1. Set $t = 0$ and start with the probabilities $\hat{u}_1(\mathfrak{x}_i,t) = u_1(\mathfrak{x}_i,t) = u(\mathfrak{x}_i)$ and $\hat{u}_2(\mathfrak{x}_i,t) = u_2(\mathfrak{x}_i,t) = u(\mathfrak{x}_i)$, $i = 1, 2, \ldots, n$, and initial observed area $a(t)$.
2. While $z(a(t)) \neq 1$ (recall that $Pr\{z(a(t)) = 1 | \hat{\mathfrak{x}}(t) \in a(t)\} = \varphi(a(t)) \leq 1$) do:
3. For all $\mathfrak{x}_i$ do: Set $\hat{u}_1(\mathfrak{x}_i,t) = u_1(\mathfrak{x}_i,t)$.
4. For all $\mathfrak{x}_i$ do: Set $\hat{u}_2(\mathfrak{x}_i,t) = u_2(\mathfrak{x}_i,t)$.

Calculate observed probabilities $\bar{u}_1(\mathfrak{x}_i,t)$ and $\bar{u}_2(\mathfrak{x}_i,t)$, $i = 1, 2, \ldots, n$:

5. For all $\mathfrak{x}_i$ do: If $\mathfrak{x}_i \in a(t)$, then set $\hat{u}_1(\mathfrak{x}_i,t) = 0$ (since $z(a(t)) \neq 1$, that is $z(a(t)) = 0$, it holds true that $z(\{\mathfrak{x}_i\})\hat{u}_1(\mathfrak{x}_i,t) = 0$ for each $\mathfrak{x}_i \in a(t)$).
6. For all $\mathfrak{x}_i$ do: Set $\bar{u}_1(\mathfrak{x}_i,t) = \hat{u}_1(\mathfrak{x}_i,t) \Big/ \sum_{j=1}^{n} \hat{u}_1(\mathfrak{x}_j,t)$.
7. For all $\mathfrak{x}_i$ do: If $\mathfrak{x}_i \in a(t)$, then set $\hat{u}_2(\mathfrak{x}_i,t) = \varphi(\mathfrak{x}_i)\hat{u}_2(\mathfrak{x}_i,t)$.
8. For all $\mathfrak{x}_i$ do: Set $\bar{u}_2(\mathfrak{x}_i,t) = \hat{u}_2(\mathfrak{x}_i,t) \Big/ \sum_{j=1}^{n} \hat{u}_2(\mathfrak{x}_j,t)$.

Calculate location probabilities $u_1(\mathfrak{x}_i,t+1)$ and $u_2(\mathfrak{x}_i,t+1)$, $i = 1, 2, \ldots, n$:

9. For all $\mathfrak{x}_i$ do: Set $u_1(\mathfrak{x}_i,t+1) = \sum_{j=1}^{n} \hat{\rho}_{ij}\bar{u}_1(\mathfrak{x}_j,t)$.
10. For all $\mathfrak{x}_i$ do: Set $u_2(\mathfrak{x}_i,t+1) = \sum_{j=1}^{n} \hat{\rho}_{ij}\bar{u}_2(\mathfrak{x}_j,t)$.

Calculate weights β_k *for the directions* $d_k \in D$:

11. For all d_k do: Set $\beta_k = \sum_{\mathfrak{x}\in a_k} u_1(\mathfrak{x},t+1)$, where a_k is an area that will be observed if the agent will move by the direction d_k.

Estimate global probabilities $\hat{u}_1(\mathfrak{x}_i,t+1)$ *if needed*, $i = 1, 2, \ldots, n$:

12. If $\beta_k = 0$ for each k, then
13. For all $\mathfrak{x}_i$ do: Set $\hat{u}_1(\mathfrak{x}_i,t+1) = \text{estimate probabilities}(\breve{\mathfrak{x}}(t), u_1(\mathfrak{x}_i,t+1), \hat{\rho})$,
14. Else
15. For all $\mathfrak{x}_i$ do: Set $\hat{u}_1(\mathfrak{x}_i,t+1) = u_1(\mathfrak{x}_i,t+1)$.
16. End if.
17. Set $\tilde{\mathfrak{x}}(t+1) = \text{estimated location}(\hat{u}_1(\mathfrak{x}_i,t+1))$.
18. If $\hat{u}_1(\tilde{\mathfrak{x}}(t+1))d(\tilde{\mathfrak{x}}(t+1), \mathfrak{x}_k(t)) > \beta_k \sum_{i=1}^{n} d(\bar{u}_{1k0}(\mathfrak{x}_i,t+1), \hat{u}_1(\mathfrak{x}_i,t+1))$ for at least one movement $d_k \in D$, then
19. For all $\mathfrak{x}_i$ do: Set $\hat{u}_1(\mathfrak{x}_i,t+1) = \hat{u}_2(\mathfrak{x}_i,t+1)$.

20. For all $\mathfrak{x}_i$ do: Set $u_1(\mathfrak{x}_i, t+1) = u_2(\mathfrak{x}_i, t+1)$.
21. End if.

Select the next observed area $a(t + 1)$ *using estimated probabilities* $\hat{u}(\mathfrak{x}_i, t+1)$, $i = 1,2,\ldots,n$:

22. Set $a(t+1) = \text{next observed area}\left(\hat{u}_1(\mathfrak{x}_i, t+1)\right)$,

Move to the area $a(t + 1)$ *and observe it*:

23. Set $t = t + 1$ and get $\varphi(a(t))$.
24. End while.
25. Return $a(t)$ (notice again that $Pr\{z(a(t)) = 1 | \hat{\mathfrak{x}}(t) \in a(t)\} = \varphi(a(t)) \leq 1$).

Similar to the notation used in Algorithm 4.8, $\bar{u}_{1k0}(\mathfrak{x}_i, t+1)$, $i = 1, 2,\ldots, n$, which appears in line 18, stands for the observed probabilities $\bar{u}_{1k}(\mathfrak{x}_i, t+1)$ of the unsuccessful observation corresponding to the movement $d_k \in D$, $t = 0, 1, 2\ldots$ Also, as in the preceding text, the target moves to its next location according to the transition probabilities matrix $\hat{\rho}$ and is not aware of the agent activities.

The actions of the Algorithm 4.9 are illustrated by the following example.

Example 4.12

Assume that the search is conducted in the same setting as in the Example 4.11. The searcher starts in the point $\breve{\mathfrak{x}}(0) = (15,1)$, and the target's initial location, which is drawn according to the target's location probabilities, is $\hat{\mathfrak{x}}(0) = (30,34)$; thus, initial distance between the searcher and the target is $d(\breve{\mathfrak{x}}(0), \hat{\mathfrak{x}}(0)) = 48$.

The detection probability $\varphi(a(t))$ is defined by a binormal distribution over the observed area $a(t)$:

$$\varphi(a(t)) = (1/2\pi\sigma)\exp\left(\frac{-(\mathfrak{x} - \mathfrak{x}^*)^2}{2\sigma^2}\right),$$

where $\mathfrak{x}^*$ is a center of the area $a(t)$ and $\sigma = (\sigma_x, \sigma_y)$ is a pair of deviation in x and y directions.

As earlier, it is assumed that $|a(t)| = 9$, $t = 0, 1, 2\ldots$ The values of the deviation $\sigma = (\sigma_x, \sigma_y)$ in both directions are specified as half of the area radius in both directions, that is, $\sigma_x = \sigma_y = 0.5$. Such values correspond to the detection error of the order 40%. In other words, the searcher does not

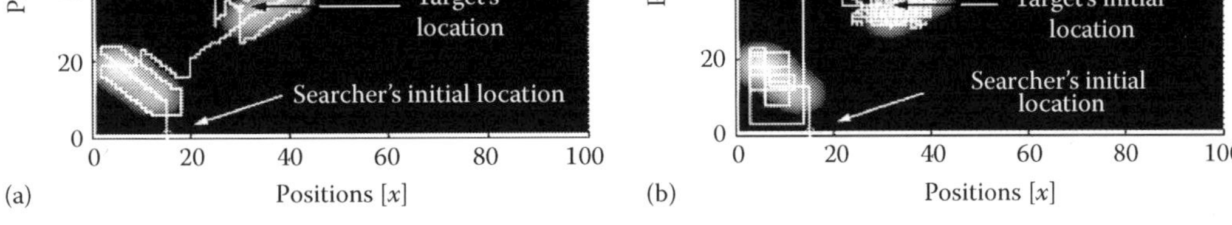

FIGURE 4.32 Trajectories of the searcher obtained by Algorithm 4.9 in (a) the search for a static target and (b) the search for a target moving according to the Brownian walk. The regions with zero target's location probabilities are depicted by black color, and nonzero probabilities are depicted by gray color such that the brighter tone corresponds to the higher location probability.

detect the existing target in nearly 40% of the observations. An example of the searcher's trajectories for the search for static and Brownian target is shown in Figure 4.32.

Certainly, the obtained trajectories have the same form as the trajectories shown in Figure 4.31. However, notice that in the considered search with erroneous detections, the search can observe the area, which includes the target, without detecting the target. In particular, the searcher's trajectory shown in Figure 4.32.a follows over the exact location of the target, but because of the erroneous detection, the target has not been detected, and the searcher continues to search. ■

Numerical analysis of Algorithm 4.9 was conducted according to the same setting as in the preceding text by 1000 trials over the domain of the size $n = 50 \times 50 = 2500$. The obtained statistics are summarized in the following table (Kagan, Goren, & Ben-Gal, 2012):

	Average Number of Search Steps		
Type of the Target	$\lvert a(t)\rvert = 3 \times 3 = 9$	$\lvert a(t)\rvert = 5 \times 5 = 25$	$\lvert a(t)\rvert = 7 \times 7 = 49$
Static target	198.8	117.7	98.0
Markovian target	637.0	258.5	174.7
Brownian target	714.6	252.9	178.7

As earlier, the average number of steps depends on the size of the observed area, and a larger area results in fewer search steps, and Markovian target, which is allowed to stay in its location, is found faster than the Brownian target.

Additional comparison of the distributions of search efforts created by the Algorithm 4.9 and the off-line Brown Algorithm 2.3 of search for a moving target demonstrates that for small total search effort, the Algorithm 4.9, in contrast to the Brown Algorithm 2.3, tends to distribute most of the search effort in the region of most probable location of the target. However, while the total search effort increases, the difference between the obtained distributions decreases. As a result, for reasonable cases, the distribution obtained online using Algorithm 4.9 is close to optimal distribution provided by the off-line Brown Algorithm 2.3.

The approach used in Algorithm 4.9 allows a direct generalization for the search with both false-positive and false-negative detection errors. The next algorithm implements such possibility.

4.4.4 Static Target Search in Noisy Environment: False-Positive and False-Negative Detection Errors

The algorithm of search with both false-positive and false-negative detection errors at most follows the same lines as previously considered in Algorithms 4.8 and 4.9 and specifies the search for a static target in a noisy environment. It is motivated by the task of planning the path of sea vessel searching for a target lying on a seabed using a wideband sonar (Chernikhovsky et al., 2012). Notice that this algorithm is not applicable in the case of search for a moving target.

Similar to the previously considered algorithms, the algorithm deals with the detection probabilities $\varphi(a(t))$ and observation results $z(a(t))$, $t = 0, 1, 2, \ldots$, but, in addition to the functions φ and z, it applies the function ψ, which specifies the probability $\psi(a(t))$ of false-positive detection of the target in the area $a(t)$. The probabilities $\varphi(a(t))$ and $\psi(a(t))$ and observation results $z(a(t))$ are defined as follows:

$$\text{if } \hat{x}(t) \in a(t), \quad \text{then } z(a(t)) = 1 \text{ with probability } \varphi(a(t)) \quad \text{and}$$

$$z(a(t)) = 0 \quad \text{with probability } \bar{\varphi}(a(t)) = 1 - \varphi(a(t)); \tag{4.72}$$

$$\text{if } \breve{x}(t) \notin a(t), \quad \text{then } z(a(t)) = 1 \text{ with probability } \psi(a(t)) \quad \text{and}$$
$$z(a(t)) = 0 \quad \text{with probability } \overline{\psi}(a(t)) = 1 - \psi(a(t)). \tag{4.73}$$

In the algorithm, the instances $u_1(x_i, t)$ and $u_2(x_i, t)$, and, respectively, $\hat{u}_1(x_i, t)$ and $\hat{u}_2(x_i, t)$, $i = 1, 2, \ldots, n$, $t = 0, 1, 2, \ldots$, of the target's location probabilities have the same meaning as in Algorithm 4.9, but the updating of the probabilities $u_2(x_i, t)$ and $\hat{u}_2(x_i, t)$ is conducted using the probabilities $\varphi(a(t))$ and $\psi(a(t))$. The outline of the algorithm is the following (as in the preceding text, in the outline, the functions are the same as in Algorithm 4.9):

Algorithm 4.10: (Search with False-Positive and False-Negative Detection Errors) (Chernikhovsky et al., 2012)

Given a discrete domain $X = \{x_1, x_2, \ldots, x_n\}$ and initial target's location probabilities $u(x_i)$, $i = 1, 2, \ldots, n$, do:

1. Set $t = 0$ and start with the probabilities $\hat{u}_1(x_i, t) = u_1(x_i, t) = u(x_i)$ and $\hat{u}_2(x_i, t) = u_2(x_i, t) = u(x_i)$, $i = 1, 2, \ldots, n$, and initial observed area $a(t)$.
2. While $z(a(t)) \neq 1$ do; recall that $Pr\{z(a(t)) = 1 \mid \breve{x}(t) \in a(t)\} = \varphi(a(t))$ and $Pr\{z(a(t)) = 1 \mid \breve{x}(t) \notin a(t)\} = \psi(a(t))$:

Calculate location probabilities $u_1(x_i, t+1)$ and $u_2(x_i, t+1)$, $i = 1, 2, \ldots, n$:

3. For all x_i do: Set $u_1(x_i, t+1) = \left[1 - z(\{x_i\})\right] u_1(x_i, t) \Big/ \sum_{j=1}^{n} \left[1 - z(\{x_i\})\right] u_1(x_j, t)$.
4. For all x_i do: Set $u_2(x_i, t+1) = \left(\varphi(\{x_i\}) \overline{\psi}(\{x_i\}) u_2(x_i, t) + \varepsilon\right) \Big/ \sum_{j=1}^{n} \left(\varphi(\{x_i\}) \overline{\psi}(\{x_i\}) u_2(x_j, t) + \varepsilon\right)$, where ε, $0 \leq |\varepsilon| \ll 1$, is a uniformly distributed random variable representing the noise if need.

Calculate weights β_k *for the directions* $d_k \in D$:

5. For all d_k do: Set $\beta_k = \sum_{x \in a_k} u_1(x, t+1)$, where a_k is an area that will be observed if the agent will move by the direction d_k.

Estimate global probabilities $\hat{u}_1(x_i, t+1)$ *if needed*, $i = 1, 2, \ldots, n$:

6. If $\beta_k = 0$ for each k, then
7. For all x_i do: Set $\hat{u}_1(x_i, t+1) = \text{estimate probabilities}\left(\breve{x}(t), u_1(x_i, t+1), \hat{\rho}\right)$
8. Else
9. For all x_i do: Set $\hat{u}_1(x_i, t+1) = u_1(x_i, t+1)$.
10. End if.
11. Set $\tilde{x}(t+1) = \text{estimated location}\left(\hat{u}_1(x_i, t+1)\right)$.
12. If $\hat{u}_1(\tilde{x}(t+1)) d(\tilde{x}(t+1), \breve{x}_k(t)) > \beta_k \sum_{i=1}^{n} d(\overline{u}_{1k0}(x_i, t+1), \hat{u}_1(x_i, t+1))$ for at least one movement $d_k \in D$, then
13. For all x_i do: Set $\hat{u}_1(x_i, t+1) = \hat{u}_2(x_i, t+1)$.
14. For all x_i do: Set $u_1(x_i, t+1) = u_2(x_i, t+1)$.
15. End if.

Select the next observed area $a(t+1)$ *using estimated probabilities* $\hat{u}(x_i, t+1)$, $i = 1, 2, \ldots, n$:

16. Set $a(t+1) = \text{next observed area}\left(\hat{u}_1(x_i, t+1)\right)$,

Move to the area $a(t+1)$ *and observe it*:

17. Set $t = t + 1$ and get $\varphi(a(t))$.
18. End while.
19. Return $a(t)$; notice again that $Pr\{z(a(t))=1|\hat{x}(t)\in a(t)\}=\varphi(a(t))$ and $Pr\{z(a(t))=1|\hat{x}(t)\notin a(t)\}=\psi(a(t))$.

As indicated in the preceding text, the algorithm implements the same idea as Algorithm 4.9. The main difference is in the updating of the target's location probabilities (lines 3 and 4 instead of lines 5 through 10 in Algorithm 4.9), which takes into account both false-positive and false-negative detection errors and the fact that the target is static. The noise is represented by the value ε, $0 \le |\varepsilon| \ll 1$, which, certainly, can be omitted.

Certainly, the search trajectories provided by Algorithm 4.10 have the same form as trajectories provided by Algorithm 4.9 in the search for a static target (see Figure 4.32a). However, notice that because of nonzero false-positive detection probabilities, Algorithm 4.10 can fail in finding the target in a finite period, but for reasonable probabilities $\varphi(a(t))$ and $\psi(a(t))$, $t = 0, 1, 2,\ldots$, and noise ε, such fails are rather rare.

Similar to the one mentioned in the preceding text, numerical analysis of Algorithm 4.10 was conducted according to the same setting by 1000 trials over the domain of the size $n = 50 \times 50 = 2500$. For observed areas $a(t)$, $t = 0, 1, 2,\ldots$, the detection probabilities are $\varphi(a(t)) = 0.9$ and probabilities of false-positive detections are $\psi(a(t)) = 0.01$; the value ε is drawn from the interval $[-0.001, 0.001]$. The obtained statistics, including percentages of false-positive detections, are as follows (Chernikhovsky et al., 2012):

Observed Area	Average Number of Search Steps	Successful Trials	False Positive Detections
$\lvert a(t)\rvert = 3\times 3 = 9$	128.3	95.8%	1.98%
$\lvert a(t)\rvert = 5\times 5 = 25$	90.4	99.6%	2.91%
$\lvert a(t)\rvert = 7\times 7 = 49$	85.0	99.7%	7.62%

The simulations demonstrate that for the tested observed areas, the algorithm terminates with finding the target in more than 95% of trials, and their number increases with the size of observed areas up to 99.7%. Notice that since observations of the points in the observed area are conducted several times while the searcher moves along its path in the area, the percentage 95% of successful trials is greater than the percentage 90%, which is provided by the detection probability $\varphi(a(t)) = 0.9$. By the same reason, the percentage of false-positive detections also increases with the size of observed areas.

As it is expected, the average number of search steps required by Algorithm 4.10 is greater than the corresponding average number of search steps required by Algorithm 4.8 with errorless detections. In addition, a comparison of the simulation results for the Algorithm 4.10 with the simulation results for Algorithm 4.9 indicates the dependence of the number of search steps on the distribution of the search efforts over the observed areas. In fact, in the simulations of Algorithm 4.10, the search efforts were distributed uniformly, while in the simulations of Algorithm 4.9, they were placed according to binormal distribution with maximum in the center of the observed areas. As a result, the trials of Algorithm 4.10 terminated in a smaller average number of search steps than the trials of Algorithm 4.9.

The considered algorithm of search with false-positive and false-negative detection errors finalizes the series of the online algorithms of search in discrete time and space based on the local decision making and global estimations using diffusion-like processes. These algorithms are applicable for different search tasks that satisfy the Problem 4.3 of search by areas for static and moving target. In general, these algorithms do not require additional assumptions regarding the observed areas as well as can be used for the search, in which the pattern of the target's motion changes during the search.

The next algorithm, in contrast, implements the search in continuous time and space. In general, it follows the optimization techniques presented in Section 2.3.3 and includes the problem of shadowing that restricts the abilities of the searcher and considers the observed areas of strongly varying size.

4.4.5 Probabilistic Search with Shadowing

Let us return to the problem of search considered in Section 2.3.3, in which the searcher and the target act in continuous time and space. Similar to the previously presented algorithms, it is assumed that

1. The search can be terminated at any time, when the target is detected with certainty. However, in contrast to the considered search processes,
2. The visibility of the target changes according to an impulse function determined by the terrain topography and searcher and target locations.

The first assumption is a general one implemented by the path-planning methods and algorithms of search with infinite horizon (Singh & Krishnamurthy, 2003), while the second assumption requires solving the problems of shadowing and visibility. In the case of search for a moving target, the strict solutions of these problems are known only for the simplest searcher and target motion (Yadin & Zacks, 1988; Zacks & Yadin, 1994), and in the case of static target, the search is considered as a problem of path planning with obstacle avoidance, which is also far from being solved.

In the considered algorithm, it is assumed that the searcher moves over a domain such that the visibility distance differs in different directions. Such search models the motion of a ground vehicle over a natural terrain with different heights of the points. Following the widely accepted model (Lukka, 1977; Mangel, 1982; Ohsumi, 1991), which was already used in Section 2.3.3, it is assumed that the target's location probabilities are governed by a diffusion process. In addition, similar to Algorithm 4.8, it is assumed that the searcher applies errorless detection function and terminates the search when the target is in the searcher's observed area. The observed areas are determined by the topography of the terrain and depend on the location of the searcher and the target. Such dependence with the errorless detection results in the changing of the detection by impulses such that an infinitely small change of the searchers' coordinates can result in certain detection of previously invisible target. In discrete time and space, the search planning in such conditions is provided by Algorithm 4.8; the next algorithm provides a heuristic solution of the problem in continuous time and space (Israel, Khmelnitsky, & Kagan, 2012).

Similar to Sections 2.2.1 and 2.3.3, assume that the searcher and the target act in the bounded 2D domain $\mathfrak{X} \subset \mathbb{R}^2$ during the period $\left[0,T\right)$, $T<\infty$. The target's location probabilities given the searcher actions are defined by the density function $u:\mathfrak{X}\times\left[0,T\right)\to\mathbb{R}^+$, and these probabilities without search are defined by the density function $\hat{u}:\mathfrak{X}\times\left[0,T\right)\to\mathbb{R}^+$. The trajectories of the searcher and the target over the domain $\mathfrak{X}$ are specified by the functions $\breve{\mathfrak{x}}:\mathfrak{X}\times\left[0,T\right)\to\mathfrak{X}$ and $\hat{\mathfrak{x}}:\mathfrak{X}\times\left[0,T\right)\to\mathfrak{X}$, correspondingly. The detection of the target is defined by the function $\psi:\mathfrak{X}\times\mathfrak{X}\to\mathbb{R}^+$ such that the value $\psi\left(\hat{\mathfrak{x}}(t),\breve{\mathfrak{x}}(t)\right)dt$ is a probability of detecting the target during the period $\left[t,t+dt\right)$, while the target's location is $\hat{\mathfrak{x}}(t)$ and the searcher's location is $\breve{\mathfrak{x}}(t)$.

In addition, let $\mathcal{T}:\mathfrak{X}\to\mathbb{R}^+$ be a smooth function over the domain $\mathfrak{X}$ such that $\mathcal{T}\left(\mathfrak{x}\right)<\infty$ for any point $\mathfrak{x}\in\mathfrak{X}$. The domain $\mathfrak{X}$ with the function $\mathcal{T}$ represents a terrain, for which $\mathfrak{X}$ is a flat basis, value $\mathcal{T}\left(\mathfrak{x}\right)$ is the height of the terrain at the point $\mathfrak{x}$, and the function $\mathcal{T}$ represents topography

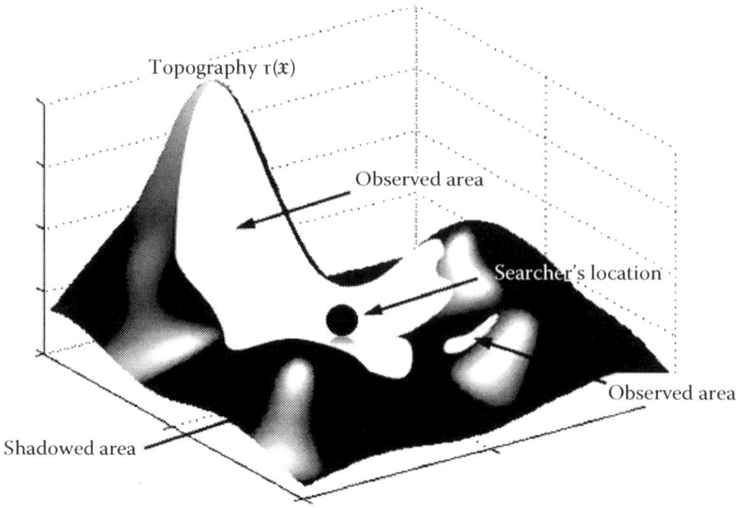

FIGURE 4.33 Scheme of the terrain and the observed area.

of the terrain. According to the topography of the terrain, for any searcher's location $\breve{\mathfrak{x}}(t)$, the observed area is defined as a part of the terrain, which can be directly observed from the point $\breve{\mathfrak{x}}(t)$ with the height $\mathcal{T}\left(\breve{\mathfrak{x}}(t)\right)$ and which is not shadowed by the other points $\mathfrak{x} \in \mathfrak{X}$ with the heights $\mathcal{T}(\mathfrak{x})$. The scheme of the observed area is shown in Figure 4.33.

The search over a terrain is conducted in two stages (Israel, Khmelnitsky, & Kagan, 2012). At the first off-line stage, the visibility analysis of the terrain is conducted, and a finite set of potential points on the terrain is created. These points are chosen in such a way that their joint observed area completely covers a terrain. At the second stage, the online search over the set of potential points is conducted.

Let us start with the visibility analysis algorithm. Assume that over a surface defined by the topography function $\mathcal{T}$ a finite measure is defined. Since the domain $\mathfrak{X}$ is bounded, the heights $\mathcal{T}(\mathfrak{x})$, $\mathfrak{x} \in \mathfrak{X}$, are finite, and $\mathcal{T}$ is smooth in the metric of $\mathfrak{X}$; such measure always exists. A measure of a set $\mathfrak{A} \subset \mathfrak{X}$ with respect to the topography $\mathcal{T}$ is denoted by $\#\mathfrak{A}$; in the simplest case, $\#\mathfrak{A}$ is an area of $\mathfrak{A}$, and in the case of discrete domain, it is a cardinality of $\mathfrak{A}$.

Let $\mathfrak{x}_1$ and $\mathfrak{x}_2$, $\mathfrak{x}_1 \neq \mathfrak{x}_2$, be two points from $\mathfrak{X}$, and let $\mathfrak{A}_1$ and $\mathfrak{A}_2$ be the sets of points visible from the points $\mathfrak{x}_1$ and $\mathfrak{x}_2$, respectively. If $\mathfrak{x}_1, \mathfrak{x}_2 \in \left(\mathfrak{A}_1 \cap \mathfrak{A}_2\right)$, then the points $\mathfrak{x}_1$ and $\mathfrak{x}_2$ are called mutually comparable; otherwise, they are mutually incomparable. For the comparable points $\mathfrak{x}_1$ and $\mathfrak{x}_2$, if $\#\mathfrak{A}_1 > \#\mathfrak{A}_2$, then it is said that the point $\mathfrak{x}_1$ dominates the point $\mathfrak{x}_2$ (or, that is the same, that the point $\mathfrak{x}_2$ is dominated by the point $\mathfrak{x}_1$). Finally, the potential point is defined as a point that is not dominated by any point, which is comparable with it. The goal of the visibility analysis is to find a finite set of potential points.

Analysis of comparability and dominance of the points and the number of resulting potential points is based on an overlap degree $a \in [0,1]$. In the algorithm, this degree is defined using the accuracy value ε, which is considered as an input parameter. Given two points $\mathfrak{x}_1$ and $\mathfrak{x}_2$, the visibility measure of $\mathfrak{x}_2$ with respect to $\mathfrak{x}_1$ is defined as follows:

$$3.\ \alpha\left(\mathfrak{x}_1, \mathfrak{x}_2\right) = \#\left(\mathfrak{A}_1 \cap \mathfrak{A}_2\right) / \#\mathfrak{A}_2, \qquad \#\mathfrak{A}_2 > 0. \tag{4.74}$$

Then, given a degree a, if for the measure it holds true that $\alpha_{12} > a$, then the observed area $\mathfrak{A}_1$ includes much enough (with respect to the factor a) of the observed area $\mathfrak{A}_2$. In other words, the point $\mathfrak{x}_1$ dominates the point $\mathfrak{x}_2$ with the overlap of at least a.

A presented below visibility analysis algorithm identifies a set of potential points with respect to a given overlap degree a, which is achieved with a certain accuracy $\varepsilon \ll 1$. The algorithm starts from an initial value a_0 of overlap degree and chooses potential points with respect to a_0. Then it increases the overlap degree up to a_1 and chooses additional potential points. The process continues up to the overlap a_n such that $0 \le a_n - a \le \varepsilon$. The algorithm is outlined as follows.

Algorithm 4.11: (Visibility Analysis) (Israel, Khmelnitsky, & Kagan, 2012)

Given a domain $\mathfrak{X}$, a topography $\mathcal{T}$, a finite measure defined by the topography $\mathcal{T}$ and an accuracy value ε do:

1. Set an initial overlap degree $a = 0$.
2. Set low and upper bounds for a as $\beta_l = 0$ and $\beta_u = 1$.
3. Do:
4. Mark all points $\mathfrak{x} \in \mathfrak{X}$ as non-dominated.
5. Initialize the set $\mathfrak{P}$ of potential points as empty.
6. For each point $\mathfrak{x} \in \mathfrak{X}$ do:
7. For each comparable point $\mathfrak{y}$ of $\mathfrak{x}$ do:
8. Calculate the visibility measures $\alpha(\mathfrak{x},\mathfrak{y})$ and $\alpha(\mathfrak{y},\mathfrak{x})$ by Equation 4.74.
9. If $\alpha(\mathfrak{x},\mathfrak{y}) > a$, then mark point $\mathfrak{y}$ as dominated by $\mathfrak{x}$ with respect to a.
10. If $\alpha(\mathfrak{y},\mathfrak{x}) > a$, then mark point $\mathfrak{x}$ as dominated by $\mathfrak{y}$ with respect to a.
11. End for.
12. End for.
13. Insert all non-dominated points into the set $\mathfrak{P}$ of potential points as empty.
14. If the observed areas of the points from $\mathfrak{P}$ cover the terrain, then set $\beta_u = a$.
15. Else set $\beta_l = a$.
16. Set $a = (\beta_l + \beta_u)/2$.
17. While the observed areas of the points from $\mathfrak{P}$ does not cover the terrain and $(\beta_u - \beta_l) > \varepsilon$.
18. Return the set $\mathfrak{P}$ of potential points.

According to the outline, the algorithm runs over the points of the domain and defines the visibility α for these points, which results in certain type of gradient flow over a domain with respect to the topography $\mathcal{T}$ and visibility measure α. Certainly, the number of potential points increases with a, and for $a = 1$, all points of the terrain are chosen as potential. Notice also that the potential points are located not only at the "peaks" of the terrain but also at the lower points so that the maximal total visibility over the terrain is reached.

The online search is conducted over the set $\mathfrak{P}$ of potential points and can be implemented using various methods; Algorithm 4.12 is based on the assumption that the target is moving randomly according to the Brownian motion in, possibly, nonhomogeneous environment (Israel, Khmelnitsky, & Kagan, 2012).

Recall (see Equation 2.107) that following this assumption, the target's location probabilities are governed by the Fokker–Planck equation

$$\frac{\partial}{\partial t}u(\mathfrak{x},t) = \frac{1}{2}\left(\hat{\sigma}_1^2(\mathfrak{x})\frac{\partial^2}{\partial x^2} + \hat{\sigma}_2^2(\mathfrak{x})\frac{\partial^2}{\partial y^2}\right)u(\mathfrak{x},t) - \left(\hat{\mu}_1(\mathfrak{x})\frac{\partial}{\partial x} + \hat{\mu}_2(\mathfrak{x})\frac{\partial}{\partial y}\right)u(\mathfrak{x},t)$$
$$-\left(\psi\left(\hat{\mathfrak{x}}(t),\breve{\mathfrak{x}}(t)\right)dt - \int_{\mathfrak{X}}\psi\left(\mathfrak{x},\breve{\mathfrak{x}}(t)\right)u(\mathfrak{x},t)\,d\mathfrak{x}\right)u(\mathfrak{x},t), \tag{4.75}$$

where two first terms in the right side define the diffusion process with varying space diffusion rates and drifts, and the second term specifies the observation process. Then, the probability of detecting the target up to the time t is defined by Equation 2.110, that is,

$$p(\breve{\mathfrak{x}},t)=1-\exp\left(-\int_0^t\int_{\mathfrak{X}}\psi\left(\mathfrak{x},\breve{\mathfrak{x}}(\tau)\right)u(\mathfrak{x},\tau)\,d\mathfrak{x}\,d\tau\right). \tag{4.76}$$

In addition, recall that the search is conducted during a finite period $\left[0,T\right)$, $T<\infty$. Then, the expected time $t(\breve{\mathfrak{x}},T)$ of detecting the target up to the time T is defined as follows (Israel, Khmelnitsky, & Kagan, 2012) (cf. Equation 2.113):

$$t(\breve{\mathfrak{x}},T)=\int_0^T t\left(1-p(\breve{\mathfrak{x}},t)\right)\int_{\mathfrak{X}}\psi\left(\mathfrak{x},\breve{\mathfrak{x}}(t)\right)u(\mathfrak{x},t)\,d\mathfrak{x}\,dt\Bigg/\int_0^T\left(1-p(\breve{\mathfrak{x}},t)\right)\int_{\mathfrak{X}}\psi\left(\mathfrak{x},\breve{\mathfrak{x}}(t)\right)u(\mathfrak{x},t). \tag{4.77}$$

In general, the algorithm of search over the set $\mathfrak{P}$ of potential points acts as follows. Assume that the searcher starts in the point $\breve{\mathfrak{x}}(t_0)$. In order to choose the next potential point, the searcher evaluates the paths from $\breve{\mathfrak{x}}(t_0)$ to all potential points $\tilde{\mathfrak{x}}\in\mathfrak{P}$. For any potential point $\tilde{\mathfrak{x}}$, this path consists of a straight line from $\breve{\mathfrak{x}}(t_0)$ to $\tilde{\mathfrak{x}}$ and of the point $\tilde{\mathfrak{x}}$ itself, in which the searcher can stay up to the time $\tilde{T}$. The value of the time $\tilde{T}$ depends on the searcher's speed v and the size of the domain $\mathfrak{X}$. For a square domain $n\times n$ with Euclidian metric, the linear part of the path is $\breve{\mathfrak{x}}(t)=\breve{\mathfrak{x}}(t_0)+\mathrm{v}t\left\|\tilde{\mathfrak{x}}-\breve{\mathfrak{x}}(t_0)\right\|\Big/\left\|\tilde{\mathfrak{x}}-\breve{\mathfrak{x}}(t_0)\right\|$ for $t_0<t\le\tilde{t}$, and the stationary part is $\breve{\mathfrak{x}}(\mathrm{t})=\tilde{\mathfrak{x}}$ over the period $\tilde{t}<t\le\tilde{T}=2n/v$, where $\tilde{t}=\left\|\tilde{\mathfrak{x}}-\breve{\mathfrak{x}}(t_0)\right\|\big/v$ is a time, which is required for reaching the potential point $\tilde{\mathfrak{x}}$ from the searcher's starting point $\breve{\mathfrak{x}}(t_0)$. Finally, given the paths to all potential points, the searcher chooses a point $\tilde{\mathfrak{x}}^*\in\mathfrak{P}$ such that the ratio $p(\tilde{\mathfrak{x}}^*,T)/t(\tilde{\mathfrak{x}}^*,T)$ is maximal over the set $\mathfrak{P}$. Certainly, instead of this ratio, any other criterion that guarantees maximum detection probability in minimal expected time of search can be applied. More formally, the actions of the algorithm are outlined as follows.

Algorithm 4.12: (Search over Potential Points) (Israel, Khmelnitsky, & Kagan, 2012)

Given a domain $\mathfrak{X}$ with the topography $\mathcal{T}$ and a set $\mathfrak{P}$ of potential points provided by Algorithm 4.11 do:

1. Start with $t=0$ and initial searcher's location $\breve{\mathfrak{x}}(t)$ at this time.
2. While target is not detected do:
3. For each potential point $\tilde{\mathfrak{x}}\in\mathfrak{P}$ do:
4. Calculate detection probability $p(\tilde{\mathfrak{x}},T)$ by Equation 4.76.
5. Calculate expected search time $t(\tilde{\mathfrak{x}},T)$ by Equation 4.77.
6. End for.
7. Set $\tilde{\mathfrak{x}}^*=\arg\max_{\mathfrak{P}}\left\{p(\mathfrak{x},T)/t(\tilde{\mathfrak{x}},T)\right\}$; for the ties, choose the point $\tilde{\mathfrak{x}}^*$ such that the distance $\left\|\tilde{\mathfrak{x}}^*-\breve{\mathfrak{x}}(t)\right\|$ is minimal.
8. If $\tilde{\mathfrak{x}}^*$ is equivalent to the current searcher's location $\breve{\mathfrak{x}}(t)$, then stay in $\breve{\mathfrak{x}}(t)$ for the time equal to the average length of the previous legs. Otherwise, move to $\tilde{\mathfrak{x}}^*$.
9. If during the leg defined by line 8 the target is detected, then terminate. Otherwise, update the current time and position of the searcher.
10. End while.

Notice that by definition of the function ψ and the target's location density u, the searcher updates the target's location probabilities with respect to observation results both while it moves over the potential points and while it stays in the current point. The actions of Algorithms 4.11 and 4.12 are illustrated by the following example.

Example 4.13: (Based on the MATLAB Scripts Developed by M. Israel)

Assume that the search is conducted over a terrain built on the square domain $\mathfrak{X}$ of the size $n = 40 \times 40$. The searcher starts in the point $\breve{x}(0) = (10,1)$ at the border of the terrain and seeks for a static target. The location of the target is drawn with respect to the target's location probabilities such that for two points $x_1 = (6,30)$ and $x_2 = (30,6)$, the probabilities are $u(x_1) = u(x_2) = 1/22$ and are zero for all other points of the domain. In the considered case, the target's location is $\hat{x}(0) = (6,30)$. The considered terrain with the searcher's and target's locations and the area, which is observed from the searcher's location $\breve{x}(0) = (10,1)$, are shown in Figure 4.34a.

The searcher starts with the visibility analysis of the terrain using Algorithm 4.11, which results in the set $\mathfrak{P}$ of potential points. The 42 potential points for the considered terrain with overlap degree $a = 0.95$ are shown in Figure 4.34b. In the figure, the terrain is shown from the top view, and potential points are depicted by white circles; the higher points of the terrain are shown in brighter gray color.

After obtaining the set $\mathfrak{P}$ of potential points, the searcher conducts online search over $\mathfrak{P}$ using Algorithm 4.12 that results in the trajectory of the searcher over a terrain. The final searcher's location $\breve{x}(142) = (14,21)$ and the area observed from this location are shown in Figure 4.34c. Notice that the search time $t = 142$ includes the considerations of all 42 potential points at each step. The length of the searcher's path over a flat domain is 50 (in Manhattan metric), and in three dimensions including stepping up and down with respect to the topography, it is 357. The trajectory of the searcher over a flat domain is shown in Figure 4.34d.

It is seen that the potential points are located both in the regions with different heights that, according to Algorithm 4.11, provides maximal visibility of the points domain. The trajectory of the searcher follows from its starting point $\breve{x}(0) = (10,1)$ to the potential point $\tilde{x} = (27,14)$, which is both the closest potential point to the searcher's starting point and the closest potential point to one of two points with nonzero target's location probability. Since the searcher have not detected the target from the potential point $\tilde{x} = (27,14)$, it again chooses the potential point, in this case, the point $\tilde{x} = (14,21)$, which is closest to the current searcher's location $\breve{x}(t) = (27,14)$ and to the second point $x = (6,30)$ with nonzero target's location probability. Since the target is visible from new searcher's location $\breve{x}(142) = (14,21)$, the search terminates. ■

The actions of Algorithms 4.11 and 4.12 were checked by numerical simulations, which were conducted for different terrains over a discrete domain $\mathfrak{X}$ of the size $n = 40 \times 40$ points with Euclidian metric (Israel, Khmelnitsky, & Kagan, 2012). For each type of target's movement, the session of the algorithm included 1000 trials, and the resulting length of the searcher's path up to the successful detection of the target was compared with the length of optimal search for static target and with the brute-force search for moving target, where such brute-force search was possible.

The simulations demonstrated that Algorithms 4.11 and 4.12 provide near-optimal solutions for the problems, in which such solutions exist and successfully find the target even in the cases, for which online solutions are unknown. In particular, the paths provided by the algorithms while solving the path-planning problem with the known destination point are in average only by 3.2% longer than the optimal paths. In the problem of tracking moving target, the algorithms provide the paths, which are in average only by 2.99% longer than optimal paths provided by the brute-force optimization. Finally, for the search for a static target hidden in one of two points of the domain with nonzero probabilities, the obtained paths are in average only by 9.96% longer than the optimal paths obtained by the brute-force optimization. For the other types of the target's motion, optimal and near-optimal solutions of the problem of search over a terrain with shadowing are unknown, while Algorithm 4.12 acting over the potential points provided by Algorithm 4.11 always terminates with detecting the target.

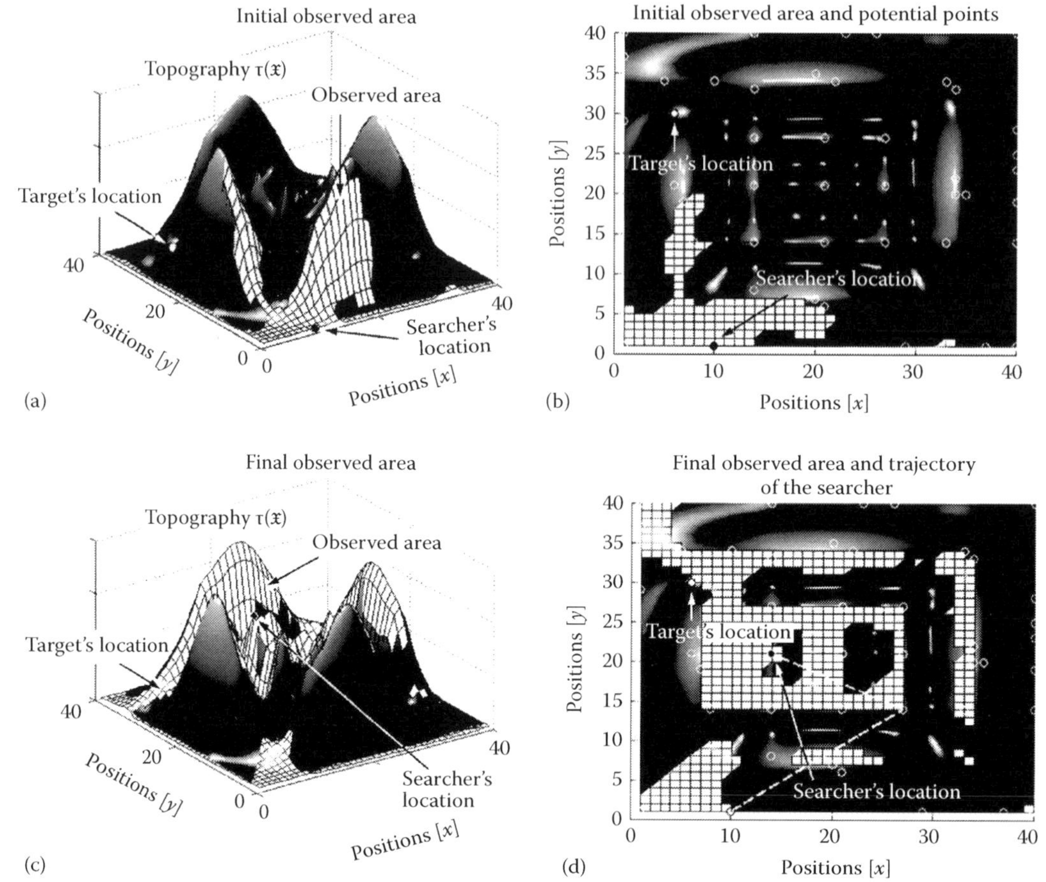

FIGURE 4.34 Search over a terrain with shadowing. Part (a) shows a terrain with the target's and the searcher's locations at the beginning of search at time $t = 0$ and an area observed (depicted in white) from the searcher's location. The target's location is $\hat{\mathfrak{x}}(0)=(6,30)$ and the searcher's location is $\breve{\mathfrak{x}}(0)=(10,1)$. A view from the top to the terrain with 42 potential points (depicted by white circles) is shown in part (b); the higher points of the terrain are shown in brighter gray color. Part (c) shows a terrain with the target's and searcher's locations at the end of search at time $t = 142$ and an area observed (depicted in white) from the final searcher's location $\breve{\mathfrak{x}}(142)=(14,21)$. A view from the top to the terrain at the end of search, including the final observed area and the trajectory of the searcher over a domain (depicted by white dashed line), is shown in part (d).

In general, the presented Algorithms (4.8 through 4.10) of search in discrete time and space and Algorithms 4.11 and 4.12 of search over the terrain with shadowing provide solutions of the path-planning problem following the approach based on the potential field (see Section 5.3). As it was mentioned at the beginning of this section, the same approach is widely used for solving the problem of obstacle avoidance. In Section 4.4.6, an application of the algorithms presented earlier for the problem of obstacle avoidance is considered briefly.

4.4.6 Remarks on Foraging with Obstacle Avoidance

The presented algorithms of search in the preceding text follow the approach to path planning and navigation of mobile robots using the potential field, where distribution of the target's location probabilities over a domain is used as a potential function, and at each step, the search agent follows maximum the field's gradient. In other words, distribution of the target' location probabilities is considered as an attractive potential (Choset et al., 2005); navigation methods, which implement such approach, are usually referred as a path planning in the belief space (Indelman, Carlone, & Dellaert, 2013a; Indelman, Carlone, & Dellaert, 2013b).

In general, the dynamics of the belief space is defined using information theory methods with respect to the considered task. In the search and foraging tasks, the changes of the agent's beliefs are specified with respect to the observation results, which are obtained during the agent's motion over a domain. In particular, in the algorithms presented earlier, the target's location probabilities in the observed areas are zeroed (for errorless detection, see Algorithm 4.8) or decreased (for erroneous detection, see Algorithms 4.9 and 4.10), and the agent tends to follow the regions with higher probabilities and avoids the areas with small or zero probabilities. This property was utilized by Cetnarowicz et al. (Cetnarowicz et al., 2013; Sosnicki et al., 2013; Zabinska et al., 2013) for constructing the algorithms of obstacle avoidance, in which obstacles correspond to regions with zero probabilities.

Notice, however, that for resolving local uncertainty in choosing the next step direction the suggested search algorithms apply additional long-distance estimations of the target's location probabilities. These estimations implement the diffusion process and use the estimated location of the target (see functions estimateprobabilities() and estimatedlocation() outlined in Section 4.4.1). In the presence of obstacles, these functions have to preserve the zero probabilities in the regions, which are indicated as obstacles, while the bounds of such regions have to be permeable. After estimating probabilities, in contrast, the bounds of these regions have to be defined as impermeable to avoid the entrance of the agent in these regions.

Additional improvements of the search algorithms can be achieved using the extensions of the well-known Bug algorithms (Choset et al., 2005), which where suggested by Shiller et al. (Shiller & Gwo, 1991; Fiorini & Shiller, 1998; Shiller et al., 2013). In fact, assume that after application of the function estimatedlocation() (see Section 4.4.1), the estimated location of the target is known. Then, the path planning in the presence of obstacles can be conducted using the Shiller algorithms, and the resulting path can be used together with the distribution of the target's location probabilities for choosing the direction of the next step.

Notice that the recently reported algorithm of path planning with obstacle avoidance (Shiller et al., 2013) builds the agent's path for both static and moving obstacles. That removes restrictions on application of obstacle avoidance methods along with the algorithms of search and foraging in dynamic environment and provides a basis for collision avoidance methods in the collective search and foraging by the groups of mobile agents. For possible implementation of such ideas, see, for example, the recent papers by Gasparri et al. (2012) and by Leccese et al. (2013); additional information about obstacle avoidance based on potential field is presented in the recent book by Fahimi (2009).

4.5 SUMMARY

In contrast to the previous chapters, which deal with the optimization problems, in this chapter, the consideration of the problem of search and foraging starts from the given trajectories of agents, which are postulated as optimal, and then addresses the methods and algorithms that provide these trajectories.

The simplest models of the searchers' motion are based on the Brownian motion (Section 4.2). These algorithms are considered following three types of Brownian motion:

1. Simple Brownian motion search (Section 4.2.1) implies that the agent follows simple random walk, and the probabilities of the agent's location are defined by diffusion equation with constant diffusion rates and zero drifts. Formally, the motion of the searcher is specified by Algorithm 4.1.
2. Search by correlated Brownian motion (Section 4.2.2) provides more realistic model of motion and implies that the agent has definite geometric size, and its displacements on

each step are provided by the movements of the legs or wheels. In addition, such model allows taking into account the preferable directions of the agent's movements. The probabilities of the agent's location are defined by the diffusion equation with nonzero drifts, and its motion is prescribed by Algorithm 4.2.

3. Search by fractional Brownian motion (Section 4.2.3) illustrates the search, in which the movements of the searcher depend on the states of the environment; in the considered tasks, these states are defined by the target's location probabilities. The probabilities of the agent's location are defined by the diffusion equation with time-dependent diffusion rates. The motion of the agent is defined by Algorithm 4.3.

Recent considerations of the foraging by animals, especially by the Viswanathan group, lead to understanding that the movements of natural foragers are rather close to the movements, defined by the Lévy flights and Lévy walks (for strict definitions, see Section 4.3.1). The Lévy flight search is considered in the following three settings:

1. Lévy flight search with constant index (Algorithm 4.4) does not consider the states of the environment, and the searcher follows the same Lévy flight during the mission.
2. Lévy flight search with varying index (Algorithm 4.5) assumes that the distribution of the steps' length changes with respect to the environmental states.
3. Lévy flight search with varying index and drifts (Algorithm 4.6) allows the dependences of both the directions of the steps and their lengths on the environment.

The intermittent search strategies (Section 4.3.3) consider the search, in which the agent changes the patterns of its motion between the search by certain process and ballistic movements with respect to the environmental states and sensing abilities of the searcher. The combination of the intermittent search and search by Lévy flights and walks is implemented in Algorithm 4.7.

Notice that the search according to the processes with environment-depending parameters allows implementation of the destructive search scenario, in which the searcher changes the environmental states in the visited points. Such scenario corresponds to the probabilistic search, where the searcher decreases the probabilities of finding the target in the unsuccessfully observed areas.

The heuristic algorithms of search inspired by the search according to Lévy walks (Section 4.4) are considered in the following several settings:

1. Search with errorless and erroneous detections (Section 4.4.1) assumes that the searcher moves in homogeneous environment with respect to the target's location probabilities and observes its close neighborhood (Algorithms 4.8 through 4.10).
2. Search with shadowing (Section 4.4.2) considers the search in the topographic terrain, in which the visibility depends on the location of the searcher in the environment (Algorithm 4.12).

In addition to effective search techniques, these algorithms form a basis for consideration of obstacle avoidance methods based on potential fields (for some remarks in this direction, see Section 4.4.3).

5 Coalitional Search and Swarm Dynamics

Mathematical modeling of collective behavior of living organisms and artificial agents can be tracked up to classical *Cybernetics* by Wiener (1948) and *Theory of Self-Reproducing Automata* by von Neumann (1966), where they suggested general framework for consideration of collective dynamics, communication, and control in the societies of active agents, and to Turing's seminal paper *The Chemical Basis of Morphogenesis* (Turing, 1952), where he suggested dynamical model and evolutionary equations, which described the formation of macroscopic spatial structures in the media of interacting elementary objects. A new impulse to the studies of collective behavior was given by the considerations of nonlinear dynamics and nonequilibrium systems conducted in parallel by Haken (1977) and by the Prigogine group (Prigogine & Nicolis, 1977), who applied the methods of statistical mechanics for analysis of the systems of different nature, including collective behavior of biological agents; nowadays these approaches are known as "synergetics" (Haken) and "self-organization" (Prigogine).

In parallel, biological observations and studies of collective behavior of animals (Bell, 1990; McElreath & Boyd, 2007; Sumpter, 2010), in particular, of ant colonies (Gordon, 2010; Holldobler & Wilson, 1990; Wilson, 1971), allowed to consider the swarm as a kind of superorganism (Holldobler & Wilson, 2009). A combination of this approach with recent advances in cybernetics and artificial intelligence led to consideration of the swarm intelligence (Kennedy, Eberhart, & Shi, 2001; Panigrahi, Shi, & Lim, 2011; Weiss, 1999), and to intensive studies of bio-inspired methods in computation (Adamatzky, Costello, & Asai, 2005; Siegelmann, 1998), optimization (Dorigo & Stutzle, 2004; Passino, 2005) and robotics (Clark, 1997; McFarland & Bösser, 1993). The progress in this direction is represented by the series of conferences on simulation and adaptive behavior (From Animals to Animats, 1991–2012) and on swarm intelligence (ANTS, 1998–2014) and in the series of symposia on distributed autonomous robotic systems (Distributed Autonomous Robotic Systems, 1992–2013).

The indicated approaches of modeling and analysis of collective behavior allow consideration of different aspects of swarm dynamics including control of the groups of mobile agents such that they are capable enough both to act independently and to communicate with other agents and to act as part of the group, which is often considered as the agents' society. Consequently, the search and foraging by the groups of mobile agents often is called social foraging (Gazi & Passino, 2011) or collective foraging (Trianni, 2008). There are several methods and techniques of specifying the activity of the mobile agents in the tasks of social foraging. This chapter considers some of these methods and presents algorithms of search by the groups of agents acting without and with communication.

5.1 SWARMING AND COLLECTIVE FORAGING

There is no direct definition of swarm, and usually it is determined as a group of active agents executing the common task that specifies the individual behavior of the agents and relations between them. In the simplest case, when the only task of the mobile agents in the group is to move together and store the geometrically distinguishable cluster of the swarm, the actions of the agents and relations between them are prescribed by three famous Reynolds rules (Reynolds, 1987):

1. *Separation*: Avoid collisions with the other agents.
2. *Alignment*: Match direction and velocity with the neighboring agents.
3. *Cohesion*: Stay close to the neighboring agents.

These rules, with corresponding definitions of distances and agent's abilities, were implemented in enormous number of simulations of the motion of animals' groups; formal presentation and analysis of these rules for particle-like agents was provided by Toner and Tu (1998).

In more complex tasks, when the collective activity of the agents is aimed to maximize definite common fitness, the rules of swarm can strongly differ from the Reynolds rules. In particular, in the tasks of collective foraging, the rules of alignment and cohesion are reconsidered, and in addition to the interagent interaction and communication, the interaction with the environment is specified (Gazi & Passino, 2011). Certainly, in the case of collective foraging, the interactions of the agents with the environment are the same as in the case of individual foraging and imply the following abilities of the agents and the environment (Kagan, Rybalov, Sela, Siegelmann, & Steshenko, 2014):

1. The agent is able to change the environment and to sign certain locations.
2. The agent is able to use the environmental states and the locations' signs in its further activity.
3. The environment is dynamic and can change its states and signs according to the internal rules.
   In addition, swarming (following the previously indicated Reynolds rules or the other) and collective activity imply that
4. The agents are able to send information about their activity and to receive information about the behavior of the other agents.
5. The agents are able to send and receive information regarding the observed environmental states and remained signs.

In the tasks of probabilistic search and foraging, the environmental states are considered as target location probabilities, and their changes are defined by the searchers' and target's activity. In Chapter 2, the search was considered in the terms of search efforts without distinguishing whether the search is conducted by a single searcher or by the group of searchers with centralized control. In contrast, Chapters 3 and 4 addressed the methods and algorithms of search and foraging by a single searcher, and the main attention was concentrated on the specification of the trajectory of the searcher.

As in individual search, the methods of collective search and foraging consider the searchers' trajectories, but the trajectories of the agents acting individually can strongly differ from the trajectories of the same agents acting in swarms. Moreover, for providing maximal fitness of the swarm, many of the individual trajectories of the agents acting in this swarm can be rather far from optimal. The difference between the trajectories obtained by individual and collective search is illustrated in Figure 5.1. The presented trajectories were obtained using the Algorithm 4.8, where in the search by multiple agents, it was assumed that at each step the agents have common information about the target's location probabilities, while the probabilities' estimation and decision making are conducted individually. Notice that in this algorithm the agents do not implement any separation techniques so that the collision avoidance is conducted automatically based on the target's location probabilities.

It is seen that the trajectories of the same agent are strongly different in the case of individual search and the search with the presence of two additional agents. In addition, notice that in the case of individual search, the searcher had not found the target in 1000 time steps, and in the case of collective search, the target was found in 687 time steps.

In the last decades, the models of swarming and of the search and foraging by multiple agents were considered in different directions, and many insights in navigation of mobile robots were obtained from the results of biological studies, especially, from the considerations of social insects (Fewell, 2003). For example, the main ideas used in navigation of the groups of mobile robots navigation follow the models of spatial movement of the foraging ants (Haefner & Crist, 1994) of the division of labor, task allocation (Robinson, Feinerman, & Franks, 2009) and leadership (Couzin, Krause, Franks, & Levin, 2005), studies of self-organization (Gaubert, Redou, Harrouet, & Tisseau, 2007), and numerous experiments with the ant's motion (see, e.g., Hayashi, Yuki, Sugawara, Kikuchi, &

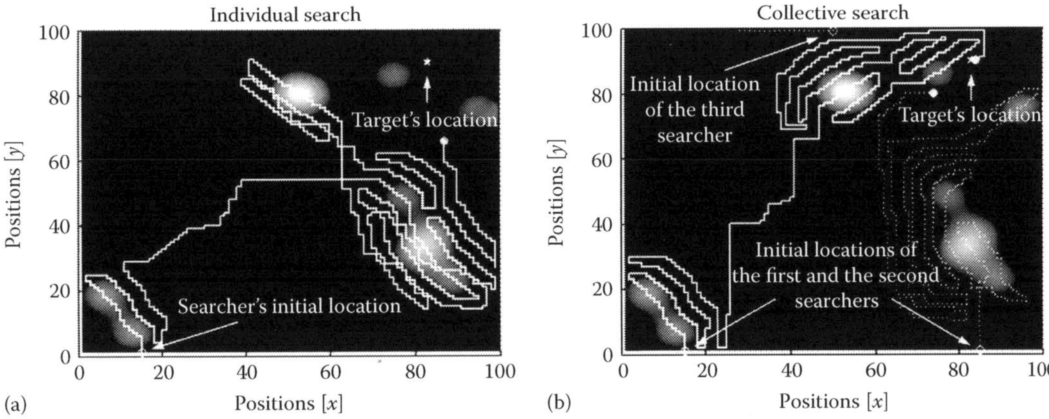

FIGURE 5.1 Probabilistic search by single and multiple agents: (a) the search trajectory of a single agent and (b) the trajectories of collective search by three agents. The trajectory of the same agent is depicted by a solid line, and the trajectories of two additional agents in figure (b) are depicted by dotted lines.

Tsuji, 2008). Some of such studies of the ants directly resulted in the algorithms of search by the group without communication (Feinerman, Korman, Lotker, & Sereni, 2012) and the models of search with restricted communication using pheromone traces (Kagan et al., 2014), while the other, for example, the studies of local coordinating in insects (Collett & Collett, 2000, 2009) are still in the stage of theoretical considerations. A detailed survey of the models of collective animals behavior and swarms dynamics was recently published by Sumpter (2010); see also the indicated aforementioned books on the animal swarms and ants colonies, which present the qualitative picture of this behavior.

Many of the biologically inspired models of collective behavior are implemented in swarm robotic systems, which, on one hand, provide a framework for verification of the models of animals behavior, and, on the other hand, allow consideration of the problems, which can be hardly solved by mobile agents acting individually (see already mentioned books by McFarland & Bösser [1993] and Clark [1997] and conference proceedings From Animals to Animats [1991–2012], ANTS [1998–2014], Distributed Autonomous Robotic Systems [1992–2013]). Some additional information regarding swarming of the robots and self-organization in the swarm robotics is presented in the papers by Baldassarre, Nolfi, and Parisi (2003), Dorigo et al. (2004), and Leccese, Gasparri, Priolo, Oriolo, and Ulivi (2013). One of the first models of foraging by swarm was suggested by Lerman and Galstyan (2002), who directly indicated the effect of collaboration between the agents. Another method of search and foraging by the robots swarm was suggested by Nouyan (2004), Liu, Winfield, and Sa (2007), and DeLima, Daniel, and Sciortino Jr (2007), and a particular consideration of aggregation of the agents in the foraging tasks is conducted by Wang, Shi, Chu, Zhang, and Lin (2005).

Following Bayesian approach to the search problems (Stone, Barlow, & Corwin, 1999) (see also Stone [2001] and Morelande, Kreucher, & Kastella [2007]), Bourgault, Furukawa, and Durrant-Whyte (2003) suggested the algorithm of search by multiple searchers and later extended it to the search for multiple targets (Wong, Bourgault, & Furukawa, 2005). Notice that Algorithm 4.8 with multiple searchers, which was applied for creating the trajectories shown in Figure 5.1, as well as mentioned in Section 4.4 methods of search based on belief space (Cetnarowicz, Turek, Zabinska, & Sosnicki, 2013; Indelman, Carlone, & Dellaert, 2013a, 2013b; Sosnicki, Turek, Cetnarowicz, & Zabinska, 2013; Zabinska, Sosnicki, Turek, & Cetnarowicz, 2013), follow the same Bayesian framework. An overview of the probabilistic methods of swarming is presented in the well-known paper by Lerman, Martinoli, and Galstyan (2005), and a detailed introduction into the world of swarm robotics is presented in excellent books by Trianni (2008) and by Hamann (2010). A brief consideration of the main ideas in the field and additional references are presented in the paper by Kagan et al. (2014).

5.2 FORAGING BY MULTIPLE FORAGERS IN RANDOM ENVIRONMENT

Collective search and foraging imply that the agents, on one hand, implement certain aggregation rules, which support the existence of the swarms as a distinguishable unit, and, on the other hand, use the environment as a shared resource and change its states with respect to their activity and results of search. This section considers the basic ideas of these processes and presents recently developed probabilistic algorithms of collective search with different models of information exchange, including shared target's location probabilities, pheromone traces, and biological signaling.

5.2.1 Information Exchange and Usage of the Environment

The simplest case of search and foraging by multiple agents is provided by the search by a number of Brownian walkers (Larralde, Trunfio, Havlin, Stanley, & Weiss, 1992a, 1992b) or of Lévy flyers (Berkolaiko & Havlin, 1997), in which the agents are acting independently according to their internal abilities. However, as it is indicated by Sumpter (2010, p. 44),

> A key benefit of being near the others is access to information. Animals often live in the environments where the resources are distributed in patches that exist only temporarily. In such environment, a single individual has a very low rate of finding a resource patch if they search independently. When large numbers of individuals search at the same time, however, the probability that one of them finds one of the patches is considerably larger. If individuals are able to monitor and use the discoveries of the others in their own search, they can increase their own rate of finding resources.

The information exchange between the search agents, both natural and artificial, can be conducted in two main manners (Iocchi, Nardi, & Salerno, 2001):

1. Direct communication between the agents according to definite rules, for example, the honeybee dances and animals' voice signals or communication protocols used in the mobile robots
2. Indirect communication via the changes of the environmental states, for example, the ants' pheromone trails or changes of the probabilities of finding the targets in the observed areas

The first type of communication is usually called signaling, and for animals' activity, it is often defined according to the bio-signaling model based on the Sir Philip Sidney game (Bergstrom & Lachmann, 1997, 1998a, 1998b; Johnstone & Grafen, 1992; Smith, 1991). For the mobile robots, such communication follows certain peer-to-peer or broadcasting protocols (Weiss, 1999); for the simple example of such protocol, see the paper by Steshenko, Kagan, and Ben-Gal (2011). The second type of communication is usually called stigmergy (see the books by Hamann [2010] and by Sumpter [2010] references therein). Often it is considered in the framework of percolation theory (De Gennes, 1976; Stauffer & Aharony, 2003) or as a usage of modified shared memory (Kohonen, 1978; Schonhage, 1980) and embodiment (Clark, 1997, 2003). In the probabilistic search, considered in the previous chapters, the changes of the environment are represented by the updates of the target's location probabilities according to the implemented detection function.

5.2.1.1 Bio-signaling in the Search and Foraging Tasks

Let us start with the direct communication between the agents and, as an example, consider the bio-signaling model based on the discrete Sir Philip Sidney game (Smith, 1991), (Bergstrom & Lachmann, 1997), and then consider its implementation in the probabilistic search and foraging

(Siegelmann, Kagan, & Ben-Gal, 2014). For historical notes regarding this game and its role in the biological studies, see the indicated paper (Bergstrom & Lachmann, 1997) and references therein; the wide context of bio-signaling is presented by Sumpter (2010).

The discrete Sir Philip Sidney game is an extensive one-stage game between two related players, in which one player, the signaler, decides whether to send a costly request for a resource that will increase the signaler's fitness and the second player, the donor, decides whether to provide the resource to the signaler according to the received request and so to decrease the own fitness or not. The relation between the players implies that even when the resource is transferred from the donor to the signaler, the donor still profits a portion from the signaler's gain, and in return, if the resource is not transferred, the signaler obtains a portion of the donor's fitness.

At the beginning of the game, the signaler is needy for the resource with probability p and not needy for the resource with probability $1-p$, and the parameters of the game are defined as follows:

1. If the signaler is needy for the resource, then without the resource has its fitness $1-a$.
2. If the signaler is not needy for the resource, then without the resource has its fitness $1-b$.
3. If the signaler sends the signal to the donor, then it pays the signaling cost c.
4. If the donor transfers the resource, then the donor's fitness decreases by d and becomes $1-d$.
5. The relativeness between the players is defined by the value k.

It is assumed that all these parameters are bounded as $0 \leq a, b, c, d, k \leq 1$ and their values are known for both the players. The extended form of the game is shown in Figure 5.2 (Bergstrom & Lachmann, 1997); for the other representation of the game, see, for example, the paper by Hutteger and Zollman (2009).

In general, the signaler and the donor can apply the following strategies (Hutteger & Zollman, 2009):

Signaler	Donor
1. Signal only if needy.	1. Donate only if no signal received.
2. Signal only if not needy.	2. Donate only if signal received.
3. Newer signal.	3. Newer donate.
4. Always signal.	4. Always donate.

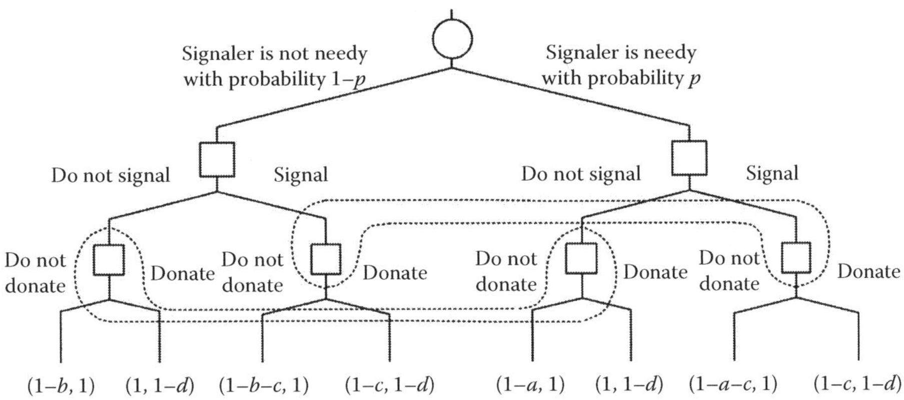

FIGURE 5.2 The Sir Philip Sidney game in the extended form. The first decision "do not signal/signal" is made by the signaler and the second decision "do not donate/donate" is made by the donor. The dotted lines depict the donor's informational sets, which represent its knowledge regarding the signaler's choice. In the pairs of outcomes, the first term denoted the fitness of the signaler and the second term the fitness of the donor.

Possible Nash equilibriums in the game are the following (Bergstrom & Lachmann, 1997; Hutteger & Zollman, 2009):

1. "Newer signal"/"Newer donate" is reached if it holds true that

$$d > k\left(pa + (1-p)b\right).$$

2. "Always signal"/"Newer donate" is reached if the backward inequality holds

$$d < k\left(pa + (1-p)b\right).$$

3. "Signal only if needy"/"Donate only if no signal" is reached if the following inequalities hold:

$$a \geq d/k \geq b \quad \text{and} \quad a \geq kd - c \geq b.$$

4. "Signal only if needy"/"Donate only if signal" is reached if the following inequalities hold:

$$a \geq d/k \geq b \quad \text{and} \quad a \geq kd + c \geq b.$$

The last equilibrium is also called signaling or separating equilibrium and its strategies are usually considered as the most natural scenario of communication between animals (Bergstrom & Lachmann, 1997; Hutteger & Zollman, 2009; Smith, 1991). Let us adopt these strategies to the tasks of search and foraging (Siegelmann et al., 2014).

Informally, the decision making is represented as follows. Assume that during the search the agent finds the patch, which contains certain amount of prey, and that both this amount and the size of the patch are known to the agent. Then, the agent has two possibilities: either to continue searching in this patch individually that will require relatively long search time or to signal to the other agents about the patch found that implies the request for help in the search, which will require less search time. Suppose that the request was sent. Then the agents, which received the request, are faced with the decision either to leave the regions, where they currently search, and to move to the patch found by the signaler or to ignore the request and to continue their individual activity.

Let us consider this situation in the terms of target's location probabilities as they used in the previous chapters. Similar to the individual search, assume that the searchers and the target act in the bounded 2D domain $\mathfrak{X} \subset \mathbb{R}^2$ and that target's location probabilities given the searcher actions are defined by the density function $u : \mathfrak{X} \times \left[0, T\right) \to \mathbb{R}^+$. Recall that at time t the searcher observes the area $a(t) \subset \mathfrak{X}$, in which it can directly detect the target. For the flat domain without obstacles, the observed areas have equal size and are defined similar to Sections 4.3 and 4.4 by the vision distance r_{vis}, that is, a maximal distance, at which the searcher can detect the target; so $a(t) = \left\{\mathfrak{x} : \left\|\breve{\mathfrak{x}}(t) - \mathfrak{x}\right\| \leq r_{\text{vis}}\right\}$, where $\breve{\mathfrak{x}}(t)$ is a location of the searcher at time t.

Similar to the foraging in patchy environment (see Section 3.3), denote by $A(t) \subset \mathfrak{X}$ a patch, which is sensed by the agent, but cannot be checked by an immediate observation. In the succeeding text, it is assumed that, like the observed areas, the patches $A(t)$ have equal size and are defined by the sensing distance r_{sense}, that is, $A(t) = \left\{\mathfrak{x} : \left\|\breve{\mathfrak{x}}(t) - \mathfrak{x}\right\| \leq r_{\text{sense}}\right\}$, where $r_{\text{sense}} > r_{\text{vis}}$. In addition, denote respectively by $r_{\text{signal}}^{\min}$ and by $r_{\text{signal}}^{\max}$ the minimal and maximal distances, to which the searcher can send information regarding the patch and the request for help. In other words, it is assumed that the

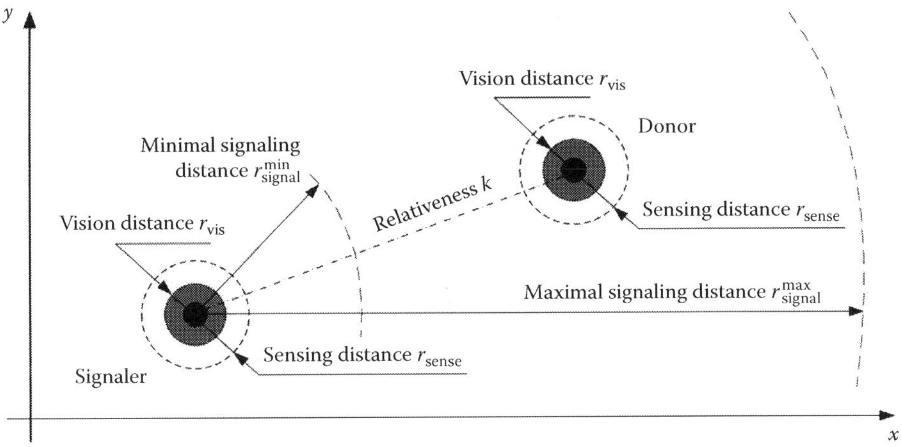

FIGURE 5.3 Parameters of the Sir Philip Sidney game. The area, which can be checked by immediate observation, is defined by the vision distance r_{vis}, and the patch A, which is sensed but requires certain effort and time for observation, is defined by the sensing distance r_{sense}. The signaler is able to send the request to the agents, which are located in the ring around the agent such that its small and large radiuses are defined by the minimal r_{signal}^{min} and maximal r_{signal}^{max} signaling distances, correspondingly. The relativeness between the agents is defined with respect to the distance between the agents.

signaler can signal to the neighboring agents such that they are not too close to its current location. The introduced parameters are illustrated in Figure 5.3.

Based on these values, parameters of the game are defined as follows. Assume that at time t, the location of the signaler is $\breve{x}_s(t)$ and the location of the donor is $\breve{x}_d(t)$, and the patches around these agents are $A_s(t) = \{x : \|\breve{x}_s(t) - x\| \leq r_{sense}\}$ and $A_d(t) = \{x : \|\breve{x}_d(t) - x\| \leq r_{sense}\}$, correspondingly. Then, the neediness a of the signaler and the fitness lost d of the donor are defined by the probabilities of finding the target in the patches $A_s(t)$ and $A_d(t)$, the signaling cost c is interpreted as an effort required for sensing the signaler's patch $A_s(t)$, and the relativeness k is defined by the distance between the signaler and the donor. Formally, given the target's probability density function u, for the search in continuous domain $\mathfrak{X}$, it follows that:

$$a \sim \int_{A_s(t)} \frac{du(x,t)}{dt} dx, \quad d \sim \int_{A_d(t)} \frac{du(x,t)}{dt} dx, \quad c \sim r_{sense}, \quad k \sim \|\breve{x}_s(t) - \breve{x}_d(t)\|, \tag{5.1}$$

and for the search in discrete domain $\mathfrak{X} = \{x_1, x_2, \ldots, x_n\}$, for the values a and d, one has:

$$a \sim \sum_{x \in A_s(t)} u(x,t), \quad d \sim \sum_{x \in A_d(t)} u(x,t). \tag{5.2}$$

In addition, it is assumed that if the signaler is not needy for the resource, its neediness is $b \ll 1$.

Certainly, since it is assumed that the search agents in the swarm are equal and there is no prescribed division to the signalers and the donors, each agent changes its role with respect to the found patches and the current target's location probabilities. The final scenario of the signaling and donating activities of the agents according to the Sir Philip Sidney game is outlined as follows.

Algorithm 5.1: (Signaling and Donating by the Sir Philips Sydney Game)

Given a domain $\mathfrak{X}$, target's probability density function u, sensing distance r_{sense} and signaling distances r_{signal}^{min} and r_{signal}^{max} do:

Signaler:

1. Being in the location $\mathfrak{x}_s(t)$, consider the patch $A_s(t)$ defined by the sensing distance r_{sense} and calculate the neediness a and the cost c by Equation 5.1 or 5.2.
2. Send the request, which includes location $\mathfrak{x}_s(t)$, neediness a and the cost c, to the agents located in the ring defined by the signaling distances r_{signal}^{min} and r_{signal}^{max} (see Figure 5.3).

Donor:

3. If the request from the signaler received, do:
4. Being in the location $\mathfrak{x}_d(t)$, consider the patch $A_d(t)$ defined by the sensing distance r_{sense} and calculate the fitness decrement d by Equation 5.1 or 5.2 and relativeness k defined by the distance between the donor's location $\mathfrak{x}_d(t)$ and the signaler's location $\mathfrak{x}_d(t)$.
5. If the signaling equilibrium (equilibrium 4) holds, then move to the signaler's location $\mathfrak{x}_s(t)$ with maximal velocity and continue search as a signaler.
6. End if.

This algorithm prescribes the activities of the search agents and results in the aggregation of the agents in the regions with high probabilities of finding the target. However, since during the search the agents decrease (or zero) the probabilities of finding the target in the already considered points, the frequency of signaling decreases with time, and at limit, the agents continue the search as non-communicating individuals. The algorithm implies that the agents are equal and, as indicated in line 5, change their roles with respect to the considered patches and the game parameters.

The signaling and donating activities of the agents are illustrated by the following example.

Example 5.1

Assume that the destructive search is conducted by two equal Brownian searchers acting in the square domain $\mathfrak{X}$ of the size $n = 100 \times 100$ with the impermeable bounds. At the beginning of search, the target's location probabilities $u(\mathfrak{x},0)$ are specified by 10 binormal distributions with random centers and the variances $\sigma_1 = \sigma_2 = 5$. The step length of each searcher is $\delta = 3$ and the vision distance is $r_{vis} = 1$. The search is conducted in discrete time $t = 0, 1, 2,..., 500$.

Consider two types of search: foraging without signaling, in which each agent acts independently on the other, and foraging with signaling, in which the agents communicate according to the Sir Philip Sidney game:

1. Search without signaling. In this case, the agents move according to the Brownian walk and zero the target's location probabilities in the observed areas $a(t)$ defined by the distance r_{vis}. An example of the trajectories of the agents is shown in Figure 5.4a.
2. Search with signaling. For such search, it is assumed that the sensing distance of both foragers is $r_{sense} = 5$, and minimal and maximal signaling distances are $r_{signal}^{min} = 2 \times r_{sense} = 10$ and $r_{signal}^{max} = \sqrt{n}/4 = 25$. Parameters of the Sir Philip Sidney game are defined according to Equations 5.1 and 5.2 as follows:

$$a(t) = \frac{1}{\pi r_{sense}^2} \sum_{\mathfrak{x} \in A_s(t)} u(\mathfrak{x},t), \quad d(t) = \frac{1}{\pi r_{sense}^2} \sum_{\mathfrak{x} \in A_d(t)} u(\mathfrak{x},t), \quad b = 0 = \text{const},$$

$$c = \frac{\pi r_{sense}^2}{n} = \text{const}, \quad k(t) = \frac{\left\| \mathfrak{x}_s(t) - \mathfrak{x}_d(t) \right\|}{r_{signal}^{max}}.$$

An example of the trajectories of the agents with signaling is shown in Figure 5.4b.

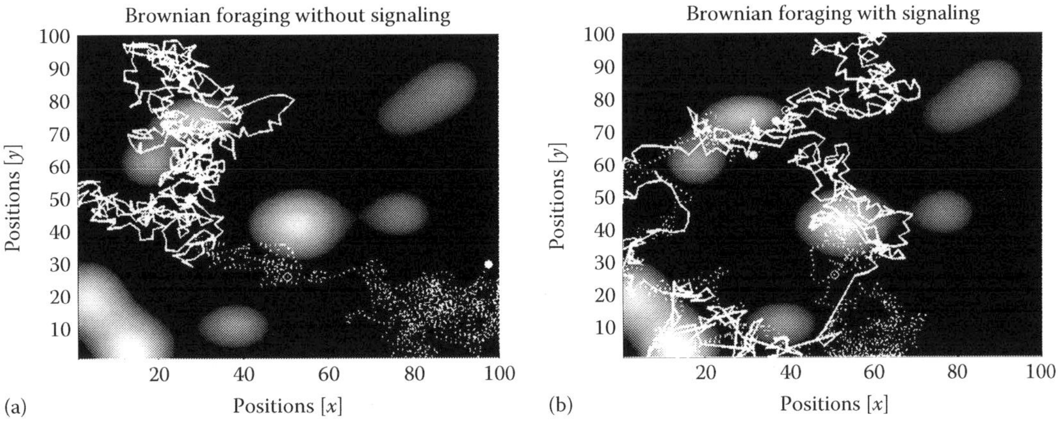

FIGURE 5.4 Trajectories of two Brownian foragers that are acting (a) individually without signaling and (b) collectively with signaling according to the Sir Philip Sidney game. The trajectory of the first agent is depicted by a solid curve and the trajectory of the second agent is depicted by a dotted curve. In both cases, the agents start in the same positions; initial locations of the agents are depicted by white circumferences and their final positions at time $t = 500$ by white circles.

It is seen that in individual foraging without signaling (Figure 5.4a) the agents move independently according to the prescribed process that in the considered case is Brownian walk. However, in the case of foraging with signaling (Figure 5.4b), the agents start by individual search (for the first agent such as the upper part of the figure), and then, after accepting request for donating, they aggregate and continue the search as a group (middle and bottom parts of the figure).

In the considered example of the individual search, the resulting weighted sum of the encountered targets, that is, the sum of the zeroed target's location probabilities, is 304.506 (243.242 for the first searcher and 61.264 for the second searcher), and the sum length of the searchers' paths is 2966.246 (1491.743 for the first search and 1474.503 for the second searcher); thus, the eat rate is 0.103. Similarly, in the considered example of the search with signaling, the resulting weighted sum of the encountered targets, that is, the sum of the zeroed target's location probabilities, is 706.566 (339.058 for the first searcher and 367.508 for the second searcher), and the sum length of the searchers' paths is 3170.405 (1627.924 for the first searcher and 1542.481 for the second searcher); thus, the eat rate is 0.223, which is more than two times greater than in the search without signaling. ■

The considered example demonstrates that the communication between the agents and their aggregation results in more effective search in the terms of encountered targets and eat rate; however, this tendency cannot be considered as general, and starting from a certain size of the group, the aggregation can decrease the search efficiency (Sumpter, 2010).

In addition, notice that in the considered scenario of foraging with signaling and donating, the searches change their behavior between Brownian search and relatively long traces. Such behavior follows the same scenario as the considered in Section 4.3.3 search with intermittent strategies, where Brownian slow motion with screening intermits with fast relocation without screening; however, in contrast to the intermittent search, the times of these motions are defined by signaling and donating and, consequently, by the number of agents in the group.

Certainly, the considered signaling and donating scenario does not exhaust possible communication and aggregation activities of the agents during the search and foraging. In the case of appropriately equipped mobile robots, communication can be organized using certain communication protocols and advanced sensors; general ideas of such communication are presented in the book (Weiss, 1999), and an illustration of simple protocol is given in the report by Steshenko, Kagan, and

Ben-Gal (2011). Nevertheless, the considered scenario stresses two important parts of communication: the peer-to-peer signaling with corresponding changes of individual behavior and dependence of aggregation and of the swarm behavior on the states of the environment. The last property is usually called stigmergy and is widely accepted as a main communication technique of the simplest animals, in particular, of ants (Sumpter, 2010), and in the last decades, this technique was introduced into the considerations of swarming in mobile robots (Clark, 1997, 2003; Hamann, 2010).

5.2.1.2 Stigmergy and Indirect Communication

Now let us consider the indirect communication. Recall that in the algorithms of destructive search presented in Sections 4.3 and 4.4, after checking the chosen observed area, the agent changes the probabilities of finding the target in this area. In the case of errorless detection, these probabilities become zero, and for erroneous detections, the probabilities of finding the target in the observed area are decreased, and the next actions of the agent are prescribed by these new probabilities. In other words, the agent changes the environment, in which it acts, and plans the actions with respect to the changes of the environment.

Assume that the agents are acting in a group and each of the agents follows the destructive search scenario and that the changes of the environment produced by one agent are observable to each other in the group. Then, during the search, each of the agents plans its actions with respect to the changes of the environment, which were done by the agent itself and by the other agents. Following Grasse (1959), such process often is considered as an indirect communication between the agents, and the resulting collective activity of the group is called stigmergy. In more complex case, it is assumed that the agents are able to change the environment both by eating the prey (or by changing the probabilities of finding the target) and by marking certain regions by the signs, which can be perceived and understood by the other agents and can be used in the further actions planning. The classical example of such signs is pheromone trails deployed by the ants; in the last time, the same mechanism was implemented in the pheromone robotics (Payton, Daily, Estowski, Howard, & Lee, 2001).

The stigmergy implies that the agent is able to change the environment either by eating the prey or by marking the regions only in the close neighborhood of its current location, while the perception abilities of the agents can be considered as a long-distance communication with the environment or the other agents. In 2003, Theraulaz, Gautrais, Camazine, and Deneubourg (2003) argued that the combination of such local and long-distance communication in the stigmergic systems results in self-organizing behavior represented by spatial or temporal ordering of the agents in swarms (Bonabeau, Dorigo, & Theraulaz, 1999; Bonabeau, Theraulaz, Deneubourg, Aron, & Camazine, 1997). The detailed description of the model in the framework of ants' behavior is presented by Sumpter (2010). The transitions from disordered states to the ordered structures in multiagent systems inspired consideration of the swarms' behavior using the percolation theory; an application of this theory for analysis of the ants' behavior was initialed by Richardson, Christensen, Franks, Jensen, and Sendova-Franks (2011a, 2011b).

To illustrate the actions of the agents with indirect communication, let us consider the activity of the group of modeling ants. The model of the ant and its correspondence with the real ant are already presented in Section 4.2.2 (see Figures 4.10 and 4.11). The consideration in the succeedings text addresses control and navigation of the modeling ant with respect to the environmental states on short and long distances and to the deployed pheromone trails (Kagan et al., 2014).

Assume that the control of the modeling ant is organized following the control of the simple Braitenberg vehicle (Braitenberg, 1984). Namely, assume that the left sensors govern the movements of the right side legs and the right sensors govern the movements of the left side legs, and in both cases the length of the step is proportional to the value of the sensed signal. Control of the legs' movements is conducted by independent left and right side controllers, which transform the sensed signals into the angles of the legs turns. Similar to Section 4.2.2, the ant's step is defined with respect to the movement of the "middle" legs at the left and right sides. The scheme of such control is depicted in Figure 5.5.

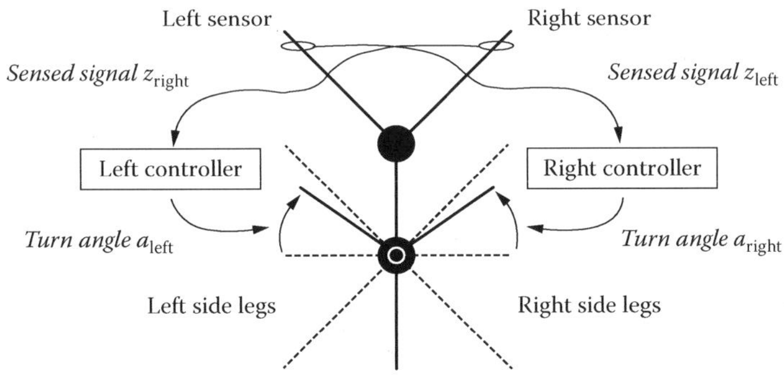

FIGURE 5.5 Control scheme of the modeling ant. The movements of the left side legs are proportional to the values of the signal sensed by the right sensor, and the movements of the right side legs are proportional to the values of the signal sensed by the left sensor.

Controllers of the modeling ant are built based on the Tsetlin automaton that is widely used as the model of biological systems acting in random environment (Tsetlin, 1973). In the simplest setup, the Tsetlin automaton is a Boolean automaton with memory such that the transitions of its internal states $\mathfrak{s}(t)$, t = 0, 1, 2, ... are defined by the "not-xor" operator with respect to the inputs $z(t)$ as follows:

$$\mathfrak{s}(t) = \overline{xor\left(z(t), \mathfrak{s}(t-1)\right)}. \tag{5.3}$$

According to the definition of the "not-xor" operator, if the automaton is in the state $\mathfrak{s}(t-1)$ and receives the input $z(t) = 0$, then it changes the state to the opposite $\mathfrak{s}(t) = \overline{\mathfrak{s}}(t-1)$, and if it receives the input $z(t) = 1$, then it stays in its current state, that is, $\mathfrak{s}(t) = \mathfrak{s}(t-1)$. Usually, the input value $z(t) = 0$ is considered as a payoff and the value $z(t) = 1$ as a reward, and the Tsetlin automaton provides the accumulation of the rewards and avoids the payoffs.

The outputs of the Tsetlin automaton are defined with respect to the considered task. In the terms of the mobile agent, which perceives the environmental states, the favorable state $\mathfrak{s}(t) = 1$ means that the agents continue its movement in the chosen direction and the unfavorable state $\mathfrak{s}(t) = 0$ means that the agent changes its direction to the opposite one.

The non-Boolean Tsetlin automaton, also called $\otimes_\vartheta$-automaton, in which inputs $z(t)$, states $\mathfrak{s}(t)$, and outputs $y(t)$ are real numbers from the interval [0, 1], is defined using the absorbing norm aggregator $\otimes_\vartheta$ that extends the "not-xor" operator to the interval [0, 1]. The absorbing norm aggregator $\otimes_\vartheta$ with the absorbing element $\vartheta \in [0, 1]$ is a function $\otimes_\vartheta$:[0, 1] × [0, 1]→[0, 1], which for any $a, b, c \in [0, 1]$ satisfies the following properties (Baturshin, Kaynak, & Rudas, 2002; Fodor, Rudas, & Bede, 2004):

1. *Commutativity*: $a \otimes_\vartheta b = b \otimes_\vartheta a$
2. *Associativity*: $(a \otimes_\vartheta b) \otimes_\vartheta c = a \otimes_\vartheta (b \otimes_\vartheta c)$
3. *Monotonicity*: $a \leq b$ implies $a \otimes_\vartheta c \leq b \otimes_\vartheta c$ for any $c \in [0, 1]$
4. *Absorbing element*: $\vartheta \otimes_\vartheta a = \vartheta$ for any $\vartheta \in [0, 1]$

and if $a, b \in \{0, 1\}$ are Boolean variables, then $a \otimes_\vartheta b = \overline{xor(a, b)}$ for any absorbing element $\vartheta \in [0, 1]$.

Transitions of the internal states of the $\otimes_\vartheta$-automaton are defined as follows (Kagan et al., 2014):

$$\mathfrak{s}(t) = z(t) \otimes_\vartheta \mathfrak{s}(t-1), \tag{5.4}$$

that is an immediate extension of the function defined by Equation 5.3.

For control of the modeled ant, the outputs of the $\otimes_\vartheta$-automaton are defined by absorbing norm aggregation of the previous and current internal states, that is,

$$y(t) = \mathfrak{s}(t) \otimes_\vartheta \mathfrak{s}(t-1). \tag{5.5}$$

Notice that for Boolean inputs and states, such transition and output functions give the equivalence $y(t) = z(t)$.

As indicated earlier, the navigation of the modeling ant is defined with respect to the environmental states and deployed pheromone trails. The fusion of these values is conducted using the uninorm aggregator $\oplus_\theta$, which generalizes the Boolean "and" and "or" operators. The uninorm aggregator $\oplus_\theta$ with the neutral element $\theta \in [0, 1]$ is a function $\oplus_\theta$:$[0, 1] \times [0, 1] \rightarrow [0, 1]$, which for any $a, b, c \in [0, 1]$ satisfies the following properties (Fodor, Yager, & Rybalov, 1997; Yager & Rybalov, 1996):

1. *Commutativity*: $a \oplus_\theta b = b \oplus_\theta a$
2. *Associativity*: $(a \oplus_\theta b) \oplus_\theta c = a \oplus_\theta (b \oplus_\theta c)$
3. *Monotonicity*: $a \leq b$ implies $a \oplus_\theta c \leq b \oplus_\theta c$ for any $c \in [0, 1]$
4. *Neutral element*: $\theta \oplus_\theta a = a$ for any $\theta \in [0, 1]$

and if $a, b \in \{0, 1\}$ are Boolean variables, then for the neutral element $\theta \in \{0, 1\}$, it holds true that $a \oplus_1 b = a \wedge b$ and $a \oplus_0 b = a \vee b$.

Denote by $z^{\text{short}}(t)$ and $z^{\text{long}}(t)$ the states of the environment detected at short and long distances, respectively, as they sensed by one of the agent's sensors at time t, $t = 0, 1, 2, \ldots$ and by $z^{\text{phm}}(t)$ the data on pheromone quantity in the close neighborhood of this sensor at this time. It is assumed that the sensed values are bounded and obtain their values from the interval [0, 1]. Then, the resulting input value $z(t)$ at this time is defined as follows (Kagan et al., 2014):

$$z(t) = z^{\text{short}}(t) \oplus_\theta z^{\text{long}}(t) \oplus_\theta z^{\text{phm}}(t). \tag{5.6}$$

Certainly, the same operation can be applied to additional sensed data, in particular, for the density of the neighboring agents, which is required by the Reynolds' swarming rules that were indicated in Section 5.1.

Finally, denote by $[]_{\text{left}}$ and $[]_{\text{right}}$ the values that correspond to the left and right sensors and controllers of the ant, respectively. Then, the processing of the sensed signals in the modeling ant with left and right sensors and controllers (see Figure 5.5) is defined as follows (Kagan et al., 2014):

1. *Inputs fusion*:

$$z_{\text{left}}(t) = z_{\text{left}}^{\text{short}}(t) \oplus_\theta z_{\text{left}}^{\text{long}}(t) \oplus_\theta z_{\text{left}}^{\text{phm}}(t), \quad z_{\text{right}}(t) = z_{\text{right}}^{\text{short}}(t) \oplus_\theta z_{\text{right}}^{\text{long}}(t) \oplus_\theta z_{\text{right}}^{\text{phm}}(t).$$

2. *States transition*:

$$\mathfrak{s}_{\text{left}}(t) = z_{\text{right}}(t) \otimes_\vartheta \mathfrak{s}_{\text{left}}(t-1), \quad \mathfrak{s}_{\text{right}}(t) = z_{\text{left}}(t) \otimes_\vartheta \mathfrak{s}_{\text{right}}(t-1).$$

3. *Outputs*:

$$y_{\text{left}}(t) = \mathfrak{s}_{\text{left}}(t) \otimes_\vartheta \mathfrak{s}_{\text{left}}(t-1), \quad y_{\text{right}}(t) = \mathfrak{s}_{\text{right}}(t) \otimes_\vartheta \mathfrak{s}_{\text{right}}(t-1).$$

Following the control scheme, the output signals $y_{\text{left}}(t)$ and $y_{\text{right}}(t)$ specify the values of the legs' turns such that the turn angles $\alpha_{\text{left}}(t)$ and $\alpha_{\text{right}}(t)$ are proportional to the values of the signals:

$$\alpha_{\text{left}}(t) \sim y_{\text{left}}(t), \quad \alpha_{\text{right}}(t) \sim y_{\text{right}}(t).$$

As a result, the modeling ant turns to the environmental state with greater values $z(t)$ and avoids the states with smaller values $z(t)$, and with time $t = 0, 1, 2, \ldots$ it tends to the favorable environmental states and follows pheromone trails. The activity of the group of modeling ants implementing the presented stigmergy techniques is illustrated by the following example.

Example 5.2: (Kagan, Rybalov, Siegelmann, & Yager, 2013)

Consider the movements of five modeling ants moving in the square domain $\mathfrak{X} = \{\mathfrak{x}_1, \mathfrak{x}_2, \ldots, \mathfrak{x}_n\}$ of the size $n = 500 \times 500$ units with the impermeable bounds. The states of the environment $u(\mathfrak{x}_i)$, $i = 1, 2, \ldots, n$, are defined by 100 circle peaks and bottoms produced by binormal distributions with random centers and random variances from the interval [5, 10], and normalized such that $\max_i u(\mathfrak{x}_i) = 1$ and $\min_i u(\mathfrak{x}_i) = 0$. In the example, it is assumed that the ants conduct nondestructive search, so the values $u(\mathfrak{x}_i)$ are constant during the search process and do not depend on the actions of the ants.

The modeling ants are defined following the scheme shown in Figure 4.10. In correspondence with the *Messor ebeninus* ant acting over a platform of the size 200 × 200 cm, which was considered in the laboratory experiments, the width of each modeling ant, as well as its length from head to tail, is $w = 15$ units, so the length of its "middle" legs is $l = w/2 = 7.5$ units. Antennas of the ant have the length $\sqrt{2}l \approx 10.61$ units and directed by 45° and −45° degrees with respect to the ant's heading.

It is assumed that each modeling ant senses the environmental states in the points lying directly under the ends of the ants' antennas and in the points located on the half circle of the radius $r_{\text{sense}} = 4w = 60$ units around each antenna. The value of deployed pheromone is sensed in the points lying directly under the ends of the antennas. The pheromone is deployed to the points around the center of the ant, and the quantity of the deployed pheromone is defined by the binormal distribution with the variances 3; the minimal quantity of the pheromone deployed in the point is 0.85. The scheme of sensing with the sizes of the modeling ant is shown in Figure 5.6a.

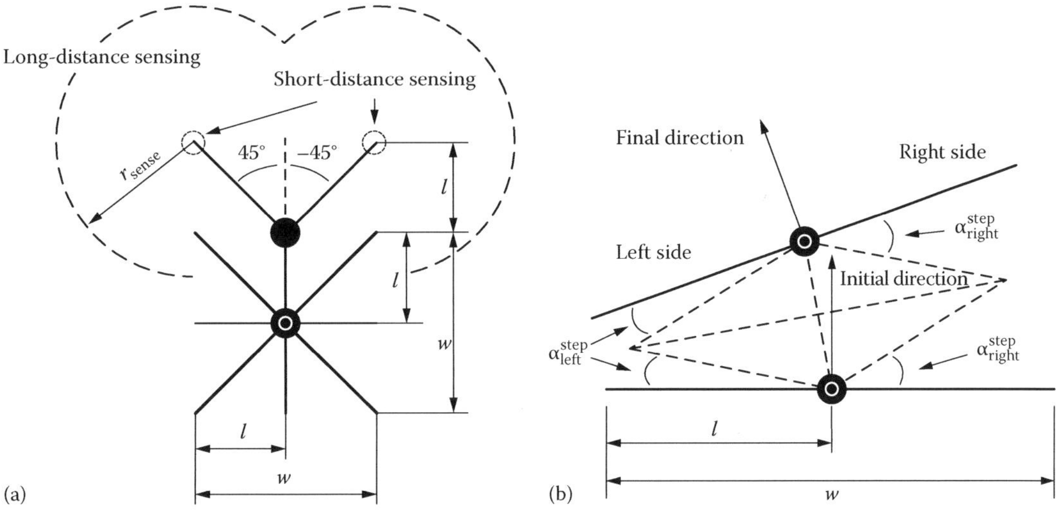

FIGURE 5.6 (a) Proportions of the modeling ant, sensing scheme, and (b) one-step motion scheme.

As indicated in Section 4.2.2, the motion of the ant follows the known odometry and stepping scheme (Wittlinger, Wehner, & Wolf, 2006, 2007; Wohlgemuth, Ronacher, & Wehner, 2001), and the turns of the ant's legs with respect to the outputs $y_{left}(t)$ and $y_{right}(t)$ of the left and right controllers are defined as follows.

The values of the outputs $y_{left}(t)$ and $y_{right}(t)$, which by definition of the $\otimes_\vartheta$-automaton are $y_{left}(t)$, $y_{right}(t) \in [0, 1]$, are transformed into the numbers l_{left}^{step} and l_{right}^{step} of units, to which the left and right sides of the ant should move. In the considered example, the values of l_{left}^{step} and l_{right}^{step} are defined as

$$l_{left}^{step} = 1.5l y_{left}(t), \quad l_{right}^{step} = 1.5l y_{right}(t).$$

The movement of the ant is conducted by the turns of the legs to the angles

$$\alpha_{left}^{step} = l_{left}^{step}\frac{360^\circ}{4\pi l}, \quad \alpha_{right}^{step} = l_{right}^{step}\frac{360^\circ}{4\pi l},$$

which correspond to the halves of the values l_{left}^{step} and l_{right}^{step}. At first, the left leg is turned to the angle α_{left}^{step}, then the right leg is turned to the angle α_{right}^{step} around the center of the ant. Then, following the rhomb properties, the ant's new center is specified. Finally, the left leg is turned to the angle α_{left}^{step}, and the right leg is turned to the angle α_{right}^{step} around the new center. The scheme of the ant's steps is shown in Figure 5.6b.

Usually, the uninorm $\otimes_\theta$ and absorbing norm $\otimes_\vartheta$ aggregators are represented using generator function that is invertible, continuous, strictly monotonously increasing function $g:(0, 1)\to(-\infty, \infty)$ such that $\lim_{a\to 0} g(a) = -\infty$ and $\lim_{a\to 1} g(a) = +\infty$. Then (Baturshin, Kaynak, & Rudas, 2002; Yager & Rybalov, 1996),

$$a \otimes_\theta b = g^{-1}\left(g_\theta(a) + g_\theta(b)\right) \quad \text{and} \quad a \otimes_\vartheta b = g^{-1}\left(g_\vartheta(a) \times g_\vartheta(b)\right).$$

In the considered example, generator function g is specified as follows (Rybalov, Kagan, & Yager, 2012):

$$g(a) = \tan\left(\pi\left(a^\beta - \frac{1}{2}\right)\right), \quad a \in (0,1), \quad \beta > 0.$$

Thus

$$a \otimes_\theta b = \frac{1}{2} + \frac{1}{\pi}\arctan\left(\tan\left[\pi\left(a^\beta - \frac{1}{2}\right)\right] + \tan\left[\pi\left(b^\beta - \frac{1}{2}\right)\right]\right),$$

$$a \otimes_\vartheta b = \frac{1}{2} + \frac{1}{\pi}\arctan\left(\tan\left[\pi\left(a^\beta - \frac{1}{2}\right)\right] \times \tan\left[\pi\left(b^\beta - \frac{1}{2}\right)\right]\right),$$

where $\theta = \vartheta = (1/2)^{1/\beta}$. Notice that if $\theta = \vartheta = 1/2$, generator function g has the same form as the Cauchy distribution $\mathcal{F}_{m,\alpha}(\xi) = \frac{1}{2} + \frac{1}{\pi}\arctan\left(\frac{\xi - m}{k}\right)$ with the median $m = 0$ and the parameter $k = 1$. In the considered example, the neutral and absorbing elements were defined as $\theta = 1/2 + \varepsilon_\theta$ and $\vartheta = 1/2 + \varepsilon_\vartheta$ where ε_θ and ε_ϑ are uniformly distributed random variables drawn from the interval $[-0.1, 0.1]$. For general definitions of the uninorm and absorbing norm based on probability distributions and their properties, see the paper by Kagan et al. (2013).

The movement of the five modeling ants is conducted in discrete time $t = 1, 2, \ldots, 1000$. An example of pheromone trails deployed by the modeling ants without and with long-distance sensing is shown in Figure 5.7.

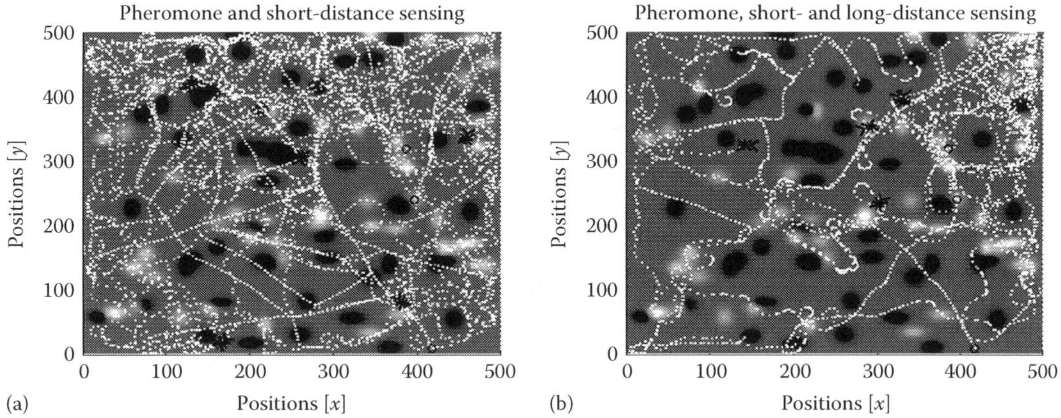

FIGURE 5.7 The pheromone trails deployed by five modeling ants. (a) Trails of the ants, which follow the pheromone trails and apply short-distance sensing of the environmental states. (b) Trails of the ants, which follow the pheromone trails and apply both short- and long-distance sensing of the environmental states, ($r_{sense} = 60$). Favorable environmental states are depicted by white, unfavorable states by black, and neutral states by gray. The trails of the ants are depicted by white dotted curves.

It is seen that the earlier presented control based on the $\otimes_9$-automaton and short-distance sensing results in avoidance of the unfavorable environmental states, and the ants tend to the favorable states following the pheromone trails. An additional application of the long-distance sensing results in more efficient avoidance of the unfavorable states and aggregation around the favorable states following the pheromone trails. However, notice that the larger values of long-distance sensing lead to less effective activity of the ants because of overabundance of information regarding the environment. As a result, the movements of the ants are shorter, and they tend to stay in their initial positions or turn around their locations. For additional considerations and examples of motion with different sensing, including the sensing of the neighboring ants, see the work by Kagan et al. (2013), and for biological inspirations, the paper by Nouvellet, Bacon, and Waxman (2009). ■

The considered implementation of stigmergy using modeling ants was simulated numerically and compared with the results of laboratory experiments with real ants. The experiments were organized in the manner that was suggested by Hayashi et al. (2008), but instead of closed circle arena, we used a square-opened platform. The ants moved on a square platform of the size 200 × 200 mm surrounded by water that prohibits them to leave the platform. The environmental states were simulated by slightly warming the platform in random points, so the warm regions simulated the favorable states and the cold regions simulated the unfavorable states. The movements of the ants were recorded by digital camera and the trajectories of the ants were analyzed, and statistical comparisons of the modeled and recorded trajectories demonstrated a high coincidence between the movements of the modeling and real ants. However, since such experiments with the ants represent rather artificial conditions, and certainly cannot support general statements regarding the ants' behavior, the problem of correct model of the ants' movement is still far from solution.

The considered examples of direct and indirect communication between the agents can be considered as reasonable illustrations of the methods of information exchange between the agents. Nevertheless, even these examples demonstrate the principles of the agents' aggregation and further collective activity. In particular, in the case of direct communication by bio-signaling, the agents follow the signaling and donating scheme and aggregate in the patches, which include a larger amount of prey or higher probability of finding the target. In the case of communication using pheromone trails, the agents aggregate in the favorable regions by following the pheromone trails deployed by themselves and by the other agents.

It is clear that in nature animals apply both direct and indirect communication that leads to the swarms' dynamics that is incomparably more complicated than the behavior provided by the considered models. General description of the swarm dynamics with such communication abilities usually follows the approach based on the nonlinear dynamical systems theory. In particular, for such purposes, Theraulaz et al. (2002) (see also the remark by Gazi and Passino in their book [Gazi & Passino, 2011]) suggested to apply the advection–diffusion–reaction equation that is the Fokker–Plank equation (see Equations 2.103, 2.104, or 3.1, and 3.6), which defines the movements of the agent in certain and may be varying potential field. This approach is considered in Sections 5.3 and 5.4 based on the dynamics of active particles.

5.2.2 Algorithms of Foraging by Multiple Agents

The algorithms of foraging by multiple agents and the resulting movements of the foragers strongly depend on the information that is circulated between the agents. In particular, in the foraging by artificial agents like mobile robots, it can be assumed that the agents have common information regarding the probabilities of the target's location over a complete domain, and each agent can transmit its own updated map of probabilities to the other agents. In contrast, in the foraging by animals, the agents can obtain only local information, while both direct and indirect communications provide the knowledge regarding the observations conducted by the agents, but the knowledge regarding the complete domain is inaccessible. The first type of communication was remarked in Section 5.1, and the examples of two last types of communication—bio-inspired signaling and stigmergy—were considered in Section 5.2.1. This section presents formal algorithms, which implement the foraging with these types of communication and which were used in the earlier considered examples.

5.2.2.1 Search with Common Target's Location Probabilities

Let us start the algorithm of search, in which the agents share complete information about the target's location probabilities and their observation results; earlier this algorithm was used for creating the trajectories shown in Figure 5.1. This algorithm extends the Algorithm 4.8 of search with errorless detection. Some remarks on such extension appear in the paper by Kagan, Goren, and Ben-Gal (2010); additional information and origins are presented in the book by Kagan and Ben-Gal (2013b).

Similar to Algorithm 4.8, the algorithm of search by multiple agents acts over a discrete domain $\mathfrak{X} = \{\mathfrak{x}_1, \mathfrak{x}_2, \ldots, \mathfrak{x}_n\}$ and can be applied for search for both static and moving targets. The motion of the target is defined by the target's transition probabilities matrix $\hat{\rho} = \rho_{ij\,n\times n}$, which for static target is a unit matrix. The pure and relative target's location probabilities are, respectively, denoted by $u(\mathfrak{x}_i, t)$ and $\hat{u}(\mathfrak{x}_i, t)$, $i = 1, 2, \ldots, n$, $t = 0, 1, 2, \ldots$; at initial time $t = 0$, it is assumed that $\hat{u}(\mathfrak{x}_i, 0) = u(\mathfrak{x}_i, 0) = u(\mathfrak{x}_i)$, and the probabilities $u(\mathfrak{x}_i)$ are known by each searcher.

In the algorithm, the searchers consider the target's location probabilities consequently and select their next observed areas according to the target's location probabilities, which are obtained after the observations and updates conducted by all the searchers. The observed area chosen by the jth searcher at time t is denoted by $a_j(t) \subset \mathfrak{X}$, and the observed probabilities, which are obtained the Bayesian update of the probabilities $\hat{u}(\mathfrak{x}_i, t)$, are denoted by $\bar{u}(\mathfrak{x}_i, t)$, $i = 1, 2, \ldots, n$, $t = 0, 1, 2, \ldots$ As mentioned earlier (see Section 4.4.1), it is assumed that if observation result of the area $a_j(t)$ is $z(a_j(t)) = 1$, then the detection probability is $\varphi(a_j(t)) = 1$, otherwise $\varphi(a_j(t)) = 0$. Finally, current location of the jth searcher at time t is denoted by $\breve{\mathfrak{x}}_j(t)$, and the set of possible directions of the searcher's next step is denoted by D. As mentioned earlier, it is assumed that $D = \{d_1, d_2, d_3, d_4, d_5\}$, where d_1 stands for "move forward"; d_2, "move backward"; d_3, "move right"; d_4, "move left"; and d_5, "stay in the current point." The decision-making regarding the next searcher's location and information-based estimation of the target's location probabilities are executed using the same as in the Algorithm 4.8 functions next observed area() and estimate probabilities().

The activities of each searcher are defined by the following functions.

calculate_observed_probabilities$\left(a(t),\hat{u}(\mathfrak{x}_i,t)\,|\,i=1,2,\ldots,n\right)$: (cf. lines 4 and 5 in Algorithm 4.8)

1. For all $\mathfrak{x}_i$ do: If $\mathfrak{x}_i \in a(t)$, then set $\hat{u}(\mathfrak{x}_i,t)=0$.
2. For all $\mathfrak{x}_i$ do: Set $\bar{u}(\mathfrak{x}_i,t)=\hat{u}(\mathfrak{x}_i,t)/\sum_{j=1}^{n}\hat{u}(\mathfrak{x}_j,t)$.
3. Return $\bar{u}(\mathfrak{x}_i,t)$, $i = 1, 2, \ldots, n$.

select_next_observed_area$\left(\breve{\mathfrak{x}}(t),\hat{\rho},\mathrm{D},u(\mathfrak{x}_i,t)\,|\,i=1,2,\ldots,n\right)$: (cf. lines 7–13 in Algorithm 4.8)

1. For all $\mathrm{d}_k \in \mathrm{D}$ do: Set $\beta_k=\sum_{\mathfrak{x}\in a_k}u(\mathfrak{x},t)$.
2. If $\beta_k = 0$ for each k, then
3. For all $\mathfrak{x}_i$ do: Set $\hat{u}(\mathfrak{x}_i,t)=$ estimate probabilities$\left(\breve{\mathfrak{x}}(t),u(\mathfrak{x}_i,t),\hat{\rho}\right)$.
4. Else
5. For all $\mathfrak{x}_i$ do: Set $\hat{u}(\mathfrak{x}_i,t)=u(\mathfrak{x}_i,t)$.
6. End if.
7. Set $a(t)=$ next observed area$\left(\hat{u}(\mathfrak{x}_i,t)\right)$.
8. Return $a(t)$.

Using these two functions, the algorithm of collective search by multiple searchers with common knowledge about the target's location probabilities is outlined as follows. In the outline, similar to Algorithm 4.8, it is assumed that at each time t the target moves according to its transition probabilities matrix $\hat{\rho}$, and the target's movements do not depend on the actions of the searchers.

Algorithm 5.2: (Collective Search with Common Target's Location Probabilities; Developed in Collaboration with G. Goren)

Given a discrete domain $\mathcal{X}=\{\mathfrak{x}_1,\mathfrak{x}_2,\ldots,\mathfrak{x}_n\}$, initial target's location probabilities $u(\mathfrak{x}_i)$, $i = 1, 2, \ldots, n$, and target's transition probabilities matrix $\hat{\rho}=\rho_{ij\,n\times n}$ do:

1. Set $t = 0$ and start with the probabilities $\hat{u}(\mathfrak{x}_i,t)=u(\mathfrak{x}_i,t)=u(\mathfrak{x}_i)$, $i = 1, 2, \ldots, n$.
2. Set initial observed areas $a_j(t)$ of the searchers, $j = 1, 2, \ldots, m$, where m is a number of searchers.
3. While all $\varphi\left(a_j(t)\right)=0, j = 1, 2, \ldots, m$, do:
4. For all $\mathfrak{x}_i$ do: Set $\hat{u}(\mathfrak{x}_i,t)=u(\mathfrak{x}_i,t)$.

   Calculate observed probabilities $\bar{u}(\mathfrak{x}_i,t)$, $i = 1, 2, \ldots, n$:
5. For all searchers $j = 1, 2, \ldots, m$ do:
6. Set $\hat{u}(\mathfrak{x}_i,t)=$ calculate_observed_probabilities$\left(a_j(t),\hat{u}(\mathfrak{x}_i,t)\,|\,i=1,2,\ldots,n\right)$, $i = 1, 2, \ldots, n$.
7. End for.
8. For all $\mathfrak{x}_i$ do: Set $\bar{u}(\mathfrak{x}_i,t)=\hat{u}(\mathfrak{x}_i,t)$.

   Calculate location probabilities $u(\mathfrak{x}_i,t+1)$, $i = 1, 2, \ldots, n$:
9. For all $\mathfrak{x}_i$ do: Set $u(\mathfrak{x}_i,t+1)=\sum_{j=1}^{n}\hat{\rho}_{ij}\bar{u}(\mathfrak{x}_i,t)$.

   Select the next observed areas $a_j(t+1)$, $j = 1, 2, \ldots, m$, *using location probabilities* $u(\mathfrak{x}_i,t+1)$, $i = 1, 2, \ldots, n$
10. For all searchers $j = 1, 2, \ldots, m$ do:
11. Set $a_j(t+1)=$ select_next_observed_area$\left(\breve{\mathfrak{x}}_j(t),\breve{\rho},D,u(\mathfrak{x}_i,t+1)\,|\,i=1,2,\ldots,n\right)$
12. End for.

Move to the new areas $a_j(t+1)$ and observe them, $j = 1, 2, \ldots, m$:
13. For all searchers $j = 1, 2, \ldots, m$ do:
14. Set $t = t + 1$ and get $\varphi(a_j(t))$.
15. End for.
16. End while.
17. Return $a_j(t)$ for which $\varphi(a_j(t)) = 1, j = 1, 2, \ldots, m$.

For comprehensibility, the presented outline follows the notation of Algorithm 4.8, while in the implementation of Algorithm 5.2, the probabilities $\bar{u}(\mathfrak{x}_i, t)$, obtained in lines 5–7, can be used in line 9 instead of the probabilities $\bar{u}(\mathfrak{x}_i, t)$, and line 8 can be dropped. In addition, the for-loops defined in lines 10–12 and in lines 13–15 can be combined into a single loop.

The sharing of information regarding the target's location probabilities is provided by the sequential calculation of the observed probabilities conducted in lines 5–7 (with the further assignment in line 8). The first searcher obtains commonly known location probabilities $\bar{u}(\mathfrak{x}_i, t)$ and updates then according to its observation result. The second searcher obtains the probabilities $\bar{u}(\mathfrak{x}_i, t)$ already updated by the first searcher and executes its own update. The third searcher in its turn obtains these probabilities updated by the second searcher (and previously—by the first searcher) and updates them according to its observations, and so on up to the last searcher. The decision-making regarding the next observed areas in lines 10–12 are conducted independently using the location probabilities updated by all the searchers.

Notice that the presented algorithm does not implement additional methods of collisions' avoidance. Since each searcher zeroes the target's location probabilities in its currently observed area, this area becomes unfavorable and is not chosen by other searchers. Such techniques are similar to the methods of obstacle avoidance noticed in Section 4.4.3, and crucially depend on the requirement of erroneous detection. This issue is clarified by the following example.

Example 5.3: (Based on the MATLAB® Scripts Developed in Collaboration with G. Goren)

Assume that the search is conducted over the square domain $\mathcal{X}$ of the size $n = 100 \times 100$ with the impermeable bounds, and that at the beginning of search, the target's location probabilities $u(\mathfrak{x}, 0)$ are specified by 10 binormal distributions with random centers and the variances $\sigma_1 = \sigma_2 = 5$. The maximal time of search is restricted by 1000 steps; however, if the target is found in a shorter time, the search will be terminated immediately at the moment of finding the target.

The search is conducted according to the destructive scenario, and according to the errorless detection, in the observed areas, the target's location probabilities are zeroed. Similar to Example 4.11, each searcher observes squire areas around its location $\mathfrak{x}(t)$ and the size of the observed areas is $|a(t)| = 9$, $t = 0, 1, 2, \ldots$ The trajectories of the searchers in individual search and in the collective search by three agents acting according to the Algorithm 5.2 are shown in Figure 5.8. This figure, contrast to Figure 5.1, shows both trajectories of the searchers and the areas with zeroed probabilities along these trajectories.

In the individual search (see Figure 5.8a), the searcher starts in the point $\mathfrak{x}(0) = (15,1)$, and in the collective search (see Figure 5.8b), the starting points of the searchers are $\mathfrak{x}_1(0) = (15,1)$, $\mathfrak{x}_2(0) = (85,1)$, and $\mathfrak{x}_3(0) = (50,99)$. In both the cases, the target is static and its location drawn with respect to the target's location probabilities is remained $\hat{\mathfrak{x}}(t) = (31,94)$ during the search time $t = 0, 1, 2, \ldots$

As it is expected, the trajectory of the first searcher in the individual search differs from its trajectory in the collective search (cf. Figure 5.1). In addition, notice that since each searcher zeroes the target's location probabilities along their trajectories, the intersections of the trajectories of different searchers are extremely rare and are led only by the movement according to the estimated probabilities. ■

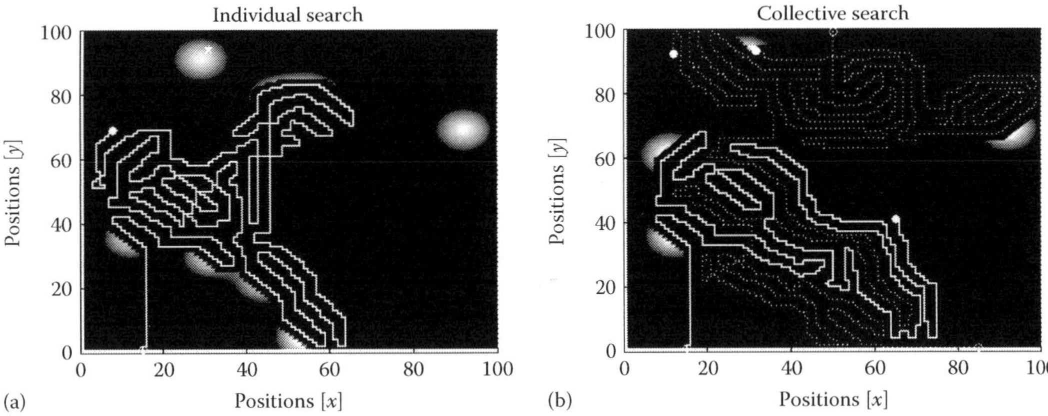

FIGURE 5.8 Destructive search by single and multiple agents with common target's location probabilities: (a) the search trajectory of a single agent and (b) the trajectories of collective search by three agents. The trajectory of the same agent is depicted by a solid line, and the trajectories of two additional agents in figure (b) are depicted by dotted lines. Higher target's location probabilities are shown in white and lower by black.

As indicated earlier, the considered search requires complete knowledge about the target's location probabilities over a domain and such communication abilities of the agents, which allow transmission of these probabilities at each search step. Such assumptions certainly can be implemented for strongly equipped artificial agents with global coordination abilities, which start with the information provided by previously activated systems. However, in the animals' foraging and in the search by reasonably equipped agents acting in noisy environment without global coordination systems, such assumptions fail and the search is conducted using local information only. In the previous section, such search was considered following an example based on the model of signaling and donating in the animals' foraging (see Algorithm 5.1); in the following, this model is combined with the mobility models and algorithms of individual considered in Section 5.4.

5.2.2.2 Foraging with Direct Communication Using Bio-signaling

Let us consider the algorithm of foraging, which implements signaling and donation model defined by Algorithm 5.1. Notice that this model can be substituted by any other signaling and donating techniques with respect to the considered search. For generality, the algorithm is defined in such a manner that supports its application for the basic types of the agents' movement considered in Section 5.4, that is, Brownian and Lévy flight searches with different parameters, and allows further comparisons of the effectiveness of different motion types.

Let $\mathfrak{X} \subset \mathbb{R}^2$ be a 2D domain with the probability density function $u : \mathfrak{X} \times \left[0, T\right] \to \mathbb{R}^+$, which defines conditional target's location probabilities $u\left(\mathfrak{x}, t\right)$ in the points $\mathfrak{x} \in \mathfrak{X}$ given that the search up to time $t \in [0, T]$ was unsuccessful. As mentioned earlier, it is assumed that the agents are equivalent and have equivalent vision, sensing, and signaling abilities. The vision distance, that is, a maximal distance, at which the agent can detect the target, is denoted by r_{vis}, and sensing distance, that is, a maximal distance, at which the agent can sense the presence of the target, is denoted by r_{sense}, $r_{\text{sense}} \geq r_{\text{vis}}$. Then, at each time $t = 0, 1, 2, \ldots$, the observed areas $a_j\left(t\right) \subset \mathfrak{X}$ of the agents are defined as $a_j\left(t\right) = \left\{\mathfrak{x} : \left\|\breve{\mathfrak{x}}_j\left(t\right) - \mathfrak{x}\right\| \leq r_{\text{vis}}\right\}$, and the patches $A_j\left(t\right) \subset \mathfrak{X}$ considered by the agents are defined as $A_j\left(t\right) = \left\{\mathfrak{x} : \left\|\breve{\mathfrak{x}}_j\left(t\right) - \mathfrak{x}\right\| \leq r_{\text{sense}}\right\}$, where $\breve{\mathfrak{x}}_j\left(t\right)$ is a location of the jth agent at time t, $t = 0, 1, 2, \ldots$ and $j = 1, 2, \ldots, m$, where m is a number of agents in the swarm. Minimal and maximal signaling distances are denoted by $r_{\text{signal}}^{\min}$ and $r_{\text{signal}}^{\max}$, respectively.

The algorithm extends the algorithm of foraging by the Lévy flyers with bio-signaling (Siegelmann et al., 2014) and is outlined as follows (for completeness, the outline includes the signaling and donating protocol defined by Algorithm 5.1 and corresponding comments).

Algorithm 5.3: (Foraging with Direct Communication Using Bio-Signaling)

Given a domain $\mathfrak{X}$, target's probability density function u, vision distance r_{vis}, sensing distance r_{sense} and signaling distances r_{signal}^{min} and r_{signal}^{max}, step length δ and parameter γ, do:

1. Set $t = 0$ and start with initial agent's locations $\breve{x}_j(t)$ and observed areas $a_j(t)$, $j = 1, 2, \ldots, m$.

   Initial observation
2. For all agents $j = 1, 2, \ldots, m$ do:
3. Observe initial observed area $a_j(t)$ and decide either to terminate foraging or to continue.
4. End for.

   Acting
5. While at least one agent continues foraging do:
6. If more than one agent continues the foraging that is signaling and donating is possible, do:

   Signaling (according to Algorithm 5.1)
7. For all agents $j = 1, 2, \ldots,$ which continue foraging, do:
8. Consider the patch $A_j(t)$ defined by the sensing distance r_{sense} and calculate the neediness a_j and the cost c_j by Equations 5.1 for continuous domain or by Equations 5.2 for discrete domain. Send the request, which includes location $\breve{x}_j(t)$, neediness a_j and the cost c_j, to the agents located in the ring defined by the signaling distances r_{signal}^{min} and r_{signal}^{max}.
9. End for (signaling by available agents; line 7).

   Donating (according to Algorithm 5.1)
10. For all agents $l = 1, 2, \ldots$ which continue the foraging do:
11. If the request from the jth agent received, then consider the patch $A_l(t)$ defined by the sensing distance r_{sense} and calculate the fitness decrement d_l by Equations 5.1 for continuous domain or by Equations 5.2 for discrete domain and the relativeness k_{lj} defined by the distance between jth and lth agents (see Equation 5.1). If the inequalities $a_j \geq d_l/k_{lj}$ and $a_j \geq k_{lj}d_l + c_j$ (signaling equilibrium with the fitness decrement $b_j = 0$) hold, then move to the location of jth agent with maximal velocity with respect to the agents' type. For example, if the agents are Lévy flyers, then such motion is direct fly, and if the agents are Brownian or Lévy walkers, then they move directly by steps, which are constant for Brownian walkers and drawn by Pareto distribution for Lévy walkers.
12. End for (donating by available agents; line 10).
13. End if (signaling and donating are possible; line 6).

    Foraging
14. For all agents $j = 1, 2, \ldots, m$ do:
15. Observe the observed area $a_j(t)$ and decide either to terminate foraging or to continue.
16. If foraging continues, do:
    Define step length (with respect to the agent's motion type)
17. If the agents are Brownian walkers, then apply the length of the step $\delta_j(t) = \delta =$ const (see Algorithm 4.1).
18. If the agents are Lévy flyers or Lévy walkers, then draw the length $\delta_j(t)$ of the step by Pareto distribution with parameter $\gamma_j = \gamma =$ const such that its probability is $p(\delta_j(t)) \sim \delta_j(t)^{-\gamma_j}$ (see Algorithm 4.4).
19. If the agents are Lévy flyers or Lévy walkers with varying index, then draw the length $\delta_j(t)$ of the step by Pareto distribution with parameter γ_j inversely proportional to $u(a_j(t))$, for example, $\gamma_j \sim 2(1 - u(a_j(t)))$ (see Algorithm 4.5).

Define step direction (with respect to the agent's sensing abilities)

20. If direction does not depend on the environment that is the sensing abilities of the agents are limited, then choose direction $\alpha_j(t)$ of the next step by random.
21. Else screen the points $\mathfrak{x}$ around $\breve{\mathfrak{x}}_j(t)$ such that $\left\|\breve{\mathfrak{x}}_j(t)-\mathfrak{x}\right\|=\delta_j(t)$ and specify direction $\alpha_j(t)$ to the point $\mathfrak{x}^*$ such that $\mathfrak{x}^*=\operatorname{argmax}\left\{u(\mathfrak{x}):\left\|\breve{\mathfrak{x}}_j(t)-\mathfrak{x}\right\|=\delta_j(t)\right\}$, ties are broken randomly (see Algorithm 4.7).

Consider observed area with respect to the foraging scenario

22. If the foraging is destructive, do:
23. For all $\mathfrak{x}\in a_j(t)$ do: Set $u(\mathfrak{x})=0$.
24. End if (destructive foraging; line 22).
25. End if (foraging continues; line 16)

Move to the next location

26. If the agents are Brownian walkers or Lévy flyers, then set the next location $\breve{\mathfrak{x}}_j(t+1)$ with respect to the current location $\breve{\mathfrak{x}}_j(t)$, step length $\delta_j(t)$ and direction $\alpha_j(t)$.
27. If the agents are Lévy walkers, do:
28. If $\left\|\breve{\mathfrak{x}}(t+1)-\tilde{\mathfrak{x}}_j\right\|<\delta$, that is the destination point $\tilde{\mathfrak{x}}_j$ is reached, then set new destination point $\tilde{\mathfrak{x}}_j$ with respect to the current location $\breve{\mathfrak{x}}_j(t)$, step length $\delta_j(t)$ and direction $\alpha_j(t)$.
29. Set the next location $\breve{\mathfrak{x}}_j(t+1)$ such that $\left\|\breve{\mathfrak{x}}(t+1)-\breve{\mathfrak{x}}_j(t)\right\|=\delta$ and direction to $\breve{\mathfrak{x}}(t+1)$ from $\breve{\mathfrak{x}}_j(t)$ is $\alpha_j(t)$. In the other worlds, make one step of the length δ towards destination $\tilde{\mathfrak{x}}_j$.
30. End if (agents are Lévy walkers; line 27).
31. End for (foraging by m agents; line 14).
32. Set $t = t + 1$.
33. End while (all agents terminated foraging; line 5).

The presented algorithm follows the most general scheme and can be extended to additional motion types, for example, to correlated or fractional Brownian searches (see Algorithms 4.2 and 4.3), and can implement different communication and sensing approaches. In addition, notice that it does not assume that the agents have long-term memory that is important for modeling of animals' foraging activity.

This algorithm with two Brownian foragers (see lines 17 and 20) and destructive foraging scenario (see lines 22–24) was already applied for depicting the trajectories shown in Figure 5.4. In the next example, the algorithm will be implemented for the similar search by the Lévy walkers.

Example 5.4

Assume that the destructive search is conducted in the same conditions as in Example 5.1, that is, the agents are acting in the square domain $\mathcal{X}$ of the size $n = 100 \times 100$ with the impermeable bounds, and at the beginning of the search, the target's location probabilities $u(\mathfrak{x},0)$ are specified by 10 binormal distributions with random centers and the variances $\sigma_1 = \sigma_2 = 5$.

The movements of the agents are defined by the Lévy walks with the following parameters: similar to the Lévy flight search with varying index (see Examples 4.6, 4.8, and 4.10), assume that the Lévy indices $a(\mathfrak{x},t)$ for the points $\mathfrak{x}\in\mathcal{X}$ are defined by $a(\mathfrak{x},t)=2(1-u(\mathfrak{x},t))$ that are normalized such that $\min_{\mathfrak{x}} a(\mathfrak{x},0)=\varepsilon>0$, $\varepsilon\ll 1$, and $\max_{\mathfrak{x}} a(\mathfrak{x},0)=2$; then the parameters $\gamma(\mathfrak{x},t)$ are $\gamma(\mathfrak{x},t)=a(\mathfrak{x},t)+1$. To stress the influence of signaling, the minimal and maximal distances to destination points (see lines 27–30 of Algorithm 5.3) are bounded as $\delta_{max} = 1$ and $\delta_{max} = 12.5$, respectively, and the length of the step, which is done towards to destination, is $\delta = 3$.

Similar to Example 5.1, for each searcher, the vision distance is $r_{vis} = 1$, and minimal and maximal signaling distances are $r_{signal}^{min}=2\times r_{sense}=10$ and $r_{signal}^{max}=\sqrt{n}/4=25$. Also, the search is conducted in discrete time $t = 0, 1, 2,\ldots, 500$.

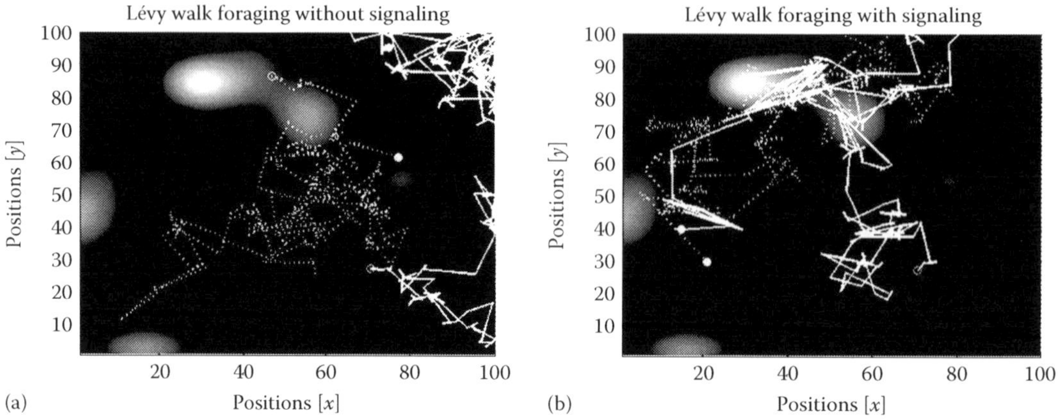

FIGURE 5.9 Trajectories of two Lévy walkers that are foraging (a) individually without signaling and (b) collectively with signaling according to the Sir Philip Sidney game (cf. Figure 5.4). The trajectory of the first agent is depicted by a solid curve, and the trajectory of the second agent is depicted by a dotted curve. In both cases, the agents start in the same positions; initial locations of the agents are depicted by white circumferences and their final positions at time $t = 500$ by white circles.

An example of the trajectories of the agents acting without and with signaling are shown in Figure 5.9.

It is seen that similar to Example 5.1, without signaling (see Figure 5.9a), the agents move independently according to the prescribed Lévy walks, and with signaling (see Figure 5.9b), the agents aggregate and continue searching by group in the region with higher target's location probabilities.

In the considered example, for the individual search, the resulting weighted sum of the encountered targets, that is, the sum of the zeroed target's location probabilities, is 228.861 (94.420 for the first searcher and 134.441 for the second searcher), and the sum length of the searchers' paths is 2311.785 (1144.521 for the first search and 1167.264 for the second searcher); so the resulting eat rate is 0.099. For the search with signaling, the resulting weighted sum of the encountered targets is 433.993 (142.089 for the first searcher and 291.904 for the second searcher) and the sum length of the searchers' paths is 2438.740 (1258.533 for the first searcher and 1180.206 for the second searcher); the resulting eat rate is 0.178, which is nearly two times greater than in the search without signaling (cf. Example 5.1).

In addition, notice that while in the case of Brownian walkers, the change of the type of trajectories can be distinguished easily; in the case of Lévy walkers, the change of the type of motion after aggregating cannot be distinguished with certainty. ■

The presented Algorithm 5.3 allows both direct communication by signaling and, in the case of destructive search, indirect communication by changing the target's location probabilities in the observed areas. However, notice that similar to the algorithms presented in Section 4, it does not consider the movement abilities of the agents and assume that each agent can turn to any direction $\alpha(t)$ at any time t. That restricts an application of the algorithm for definition of the movement "in small" and requires accurate definition of the dependence between agent's current and the next headings (see Section 4.2.2), which in the simplest case result in the correlated Brownian search defined by Algorithm 4.2.

On the other hand, consideration of the turns of the agents in the terms of models with certain geometry and distinguished sides, as it is done in the previous section using the modeling ant (see Figure 5.5 and Example 5.2), can be conducted by defining the controls of each agent with respect to the sensed signals. In Example 5.2, such controls were defined based on the $\otimes_9$-automatons. The next subsection presents this activity of the agents with the stigmergy

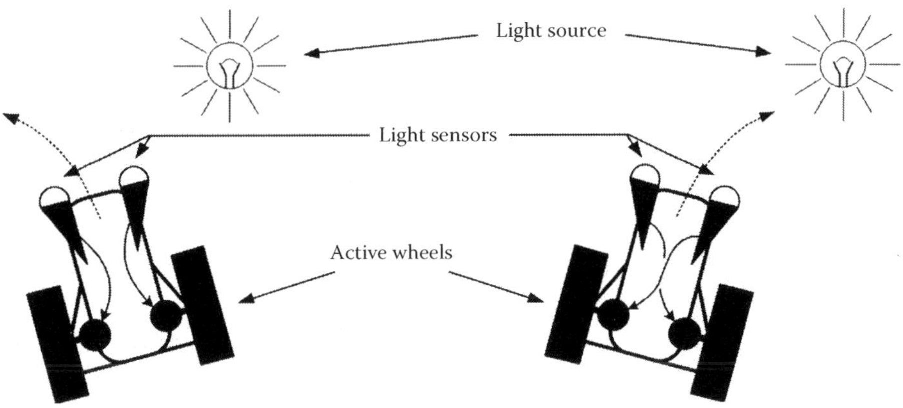

FIGURE 5.10 Movements of the Braitenberg vehicle with different connections between its sensors and motors.

defined by deploying the pheromone trail in algorithmic form. In addition, the subsection considers the algorithm of stigmergic activity of the agents executing collective task (Beckers, Holland, & Deneubourg, 1994).

5.2.2.3 Foraging with Indirect Communication

In the definition of the modeling ant, the algorithm of control of the agent, which has definite geometry and distinguished left and right side sensors and locomotion equipment, can be easily defined based on the principles of the Braitenberg vehicles (Braitenberg, 1984). An illustration of the behavior of such vehicle with two different types of connection between the sensors and the motors is shown in Figure 5.10.

Following the figure, for the left-to-left/right-to-right connections, the vehicle turns away from the light source, while for the left-to-right/right-to-left connections, the vehicle turns toward the light source. The movements of the modeling ant shown in Figure 5.5 are defined according the same principles.

The stigmergic principle that prescribes to the agent to use the changes of the environment, which were done by this agent or by the other agents of the group, can be implemented in different ways with respect to the considered task. In the earlier considered case of the ants' motion, it was required to follow over maximal number of favorable environmental states and avoid unfavorable states using both the environmental states and the deployed pheromone trails. In the other case, which is considered in the succeeding text, it is required to collect the objects over the domain using local information only. Notice that the probabilistic search and foraging can be considered as a kind of the first task.

Let us start with general algorithm of control of the agents' motion with signing the already considered regions, for example, by deploying pheromone trails. The algorithm follows the principles of the Braitenberg vehicles (Braitenberg, 1984) and pheromone robotics (Payton et al., 2001). As mentioned earlier, it is assumed that each agent is equipped by left and right side sensors implementing the observation functions $z^{\text{env}} : \mathbb{R} \to [0,1]$ and $z^{\text{sgn}} : \mathbb{R} \to [0,1]$, which, respectively, transform the observed environmental states and signs into the observation results. The results of observation of the environment by the left and right sensors at time t, $t = 0, 1, 2,\ldots$ are denoted by $z_{\text{left}}^{\text{env}}(t)$ and $z_{\text{right}}^{\text{env}}(t)$, correspondingly. The outputs of the control units, which are sent directly to the left and right locomotion devices, are denoted by $y_{\text{left}}(t)$ and $y_{\text{right}}(t)$, respectively, and it is assumed that the velocity of the movement of each side is proportional to the value of corresponding output. Finally, it is assumed that each agent can sign the environment by certain signs $\zeta(\mathfrak{x},t)$, $\mathfrak{x} \in \mathfrak{X}$, that can be sensed by specific sensors with observation results $z_{\text{left}}^{\text{sgn}}(t)$ and $z_{\text{right}}^{\text{sgn}}(t)$, or by the sensors that are used for sensing environmental states. The algorithm is outlined as follows.

Algorithm 5.4: (Foraging with Indirect Communication)

Given a domain $\mathfrak{X}$, the real valued environmental states over $\mathfrak{X}$ and observation functions z^{env} and z^{sgn} for the left and right sensors, do:

1. Set $t = 0$ and start with initial agent's locations $\breve{\mathfrak{x}}_j(t)$, $j = 1, 2, \ldots, m$.

Initial observation

2. For all agents $j = 1, 2, \ldots, m$ do:
3. Obtain results $z^{\text{env}}_{\text{left}(j)}(t)$ and $z^{\text{env}}_{\text{right}(j)}(t)$ of observation of the environmental states and decide either to terminate foraging or to continue.
4. End for.

Acting

5. While at least one agent continues foraging do:

Foraging

6. For all agents $j = 1, 2, \ldots$ which continue foraging, do:

Observe environment and signs

7. Obtain results $z^{\text{env}}_{\text{left}(j)}(t)$ and $z^{\text{env}}_{\text{right}(j)}(t)$ of observation of the environmental states and results $z^{\text{sgn}}_{\text{left}(j)}(t)$ and $z^{\text{sgn}}_{\text{right}(j)}(t)$ of observation of the remained signs and decide either to terminate foraging or to continue.
8. If jth agent terminates foraging, then continue with the next agent in the loop (line 6).
9. Set $z_{\text{left}(j)}(t) = f\left(z^{\text{env}}_{\text{left}(j)}(t), z^{\text{sgn}}_{\text{left}(j)}(t)\right)$ and $z_{\text{right}(j)}(t) = f\left(z^{\text{env}}_{\text{right}(j)}(t), z^{\text{sgn}}_{\text{right}(j)}(t)\right)$ using certain sensor fusion function $f{:}[0, 1] \times [0, 1] \to [0, 1]$.

Specify motion

10. Set $y_{\text{left}(j)}(t) \sim z_{\text{left}(j)}(t)$ and $y_{\text{right}(j)}(t) \sim z_{\text{right}(j)}(t)$.

Sign the environment if need

11. Decide whether to sign the current location $\breve{\mathfrak{x}}_j(t)$, and if so—remain the sign of the value $\zeta(\breve{\mathfrak{x}}_j(t), t)$.

Move to the next location

12. Move the left side proportionally to $y_{\text{left}(j)}(t)$ and the right side proportionally to $y_{\text{right}(j)}(t)$.
13. End for (foraging; line 6)
14. Set $t = t + 1$.
15. End while (all agents terminated foraging; line 5).

It is clear that the outlined algorithm defines the actions, which are the same as the actions of the modeling ant considered in Section 5.2.1, and can be implemented for any appropriate observation, decision making, and fusion functions. The next example illustrates the actions of the algorithm and stresses an influence of the indirect communication.

Example 5.5: (Kagan et al., 2013)

Similar to Example 5.2, consider the movements of five modeling ants conducting nondestructive search in the square domain $\mathfrak{X} = \{\mathfrak{x}_1, \mathfrak{x}_2, \ldots, \mathfrak{x}_n\}$ of the size $n = 500 \times 500$ units with the impermeable bounds. However, in contrast to Example 5.2, the environmental states $u(\mathfrak{x}_i)$, $i = 1, 2, \ldots, n$, are defined by two circle peaks and bottoms produced by binormal distributions with random centers and random variances from the interval [5, 10] and normalized such that $\max_i u(\mathfrak{x}_i) = 1$ and $\min_i u(\mathfrak{x}_i) = 0$. Such states provide a mostly homogeneous environment with one favorable and one unfavorable regions.

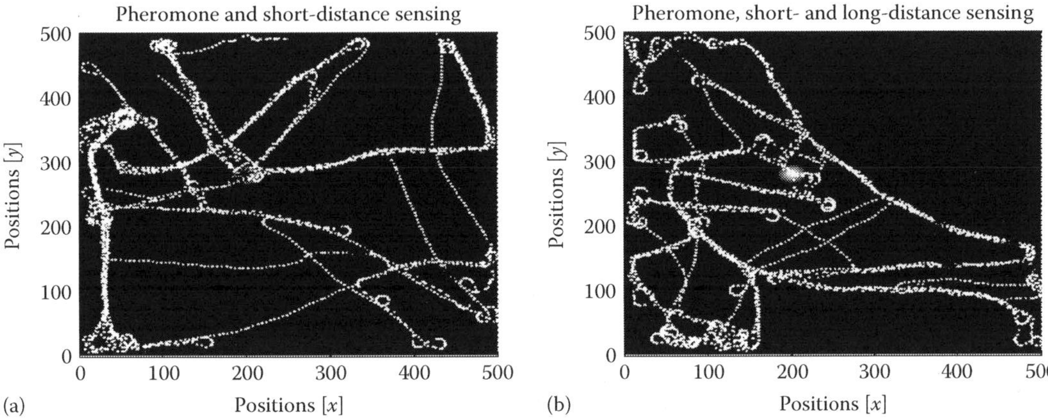

FIGURE 5.11 The pheromone trails deployed by five modeling ants. (a) Trails of the ants, which follow the pheromone trails and apply short-distance sensing of the environmental states. (b) Trails of the ants, which follow the pheromone trails and apply both short- and long-distance sensing of the environmental states, $r_{\text{sense}} = 60$. Favorable environmental state is depicted by white, unfavorable state by black, and neutral states by gray. The trails of the ants are depicted by white dotted curves.

The movement of the ants is conducted in discrete time $t = 1, 2, \ldots, 1000$. An example of pheromone trails deployed by the modeling ants without and with long-distance sensing is shown in Figure 5.11.

It is seen that similar to Example 5.2, the ants tend to follow the pheromone trails deployed by themselves and by the other ants and avoid passing over the unfavorable region. In addition, notice that in the case of short-distance sensing (Figure 5.11a), the trajectories of the ants are distributed over a domain, while if the ants apply both short- and long-distance sensing, their trajectories are concentrated in the half of the domain, in which the favorable region is located. ■

In the considered algorithm, the stigmergy was implemented by signing the regions of the domain by pheromone trails and using these trails in specification of the further behavior of the agents.

Finally, let us consider an example of the algorithm with stigmergy, in which the agents are executing collective task and are prescribed to collect the objects distributed over a domain. This algorithm formalizes the activity of mobile robots suggested and simulated in 1994 by Beckers et al. (1994).

Assume that each agent is equipped by devices, which allow to collect and to carry a certain number k of objects. In addition, each agent is equipped with the sensor for detecting the presence of the object in the close neighborhood and a sensor for detecting the obstacles such that the heap of objects, which includes more than k objects, is also considered as obstacle. Communication between the agents is provided by changing the number and locations of the objects in the environment, which is sensed by the agents and influenced on their activity.

The succeeding outline presents the algorithm for the simplest behavior of the agents acting in homogeneous environment.

Algorithm 5.5 (Collecting Objects in Homogeneous Environment; Following the Simulation by Beckers et al. [1994])

Given domain $\mathfrak{X}$ with the impermeable bounds, objects distributed objects over the domain and a heap size k, do:

1. Start with initial agent's locations $\breve{x}_j(t), j = 1, 2, \ldots, m$.
2. While exists a heap with less than k objects, do:
3. For all agents $j = 1, 2, \ldots, m$ do:

Ballistic motion (cf. search with intermittent strategies considered in Section 4.3.3)
4. Move forward up to detecting obstacle (or heap of objects).

Obstacle avoidance
5. If obstacle detected, then turn by a random angle away from obstacle and continue with forward movement.

Objects collecting
6. If object detected, do:

Pick up object
7. If the number of collected objects is less than the heap size k then pick up the object,

Collect objects into heap
8. Else put the already collected objects near the detected object.
9. End if (object detected; line 6).
10. End for (activity of m agents; line 3).
11. End while (all objects are collected into heaps with k or more objects; line 2).

In the presented algorithm, the communication between the agents is provided by changes of the environment, that is, the changes in distribution of the objects over a domain, which lead to the changes in the agents' behavior. Certainly, the termination condition appearing in line 2, which assumes a knowledge about the state of the objects over a domain, can be substituted by the condition, which can be implemented for each agent and checked locally, for example, based on the time of forward movement without meeting objects, or on the number of already collected objects.

As indicated earlier, Beckers et al. (1994) implemented Algorithm 5.5 using mobile robots acting; the succeeding example illustrates the actions of the algorithm using numerical simulations.

Example 5.6

Assume that the agents act in the square domain $\mathcal{X}$ of the size $n = 100 \times 100$ with the impermeable bounds, over which there are distributed 400 objects. Following the simulation by Beckers et al. (1994), at the beginning, the objects are located regularly in rows and columns. Each object has a unit size, that is, it occupies a single point of the domain, and the heap size is $k = 3$. Obstacle avoidance is implemented as follows. If the agent detects a heap or a bound, it moves three steps backward and then turns by random angle and continues forward motion. The step's length in forward and backward motion is $l = 3$. Examples of the trajectories of a single agent ($m = 1$) and of five agents ($m = 5$) are shown in Figure 5.12.

It is seen that during the motion, the agents change the distribution of the objects, which in its turn determines the motion. At the beginning, when the heaps were not yet collected, the agents move directly, and with increasing the number of collected heaps, the direct movements become shorter and the turns become more frequent. Notice that such motion is rather similar to the Lévy flights (see Section 4.3.1) with the parameter depending on the objects distribution and decreasing with time; however, this issue requires additional considerations. ■

The considered example finalizes the consideration of aggregation and communication between the agents in the tasks of multiagent search and foraging. Certainly, the presented algorithms do not exhaust numerous techniques of the agents' aggregation and communication protocols (for the progress in the field, see earlier indicated books and, in particular, proceedings like ANTS [1998–2014], From Animals to Animats [1991–2012], Distributed Autonomous Robotic Systems [1992–2013]. However, they clarify the main ideas of such methods and provide a basis for consideration of collective search and foraging and swarm dynamics.

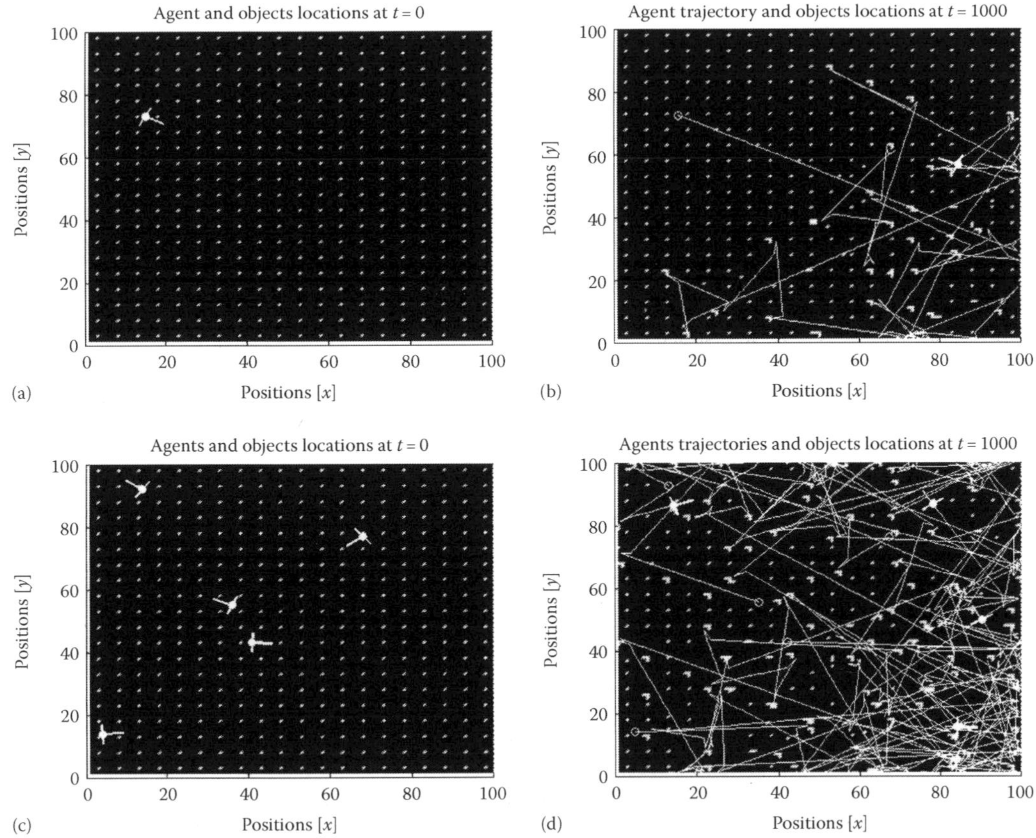

FIGURE 5.12 Collecting objects in homogeneous environment by single agent ($m = 1$) and by the group of five agents ($m = 5$): (a) initial location of the agent and locations of the objects at time $t = 0$ and (b) agent's trajectory and locations of the objects at time $t = 1000$; (c) initial location of $m = 5$ agents and locations of the objects at time $t = 0$ and (d) trajectories of the agents and locations of the objects at time $t = 1000$. The agents are depicted by white circles; the heading of the agents is indicated by the longer line. In figures (b) and (d), trajectories of the agents are depicted by the white lines. Locations of the objects are depicted by white points.

5.2.3 Simulation Results of Group Foraging with Direct Communication

This short section presents results of simulations of group foraging with direct communication, which support certain conclusions regarding animals' behavior. All over the section, it is assumed that $m = 10$ agents act according to Algorithm 5. 3 of destructive foraging (see Section 4.3.2) with direct communication using the bio-signaling defined in Section 5.2.1. The activity of the agents is considered in the discrete domain $\mathcal{X}$ of the size $n = 100 \times 100$ with impermeable bounds. Similar to previous examples, parameters of signaling and sensing are specified as follows: minimal and maximal distances are $\delta_{min} = 1$ and $\delta_{max} = 12.5$, respectively, the length of the step is $\delta = 3$, vision distance is $r_{vis} = 1$, sensing distance is $r_{sense} = 5$, and minimal and maximal signaling distances are $r_{signal}^{min} = 10$ and $r_{signal}^{max} = 25$. In all cases, the search is conducted in discrete time $t = 0, 1, 2, \ldots, 100$.

The foraging was considered for three cases of prey distribution (target's location probabilities). In the first case, it was assumed that the most part of the domain does not contain prey, and it is distributed only over a fourth part of the points. In the second case, the prey was distributed over a half of the points, and in the third case, it was assumed that each point of the domain contains a nonzero quantity of prey. For each case, the simulations included 1000 sessions. For the first prey distribution, examples of the trajectories of the agents and distributions of the eaten prey over the sessions are shown in Figure 5.13 (Brownian foraging), Figure 5.14 (Lévy flight foraging), and Figure 5.15 (Lévy walk foraging).

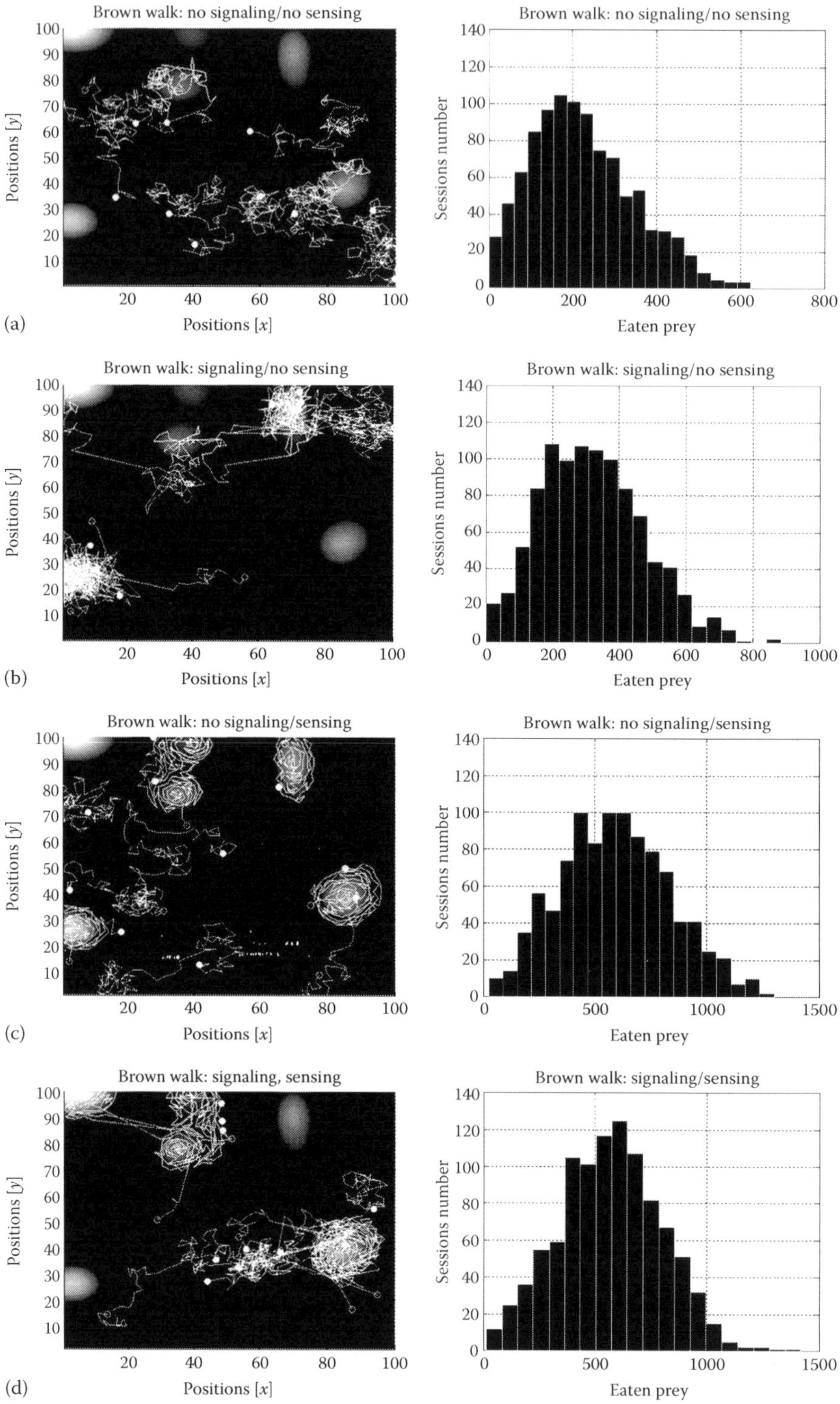

FIGURE 5.13 Brownian foraging without and with sensing and signaling: left column—examples of the trajectories of ten agents and right column—distributions of eaten prey over 1000 sessions. (a) Agents without signaling and sensing, (b) agents with signaling, but without sensing, (c) agents without signaling, but with sensing, and (d) agents with both signaling and sensing.

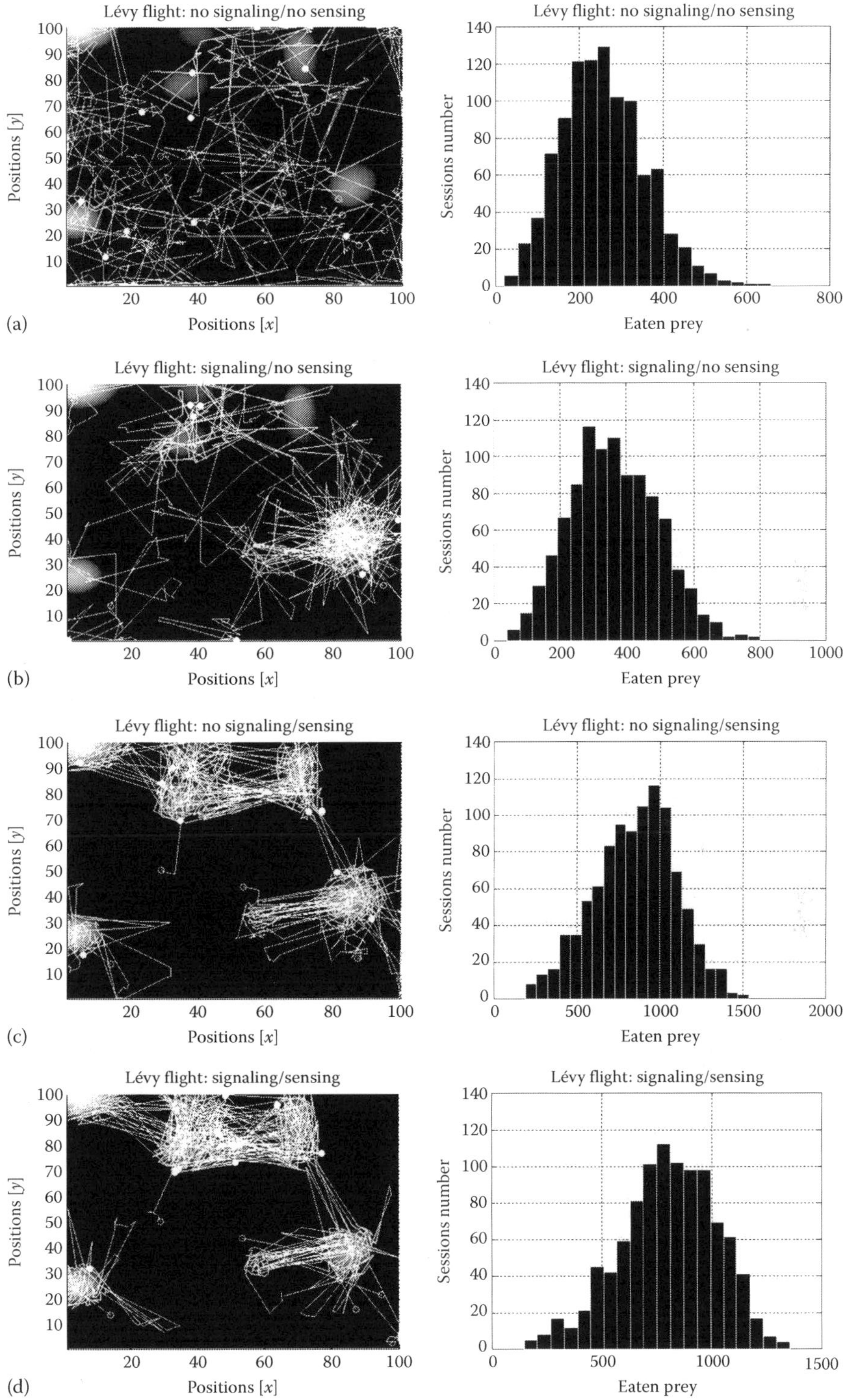

FIGURE 5.14 Lévy flight foraging without and with sensing and signaling: left column—examples of the trajectories of 10 agents and right column—distributions of eaten prey over 1000 sessions. (a) Agents without signaling and sensing, (b) agents with signaling, but without sensing, (c) agents without signaling, but with sensing, and (d) agents with both signaling and sensing.

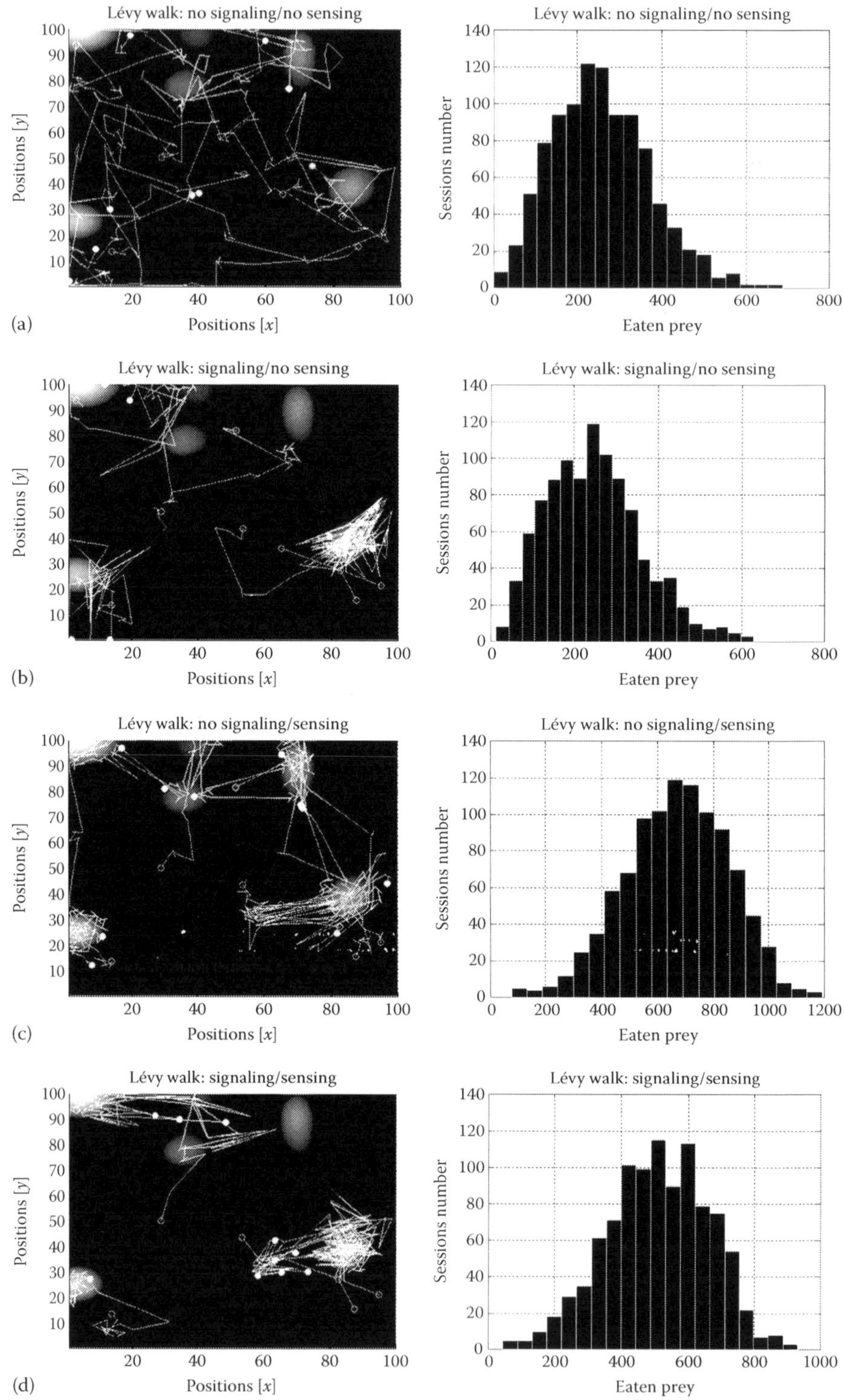

FIGURE 5.15 Lévy walk foraging without and with sensing and signaling: left column—examples of the trajectories of 10 agents and right column—distributions of eaten prey over 1000 sessions. (a) Agents without signaling and sensing, (b) agents with signaling, but without sensing, (c) agents without signaling, but with sensing, and (d) agents with both signaling and sensing.

In the remaining two cases of prey distribution, the trajectories and distributions of the eaten prey are the same.

In addition, for each session, the average path length and average eat rate for the agents were calculated. The results of simulations for the indicated distributions of prey and three types of foragers' motion are the following (Siegelmann et al., 2014) (maximal values are stressed by bold):

1. The prey is distributed over a fourth part of the points of the domain:

	Brownian Walk		Lévy Flight		Lévy Walk	
	Eaten Prey	Eat Rate	Eaten Prey	Eat Rate	Eaten Prey	Eat Rate
No signaling/no sensing	107.00	0.037	120.72	0.019	116.16	0.047
Signaling/no sensing	160.98	0.055	187.84	0.026	127.58	0.048
No signaling/sensing	262.83	0.090	398.13	0.062	355.16	0.144
Sensing and signaling	253.66	0.086	404.70	0.057	310.36	0.117

2. The prey is distributed over a half of the points of the domain:

	Brownian Walk		Lévy Flight		Lévy Walk	
	Eaten Prey	Eat Rate	Eaten Prey	Eat Rate	Eaten Prey	Eat Rate
No signaling/no sensing	226.71	0.077	257.47	0.041	256.35	0.104
Signaling/no sensing	320.17	0.109	361.82	0.048	247.42	0.091
No signaling/sensing	591.11	0.201	849.63	0.129	668.94	0.272
Sensing and signaling	567.29	0.192	800.79	0.102	510.45	0.191

3. The prey is distributed over all points of the domain:

	Brownian Walk		Lévy Flight		Lévy Walk	
	Eaten Prey	Eat Rate	Eaten Prey	Eat Rate	Eaten Prey	Eat Rate
No signaling/no sensing	341.34	0.116	393.07	0.063	372.71	0.151
Signaling/no sensing	431.01	0.146	482.56	0.062	330.07	0.120
No signaling/sensing	830.67	0.280	944.95	0.143	758.35	0.309
Sensing and signaling	730.20	0.246	887.05	0.112	600.29	0.224

From the obtained results, it follows that the maximum both of the quantity of eaten prey and of the eat rate is reached for the agents with strong sensing abilities acting individually and using indirect communication based on the changes of the environmental states that are the destroyed prey or zeroed target's location probabilities. If such sensing and indirect communication is combined with the direct communication by bio-signaling, then the foraging becomes less effective but still strongly outperforms the foraging without sensing and without both signaling and sensing. The only exception is the foraging by the Lévy flight over the domain with large empty regions (Case 1), where sensing and bio-signaling result in slightly larger quantity of eaten prey; however, notice that statistically this difference is insignificant. If the agents have no sensing abilities, then, as it is expected, bio-signaling leads to the larger quantities of eaten prey than in the case of free motion without sensing and signaling. However, the relation between eat rates in the foraging without sensing and with and without signaling depends on the distribution of the prey over a domain. If most of the regions do not contain prey (Case 1), then application of bio-signaling results in greater eat rates for all types of the agent's motion, but if the domain is rich enough with prey (Cases 2 and 3), then the greater eat rate for the Lévy walk foraging is provided by the motion without sensing and signaling.

In addition, notice that, in opposite to the Lévy flight foraging hypothesis (see Section 4.3.2), for some cases, the Brownian foraging, outperforms the Lévy walk foraging both in the quantities of eaten prey and in the eat rates. Nevertheless, as it follows from the histograms shown in Figure 5.13, in the Brownian foraging the search often results in small or even zero eaten prey, while in the Lévy walk foraging such results are rather rare (see histograms shown in Figure 5.15). Hence, from the evolutional point of view, such observation means that the Brownian foraging will result in greater death rate of the agents than the Lévy walk foraging, and the Lévy walk foraging can be considered as evolutionarily preferable. Certainly, since Lévy flight foraging outperforms both the Brownian and the Lévy walk foraging, for the flying agents this motion type is preferable in complete coincidence with the Lévy flight foraging hypothesis.

The indicated observation finalizes the consideration of the algorithms of group foraging and stigmergy, and the rest of the section addresses general models of swarm dynamics based on active Brownian motion and corresponding Fokker–Planck equations.

5.3 MODELING BY ACTIVE BROWNIAN MOTION

Let us return to the general description of the agents' motion that specifies the agents' trajectories and probability distributions of the agents' location in the points of the domain rather than the reactions of the agents on the observed states of the environment. Such description is provided by different stochastic processes; the main types of these processes, namely, Brownian motion and its extensions and Lévy flight, were presented in Section 4. This section continues the consideration of the Brownian motion that is exposed in more physical manner and in the form of active Brownian motion is implemented for modeling of the agents' motion and swarm dynamics.

Generally speaking, the difference between classical Brownian motion and active Brownian motion is the following (Schweitzer, 2000, pp. 161–162):

> The motion of "simple" Brown particle is due to fluctuations of the surrounding medium, i.e. the result of random impacts of the molecules or atoms of the liquid or gas, the particle is immersed in… Active motion, on the other hand, relies on the supply of energy… [such that the active particles have] the ability to take up energy from the environment, to store it in an internal depot and to convert internal energy to perform different activities, such as metabolism or motion.

In other words, active Brownian particles act as autonomous mobile agents, while their mobility and locomotion are defined by the laws of the dynamics of particles and allow direct application of the methods of classical and statistical mechanics for specifying the agents' motion. In addition, an application of energy supply provides flexible techniques for the definition of swarms and consideration of swarm dynamics using well-established physical methods (Ebeling, 2007; Erdmann, Ebeling, Schimansky-Geier, & Schweitzer, 2000; Klimontovich, 1991, 1994; Schweitzer, 2003).

5.3.1 Active Brownian Particles

Let us start with the consideration of active Brownian particles. As mentioned earlier, the discourse concentrates on the motion in 2D space; however, an extension of the presented methods of 3D case in most cases is straightforward.

Let $\mathfrak{X} \subset \mathbb{R}^2$ be a 2D domain with the points $\mathfrak{x} = (x, y)$ and assume that the simple Brownian particle moves in the domain $\mathfrak{X}$. In Section 4, the motion of this particle was defined starting from the diffusion Equation 4.22 and using the corresponding Wiener processes $\mathcal{W}_t$. The other approach suggested by Langevin (1908) follows classical mechanics and defines the motion of the Brownian particle as follows.

5.3.1.1 Langevin Equation for Simple Brownian Motion

As mentioned earlier, denote by $\mathfrak{x}(t)$ the location of the particle at time $t \in [0, T]$ and by $v(t) = \frac{d}{dt}\mathfrak{x}(t)$ its velocity at this time. Then, by the Newtonian law, the motion of the particle is defined as follows (Klimontovich, 1994; Schweitzer, 2003):

$$\frac{d}{dt}\mathfrak{x}(t) = v(t), \quad \frac{d}{dt}v(t) = \frac{1}{m}F(t), \tag{5.7}$$

where

m is the mass of the particle

F(*t*) is a sum of the forces applied to the particle at time *t*.

For the Brownian particle, the value *F*(*t*) is combined from two forces: the velocity-dependent dissipative force

$$F^{\text{diss}}(t) = -\gamma_0 v(t),$$

where $\gamma_0 \geq 0$ has a meaning of friction coefficient, and the stochastic force

$$F^{\text{stoch}}(t) = \mathcal{F}_t,$$

where $\mathcal{F}_t$ is a stochastic process with zero mean $\mathcal{E}(\mathcal{F}_t) = 0$ and δ-correlated time dependence $\mathcal{E}(\mathcal{F}_{t_2}\mathcal{F}_{t_1}) = 2S\delta(t_2 - t_1)$ with the strength $S = \gamma_0 \frac{k_B \mathcal{T}}{m}$, where $k_B \approx 1.38 \times 10^{-23}\left[\frac{\text{m}^2\text{kg}}{\text{s}^2\text{K}}\right]$ is the Boltzmann constant and $\mathcal{T}$ is a temperature.

The substitution of the indicated values into Equation 5.7 results in the following system:

$$\frac{d}{dt}\mathfrak{x}(t) = v(t), \quad \frac{d}{dt}v(t) = -\frac{\gamma_0}{m}v(t) + \sqrt{2\gamma_0 k_B \mathcal{T}}\,\mathcal{G}_t, \tag{5.8}$$

where $\mathcal{G}_t$ is the Gaussian white noise, that is, $\mathcal{E}(\mathcal{G}_t) = 0$ and $\mathcal{E}(\mathcal{G}_{t_2}\mathcal{G}_{t_1}) = \delta(t_2 - t_1)$. The second equation in this system is called the Langevin equation for the Brownian motion. Notice that since the Gaussian white noise $\mathcal{G}_t$ is a generalized derivative of the Wiener process $\mathcal{W}_t$, that is, $\mathcal{G}_t = f(t)\mathcal{W}_t - \int_0^t f(\tau)\frac{d\mathcal{W}_\tau}{d\tau}d\tau + \int_0^t \frac{df}{d\tau}\mathcal{W}_\tau d\tau$, the definition given by Equation 5.8 is equivalent to the direct definition of the Brownian motion based on the Wiener process (see also Klimontovich, 1994; Romanczuk, 2012).

Assume that at time $t = 0$, the particle starts in the point $\mathfrak{x}(0)$. Then, the mean squared displacement $(\mathfrak{x}(t) - \mathfrak{x}(0))^2$ of the particle given by Equation 5.8 is the following (Romanczuk, 2012; Schweitzer, 2003):

$$(\mathfrak{x}(t) - \mathfrak{x}(0))^2 = 2\sigma^2\left\{t - \frac{m}{\gamma_0}\left[1 - \exp\left(-\frac{\gamma_0}{m}t\right)\right]\right\}, \tag{5.9}$$

where $\sigma^2 = \frac{k_B \mathcal{T}}{\gamma_0}$ stands for the diffusion rate (cf. Chapter 4) and γ_0 represents the Stokes friction.

For the times $t \ll \frac{m}{\gamma_0}$, this solution corresponds to the free motion of the particle with mean velocity

$$\left(\mathfrak{x}(t)-\mathfrak{x}(0)\right)^2 = v^2(t)t^2,$$

and for the times $t \gg \frac{m}{\gamma_0}$, this describes the motion of the particle with respect to friction and stochastic forces

$$\left(\mathfrak{x}(t)-\mathfrak{x}(0)\right)^2 = 2\sigma^2\left(t-\frac{m}{\gamma_0}\right),$$

that agrees with the motion of Brownian particle in micro- and macroscales (Schweitzer, 2003).

5.3.1.2 Brownian Motion in External Potential Field

Now assume that the particle moves in the external potential field defined by the function $\mathcal{U}:\mathfrak{X}\to\mathbb{R}$ that specifies the states $\mathcal{U}(\mathfrak{x})$ of the environment, into which the particle is immersed. Then, in addition to dissipative force F^{diss} and stochastic force F^{stoch}, the forces F appearing in Equation 5.7 include additional force

$$F^{\text{poten}} = -\nabla\mathcal{U} = -\frac{\partial}{\partial x}\mathcal{U}-\frac{\partial}{\partial y}\mathcal{U},$$

where it is assumed that the particle is attracted to the points with minimal potential $\mathcal{U}(\mathfrak{x})$. The Langevin equation, which governs the movement of the particle, is written as follows (Romanczuk, 2012; Schweitzer, 2003):

$$\frac{d}{dt}\mathfrak{x}(t)=v(t),\quad \frac{d}{dt}v(t)=-\frac{\gamma_0}{m}v(t)-\frac{1}{m}\nabla\mathcal{U}(\mathfrak{x})+\sqrt{2\gamma_0 k_B\mathcal{T}}\mathcal{G}_t, \tag{5.10}$$

that is direct extension of Equation 5.8.

Consider the probability $v(\mathfrak{x},v,t)$ that at time t the particles are located in the point $\mathfrak{x}=\mathfrak{x}(t)$ and move with the velocity $v = v(t)$. The Fokker–Planck equation, which governs the distribution of the probability $v(\mathfrak{x},v,t)$ with respect to the Langevin equation (5.10) is the following (Erdmann et al., 2000; Romanczuk, Bar, Ebeling, Lindner, & Schimansky-Geier, 2012; Schweitzer, 2003):

$$\frac{\partial}{\partial t}v(\mathfrak{x},v,t)=\frac{\gamma_0}{m}\frac{\partial}{\partial v}\left(\gamma_0\sigma^2\frac{\partial}{\partial v}v(\mathfrak{x},v,t)+v(t)v(\mathfrak{x},v,t)\right)-v(t)\frac{\partial}{\partial \mathfrak{x}}v(\mathfrak{x},v,t)-\frac{1}{m}\nabla\mathcal{U}(\mathfrak{x})\frac{\partial}{\partial v}v(\mathfrak{x},v,t). \tag{5.11}$$

It is clear that the solution of Equation 5.11 strongly depends on the potential $\mathcal{U}$. In particular, for constant potential, this equation is reduced to the diffusion Equation 4.29, and its solution, with respect to constant coefficients, is given by the Gaussian distribution (4.30). Similarly, for periodic potential $\mathcal{U}(\mathfrak{x}+|\mathfrak{X}|)=\mathcal{U}(\mathfrak{x})$, where $|\mathfrak{X}|$ is a size of the domain $\mathfrak{X}$, stationary solution $v_0(\mathfrak{x},v)$ of Equation 5.11 is provided by the Boltzmann distribution (Erdmann et al., 2000; Romanczuk et al., 2012; Schweitzer, 2003):

$$v_0(\mathfrak{x},v)=C\exp\left(-\frac{1}{\gamma_0\sigma^2}\mathcal{U}(\mathfrak{x})\right)\exp\left(-\frac{m}{2\gamma_0\sigma^2}v^2\right), \tag{5.12}$$

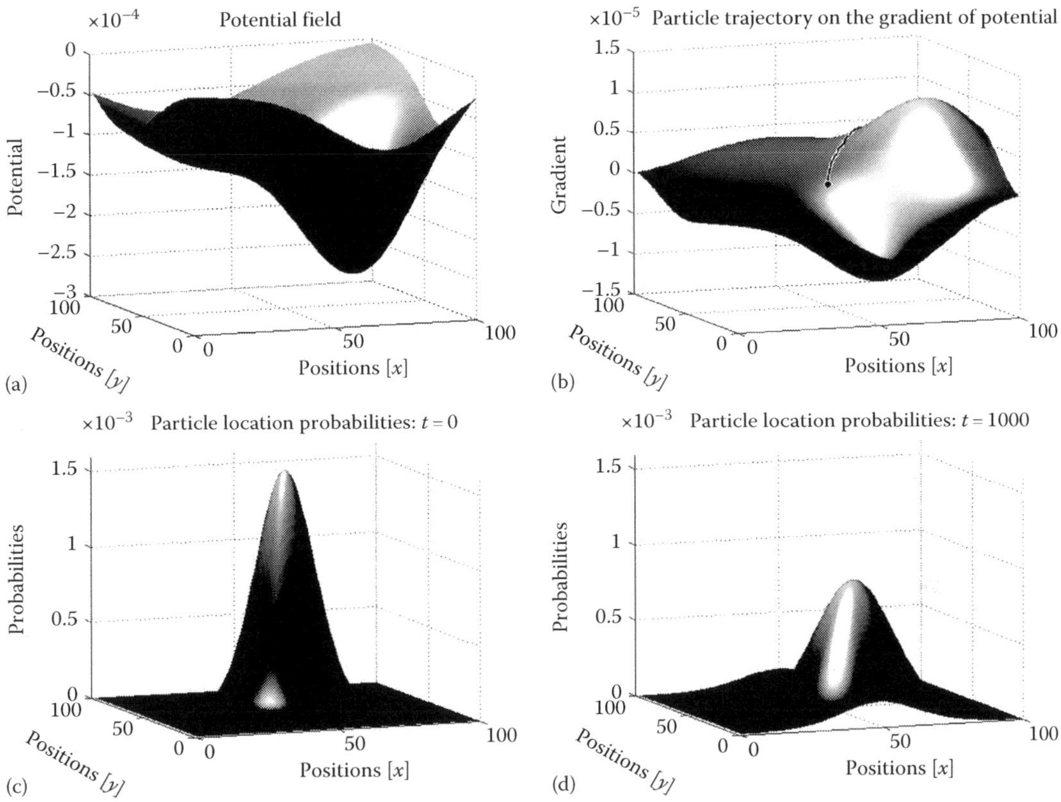

FIGURE 5.16 Example of the trajectory of active Brownian particle governed by the Langevin Equation 5.10 and distribution of the particle's location probabilities specified by the Fokker–Planck Equation 5.11 with respect to the potential field shown in figure (a). Figure (b) shows an example of the trajectory of the particle starting from the point $\mathfrak{x}(0) = (50, 50)$ on the gradient of potential field, figure (c) initial distribution of the particle's location probabilities at $t = 0$, and figure (d) distribution of the location probabilities at $t = 1000$. In the example, it is assumed that mass is $m = 1$, friction is $\gamma_0 = 0.5$, diffusion rate is $\sigma^2 = 0.25$, and the velocities are $v(t) = 0.001$ for all $t = 0, 1, \ldots, 1000$.

where C is a normalization coefficient, which provides the equivalence $\int v_0(\mathfrak{x}, \mathrm{v})\, d\mathfrak{x} d\mathrm{v} = 1$ over $\mathfrak{X}$. Notice that, as it is expected, for constant potential $\mathcal{U}$, this solution is equivalent to the Gaussian distribution. Finally, for arbitrary potential $\mathcal{U}$, the solution of Equation 5.11 cannot be obtained in analytic form; however, for any reasonable potential, numerical solutions are always available.

The trajectory of the particle governed by Equation 5.10 and probability distributions specified by Equation 5.11 given potential field are shown in Figure 5.16. The implemented potential field shown in Figure 5.16a is defined with respect to the target's location probabilities used in the examples in Chapter 2 (see, e.g., Figures 2.2a and 2.7a). Initial distribution of particle location probabilities is defined similarly to the examples in Section 4.2 (see, e.g., Figures 4.8a and 4.14a, and further) by the binormal distribution with the central point $\breve{\mathfrak{x}}(0) = (50, 50)$ and equal variances over the axes $\sigma_1 = \sigma_2 = 10$. The applied parameters are mass $m = 1$, friction $\gamma_0 = 0.5$, and diffusion rate $\sigma^2 = 0.25$. In addition, it is assumed that the velocities are $v(t) = 0.001$ for all $t = 0, 1, \ldots, 1000$.

It is seen that, as it is expected, the particle moves, respectively, to the potential field and tends to the regions where the gradient of the potential is higher. Correspondingly, the distribution of the particle's location probabilities tends to follow the gradient of potential field with certain disturbance led by friction and drift. Additional examples and informative simulation results appear in the book by Schweitzer (2003) and the later overview by Romanczuk et al. (2012).

In the Langevin equation (5.10) and corresponding Fokker–Planck equation (5.11), it is assumed that the medium, in which the particle moves, is rarefied enough for significant influence of the inertia effects. If, in contrast, the particle moves in viscous medium with high friction, then the influence of inertia is neglected and the Langevin equation in this case is defined as follows (Romanczuk et al., 2012; Schweitzer, 2003):

$$\frac{d}{dt}\mathfrak{x}(t) = v(t) = -\frac{1}{m}\nabla\mathcal{U} + m\sqrt{2\sigma^2}\mathcal{G}_t. \tag{5.13}$$

The Fokker–Plank equation in this case is

$$\frac{\partial}{\partial t}v(\mathfrak{x}, v, t) = \frac{1}{m}\sigma^2\frac{\partial^2}{\partial \mathfrak{x}^2}v(\mathfrak{x}, v, t) - \frac{1}{m}\nabla\mathcal{U}(\mathfrak{x})\frac{\partial}{\partial \mathfrak{x}}v(\mathfrak{x}, v, t), \tag{5.14}$$

and its stationary solution is

$$v_0(\mathfrak{x}) = C\exp\left(-\frac{1}{\sigma^2}\mathcal{U}(\mathfrak{x})\right), \tag{5.15}$$

where normalization constant C has the same meaning as in Equation 5.12.

Brownian motion with high friction often called overdamped motion, and Equations 5.13 and 5.14 are referred as equations of "life at low Reynolds number" and are considered as a model of bacterial movements (Berg, 1993; Romanczuk et al., 2012).

5.3.1.3 Active Brownian Motion

The earlier considered Langevin and Fokker–Planck equations with constant coefficients govern the "passive" Brownian motion, which imply that the particles have no energy supply. In opposite, in the active Brownian motion, it is assumed that the particles are able to obtain and accumulate energy and transform it into movements, which are not governed directly by the environmental states. Formally, such mechanism of energy support implies that the dissipative force $F^{\text{diss}}(t)$, which is included into the overall forces $F(t)$ applied to the particle at time t, depends on the particle position $\mathfrak{x}(t)$ and velocity $v(t)$ and can change its sign and value with respect to $\mathfrak{x}(t)$ and $v(t)$.

The dependence of the dissipative force F^{diss} on the position and velocity of the particle is represented by the use of varying friction coefficient $\gamma(\mathfrak{x}, v) \in \mathbb{R}$, and the force F^{diss} is defined as follows:

$$F^{\text{diss}}(t) = -\gamma(\mathfrak{x}, v)v(t).$$

In other words, it is assumed that the friction in the medium, in which the particle moves, is nonlinear and can both dissipate the particle's energy and activate the particle with additional energy support. Such a behavior is inherent to living organisms and artificial mobile agents; that supports the usage of active Brownian motion for description of biological and robotic systems (Schweitzer, 2003).

The Langevin equation, which governs the motion of active Brownian particle, is the following (Condat & Sibona, 2002; Klimontovich, 1994; Romanczuk et al., 2012; Schimansky-Geier, Ebeling, & Erdmann, 2005; Schweitzer, 2003):

$$\frac{d}{dt}\mathfrak{x}(t) = v(t), \quad \frac{d}{dt}v(t) = -\frac{1}{m}\gamma(\mathfrak{x}, v)v(t) - \frac{1}{m}\nabla\mathcal{U}(\mathfrak{x}) + \sqrt{2\gamma_0 k_B \mathcal{T}}\mathcal{G}_t. \tag{5.16}$$

The corresponding Fokker–Planck equation for the distribution of the probability $v(\mathfrak{x},v,t)$ is

$$\frac{\partial}{\partial t}v(\mathfrak{x},v,t)=\frac{1}{m}\frac{\partial}{\partial v}\left(\gamma_0^2\sigma^2\frac{\partial}{\partial v}v(\mathfrak{x},v,t)+\gamma(\mathfrak{x},v)v(t)v(\mathfrak{x},v,t)\right)-v(t)\frac{\partial}{\partial \mathfrak{x}}v(\mathfrak{x},v,t)-\frac{1}{m}\nabla\mathcal{U}(\mathfrak{x})\frac{\partial}{\partial v}v(\mathfrak{x},v,t), \quad (5.17)$$

where, as mentioned earlier, $\sigma^2=\dfrac{k_B\mathcal{T}}{\gamma_0}$. It is clear that if $\gamma(\mathfrak{x},v)\equiv\gamma_0$, then this equation is equivalent to Equation 5.11.

Equations 5.16 and 5.17 provide a general description of the motion of Brownian particle in the medium with nonlinear friction γ or, that is the same, the motion of active Brownian particle, which is able both to dissipate energy and to obtain energy from the environment. An energy E averaged over the Gaussian white noise $\mathcal{G}_t$ is (Romanczuk et al., 2012):

$$\frac{d}{dt}E=-\frac{1}{m}\gamma(\mathfrak{x},v)v^2(t)+\gamma_0 k_B\mathcal{T}.$$

From this equation, it follows that negative friction leads to increasing an energy of the particle, while positive friction results in dissipating the energy.

There are several standard models, which specify function γ and lead to different types of the active Brownian particle behavior. The most popular are quadratic Rayleigh–Helmholtz model and linear Scheinbein–Gruler model (Romanczuk et al., 2012).

In the Rayleigh–Helmholtz model, friction is defined by the following quadratic function (see, e.g., Ebeling, 2007; Erdmann et al., 2000; Romanczuk et al., 2012; Schimansky-Geier, Ebeling, & Erdmann, 2005):

$$\gamma(\mathfrak{x},v)=-\alpha+\beta v^2=\alpha\left(\frac{|v|^2}{|v_0|^2}-1\right)=\beta\left(|v|^2-|v_0|^2\right), \quad (5.18)$$

where

$\alpha,\beta>0$

v_0 is a critical velocity such that $|v_0|^2=\dfrac{\alpha}{\beta}$ for which the friction is zero. The corresponding stationary distribution of the velocities is

$$v_0(v)=C\exp\left(\frac{m}{\gamma_0\sigma^2}\left(\frac{1}{2}\alpha v^2-\frac{1}{4}\beta v^4\right)\right), \quad (5.19)$$

where C is a normalization coefficient, which provides $\int v_0(v)\,dv=1$ over $\mathfrak{X}$.

The Scheinbein–Gruler model, in contrast, implements the friction specified by the linear function (Romanczuk et al., 2012)

$$\gamma(\mathfrak{x},v)=-\gamma_0\frac{|v|-|v_0|}{|v|}=-\gamma_0\left(1-\frac{|v_0|}{|v|}\right), \quad (5.20)$$

where

$\gamma_0>0$

$|v_0|$, as mentioned above, is critical velocity

The stationary distribution of velocities in this case is

$$v_0(v) = C \exp\left(-\frac{m}{2\sigma^2}(v - v_0)^2\right), \tag{5.21}$$

where C, as mentioned above, is a normalization coefficient.

For a detailed consideration of these models and of the velocity-dependent friction in general, see the papers by Klimantovich (1994) and by Ebeling and his collaborators (Ebeling, 2007; Erdmann et al., 2000; Romanczuk et al., 2012; Schimansky-Geier, Ebeling, & Erdmann, 2005).

In the framework of search and foraging, the nonlinear friction can be interpreted as follows. Notice that in both indicated models of velocity-dependent friction, at low velocities $|v|^2 < |v_0|^2$, the friction is negative, and the particle gains energy from the environment. In contrast, if the particle moves with high velocity $|v|^2 > |v_0|^2$, then the friction coefficient is positive and the energy decreases. Such activity is similar to the activity prescribed by the intermittent strategies of search considered in Section 4.3.3. In fact, the movement with low velocity $|v|^2 < |v_0|^2$ and gaining energy from the environment is closely related to the slow motion with screening (Phase 1 of intermittent search), during which the agent screens its close environment and obtains information whether the target is in this environment or not. In the case of search using the target's location probabilities, this phase results in zeroing the probabilities in the regions, which were observed by the agent and in which the target was not detected. Similarly, the movement with high velocity $|v|^2 > |v_0|^2$ with dissipation corresponds to the fast relocation of the agent without screening (Phase 2 of intermittent search). However, notice that in the intermittent search as well as in the Lévy flight search, fast relocation is conducted with constant speed, while in the active Brownian motion, this speed can be changed with respect to the friction.

The indicated models of nonlinear friction describe the mechanisms of obtaining energy from the environment and transforming it into self-propelling movements. Now let us consider a widely accepted model of accumulating energy in the internal depot and the friction function, which follows from this model (Romanczuk et al., 2012; Schweitzer, 2000, 2003; Schweitzer, Ebeling, & Tilch, 1998).

Denote by $r_e(\mathfrak{x}) \in \mathbb{R}^+$ the nonnegative space-dependent energy intake rate or, in the terms of the models of foraging considered in Section 3.3, the reward rate that characterizes the energy taken by the agent from the environment in the point $\mathfrak{x} \in \mathfrak{X}$ and by $c_e(v) \in \mathbb{R}^+$ the nonnegative rate of conversion of internal energy into motion with respect to the agent's velocity v. In addition, denote by $\gamma_e \geq 0$ the rate of internal dissipation or energy lost. Then, the energy balance equation of the internal energy depot is specified as follows (Romanczuk et al., 2012; Schweitzer, 2003):

$$\frac{d}{dt}e(t) = r_e(\mathfrak{x}) - \gamma_e e(t) - c_e(v)e(t)$$

In other words, the overall energy $e(t)$ in the depot is supplied by the energy taken from the environment with the rate $r_e(\mathfrak{x})$, which then is dissipated with the rate γ_e and transformed into motion with the rate $c_e(v)$.

Usually (Schweitzer, 2000, 2003), the energy conversion rate $c_e(v)$ is defined as $c_e(v) = c_0 v^2$, and it is assumed that the energy intake is constant $r_e(\mathfrak{x}) \equiv r_0$, where both $c_0 > 0$ and $r_0 > 0$. Then, for constant friction γ_0 and stationary energy depot with $\frac{d}{dt}e(t) = 0$, the dissipative force F^{diss} is defined the adiabatic elimination as follows

$$F^{\text{diss}}(t) = -\left(\gamma_0 - \frac{r_0 c_0}{\gamma_e + c_0 |v|^2}\right) v(t);$$

This dissipative force results in the following Langevin equation (cf. Equation 5.10):

$$\frac{d}{dt}\mathfrak{x}(t)=v(t),\quad \frac{d}{dt}v(t)=\frac{1}{m}\left(\frac{r_0c_0}{\gamma_e+c_0|v|^2}-\gamma_0\right)v(t)-\frac{1}{m}\nabla\mathcal{U}(\mathfrak{x})+\sqrt{2\gamma_0k_B\mathcal{T}}\mathcal{G}_t, \tag{5.22}$$

where the term $\frac{1}{m}\left(\frac{r_0c_0}{\gamma_e+c_0|v|^2}-\gamma_0\right)v(t)$ specifies the energy intake and dissipation with respect to the velocity-dependent friction (cf. Equations 5.10 and 5.16)

$$\gamma(v)=\gamma_0-\frac{r_0c_0}{\gamma_e+c_0|v|^2}. \tag{5.23}$$

Dependence of the value $\gamma(v)$ on the velocity v and the corresponding probability distribution defined by the Fokker–Planck equation (5.17) with $\gamma(\mathfrak{x},v)\equiv\gamma(v)$ are illustrated in Figure 5.17. The parameters used for definition of friction $\gamma(v)$ are $\gamma_0=0.5$, $r_0=1$, $c_0=1$, and $\gamma_e=1$, and Figure 5.17a

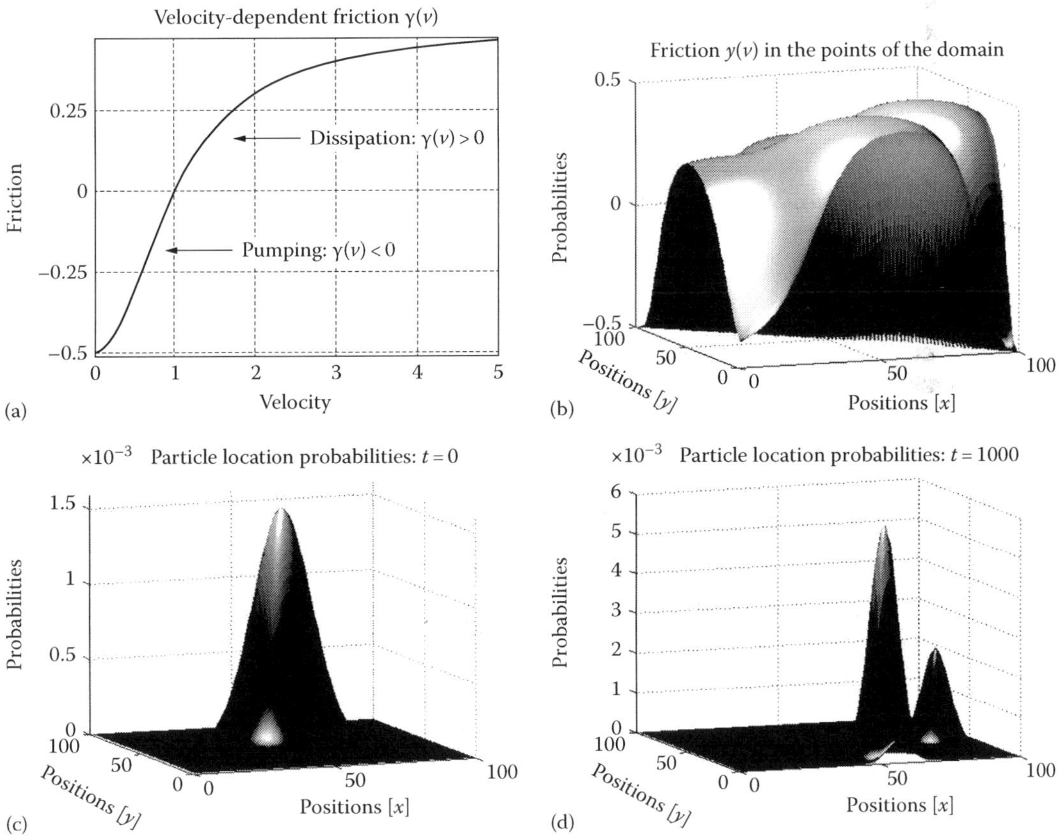

FIGURE 5.17 Example of the dependence between friction on the velocity and distribution of the particle's location probabilities with respect to the nonlinear friction. Figure (a) shows the dependence of friction $\gamma(v)$ on v, $|v| \in [0, 5]$ with the parameters $\gamma_0 = 0.5$, $r_0 = 1$, $c_0 = 1$, and $\gamma_e = 1$. Figure (c) shows initial distribution of the particle's location probabilities at $t = 0$, and figure (d) the distribution of the location probabilities at $t = 1000$ with respect to the distribution of friction shown in figure (b). As mentioned earlier, it is assumed that the mass of the particle is $m = 1$ and the diffusion rate is $\sigma^2 = 0.25$.

shows the relation between $\gamma(v)$ and v such that $|v| \in [0, 5]$. In the example, it is assumed that in each point $\mathfrak{x} \in \mathcal{X}$, the value $|v(\mathfrak{x})|$ is proportional to the gradient of potential field $\nabla\mathcal{U}(\mathfrak{x})$. The corresponding values $\gamma(v(\mathfrak{x}))$ for the previously defined potential field (see Figure 5.16a) are shown in Figure 5.17b. As mentioned earlier, it is assumed that the particle's mass is $m = 1$ and diffusion rate is $\sigma^2 = 0.25$. As mentioned previously, initial probability distribution of the particle's location at $t = 0$ is defined by the binormal distribution with the central point $\mathfrak{x}(0) = (50, 50)$ and variances over the axes $\sigma_1 = \sigma_2 = 10$; this distribution is shown in Figure 5.17c. The resulting probability distribution at $t = 1000$ defined by the Fokker–Planck equation (5.17) with respect to the friction $\gamma(v)$ shown in Figure 5.17b is presented in Figure 5.17d.

In contrast to the previously considered distribution with constant friction (see Figure 5.16d), where the particle is led by the potential field, in the case of nonlinear friction, the particle leaves the regions with negative friction, where it obtains the energy from the environment, and moves slowly in the regions with positive friction. As it is expected, the obtained distribution (see Figure 5.17d) demonstrates that the probability of the particle's location is higher the larger the friction in the corresponding points, and with time, maximal probabilities are obtained in the points with maximal friction.

It is clear that the model active Brownian particle with energy depot is closely related with the considered methods of optimal foraging in Section 3.3, which implement the gain or reward rate functions that allow an application of the optimization techniques used in foraging algorithms to specification of the parameters of active Brownian motion. However, notice that in addition to energy rate used in the foraging algorithms, the model of energy depot includes both pumping and dissipation modes that provide more flexible mechanism of control of the agent motion.

The earlier considered models describe the motion of a single active Brownian particle in the medium with nonlinear friction such that the particle is allowed both to obtain the energy from the environment and to dissipate it during its motion. In the next section, these models are extended to the motion of several particles considered as a multiagent system.

5.3.2 Multiagent Systems and Canonical Active Brownian Motion

In the model of active Brownian motion, the dynamics of a single particle was defined by direct implementation of energy balance equations. However, for consideration of multiagent system, it is more convenient to apply classical mechanics approach and to obtain the dynamical equations from the Hamilton equations specifying the behavior of each particle in the system and interactions between them (Glück, Hüffel, & Ilijić, 2009; Schweitzer, 2000, 2003; Schweitzer, Ebeling, & Tilch, 2001). This section presents the model of swarm dynamics, including attraction and repulsion of the agents, using the Langevin equation, and then provides the Hamilton equations for the swarm.

Since in the succeeding discourse the mobile entities are considered as mobile agents rather than as particles, the notation of the agent's coordinates, velocities, location probabilities and so on follows the manner used in previous parts of the book such that it stresses the difference between the search agent and the target. In addition, it is assumed that the mass of the agents is $m = 1$ and that they are equipped with appropriate sensors and information processing units, which provide the required perception and operations.

Assume that the agents are acting in group of m units, and denote by $\breve{\mathfrak{x}}_j(t)$ the location of the jth agent at time t and by $\breve{v}_j(t)$ its velocity as this time. Following general dynamical systems theory (see, e.g., Opoytzev, 1977), the behavior of such group is defined by the following system of Langevin equations (Romanczuk et al., 2012):

$$\frac{d}{dt}\breve{\mathfrak{x}}_j(t) = \breve{v}_j(t), \quad \frac{d}{dt}\breve{v}_j(t) = F_j\left(\breve{\mathfrak{x}}_1(t), \ldots, \breve{\mathfrak{x}}_m(t), \breve{v}_1(t), \ldots, \breve{v}_m(t), t\right), \quad j = 1, 2, \ldots, m, \tag{5.24}$$

where $F_j\left(\breve{\mathfrak{x}}_1(t), \ldots, \breve{\mathfrak{x}}_m(t), \breve{v}_1(t), \ldots, \breve{v}_m(t), t\right)$ is a sum of the forces applied to the jth agent at time t (cf. Equation 5.7).

Following the model of active Brownian motion (see Section 5.3.1), usually it is assumed that the force F_j consists of the velocity-dependent dissipative force, location-dependent force resulted by potential field $\mathcal{U}$, and stochastic force resulted by the influence of Gaussian white noise. Namely,

$$F_j\left(\breve{x}_1(t),\ldots,\breve{x}_m(t),\breve{v}_1(t),\ldots,\breve{v}_m(t),t\right)=F_j^{\text{diss}}\left(\breve{v}_1(t),\ldots,\breve{v}_m(t)\right)+F_j^{\text{poten}}\left(\breve{x}_1(t),\ldots,\breve{x}_m(t)\right)+F_j^{\text{stoch}}(t),$$

for each $j = 1, 2,\ldots, m$. It is clear that if $F_j^{\text{stoch}} \equiv 0$, system (5.24) defines a well-known deterministic multiagent system (Opoytzev, 1977).

Similar to the dynamical systems theory, the forces F_j^{diss} and F_j^{poten}, $j = 1, 2,\ldots, m$, can be specified in different ways with respect to the considered task and desired behavior of the resulting system. In particular, in the models of swarm dynamics, these forces, in addition to the individual dynamics and environmental influences, include the influence of the neighboring agents that provide the activity of the agents as a united group or swarm. In the succeeding text, the forces are defined in coincidence with the indicated Reynolds swarming rules in Section 5.1 (Reynolds, 1987); for additional definitions, see the book by Schweitzer (2003) and the review by Romanczuk et al. (2012) and references therein.

At first, let us consider the alignment of the agents, that is, the direction and velocity matching (the Reynolds rule 2 in Section 5.1). Usually, such alignment is provided by the velocity-dependent dissipation such that

$$F_i^{\text{diss}}\left(\breve{v}_1(t),\ldots,\breve{v}_m(t)\right)=-\gamma\left(\breve{v}_i\right)\breve{v}_i(t)-\sum_{j=1}^{m}k^{\text{diss}}\left(\breve{x}_i(t),\breve{x}_j(t)\right)\left(\breve{v}_i(t)-\breve{v}_j(t)\right), \tag{5.25}$$

where $\gamma\left(\breve{v}_i\right)$ is the earlier considered velocity-dependent friction for the ith agent and $k^{\text{diss}}\left(\breve{x}_i(t),\breve{x}_j(t)\right)$ is the value, which defines the strength of interaction between the ith and jth agents, $i,j = 1, 2,\ldots, m$, and has the same meaning as relativeness used in the definition of bio-signaling by the Sir Philip Sidney game (see Section 5.2.1). In the case of distance-based interactions, the relativeness is defined with respect to the distance $r\left(\breve{x}_i,\breve{x}_j\right)=\left\|\breve{x}_i(t)-\breve{x}_j(t)\right\|$ between the agents such that the values $k^{\text{diss}}\left(\breve{x}_i(t),\breve{x}_j(t)\right)$ decrease with increase of the distances $r\left(\breve{x}_i,\breve{x}_j\right)$ (Romanczuk et al., 2012).

In the models of motion of artificial swarms with common maps and shared information regarding the coordinates and velocities of all m agents, the velocity-based alignment is defined with respect to the velocity $\langle\breve{v}\rangle$ of the center mass of swarm, that is, the average velocity of all the agents at time t:

$$\left\langle\breve{v}(t)\right\rangle=\frac{1}{m}\sum_{j=1}^{m}\breve{v}_j(t). \tag{5.26}$$

Then for distance-based interactions (Romanczuk et al., 2012),

$$F_i^{\text{diss}}\left(\breve{v}_1(t),\ldots,\breve{v}_m(t)\right)=-\gamma\left(\breve{v}_i\right)\breve{v}_i(t)-k^{\text{diss}}\left(\breve{x}_i(t),\left\langle\breve{x}(t)\right\rangle\right)\left(\breve{v}_i(t)-\left\langle\breve{v}(t)\right\rangle\right), \tag{5.27}$$

where

$$\left\langle\breve{x}(t)\right\rangle=\frac{1}{m}\sum_{j=1}^{m}\breve{x}_j(t) \tag{5.28}$$

is a location of the center mass of the swarm. However, notice that for natural agents and for the agents with restricted information exchange, the usage of the swarm's center mass and its velocity is problematic or even impossible.

Now, consider the cohesion rule that prescribes to the agents to stay close to the neighbors (the Reynolds rule 3 in Section 5.1). For each agent j, $j = 1, 2,\ldots, m$, this rule is implemented using the force F_j^{poten}, which combines two potential fields: the earlier indicated external potential field is defined by the function $\mathcal{U}:\mathfrak{X}\to\mathbb{R}$ and the pairwise interactions potential field by the function $\mathcal{V}:\mathfrak{X}\times\mathfrak{X}\to\mathbb{R}$. Then for each $i = 1, 2,\ldots, m$, it follows that

$$F_i^{\text{poten}}\left(\breve{\mathfrak{x}}_1(t),\ldots,\breve{\mathfrak{x}}_m(t)\right)=-\nabla\mathcal{U}\left(\breve{\mathfrak{x}}_i(t)\right)-\sum_{j=1}^{m}k^{\text{poten}}\left(\breve{\mathfrak{x}}_i(t),\breve{\mathfrak{x}}_j(t)\right)\nabla\mathcal{V}\left(\breve{\mathfrak{x}}_i(t),\breve{\mathfrak{x}}_j(t)\right), \tag{5.29}$$

where $k^{\text{poten}}\left(\breve{\mathfrak{x}}_i(t),\breve{\mathfrak{x}}_j(t)\right)$ is also the value that, as the value $k^{\text{diss}}\left(\breve{\mathfrak{x}}_i(t),\breve{\mathfrak{x}}_j(t)\right)$, defines the strength of integration between the ith and jth agents, but the form of the function k^{poten} can be strongly different than the form of the function k^{diss}.

For the distance-based interactions, the field $\mathcal{V}$ is usually defined as follows (Romanczuk et al., 2012):

$$\mathcal{V}\left(\breve{\mathfrak{x}}_i(t),\breve{\mathfrak{x}}_j(t)\right)=\frac{1}{2}\left(r\left(\breve{\mathfrak{x}}_i(t),\breve{\mathfrak{x}}_j(t)\right)\right)^2, \tag{5.30}$$

where, as mentioned earlier, $r\left(\breve{\mathfrak{x}}_i,\breve{\mathfrak{x}}_j\right)$ is the distance between ith and jth agents, $i, j = 1, 2,\ldots, m$. Then for each i and j, one obtains

$$\nabla\mathcal{V}\left(\breve{\mathfrak{x}}_i(t),\breve{\mathfrak{x}}_j(t)\right)=r\left(\breve{\mathfrak{x}}_i(t),\breve{\mathfrak{x}}_j(t)\right),$$

and the force F_i^{poten} influencing on the ith agent is

$$F_i^{\text{poten}}\left(\breve{\mathfrak{x}}_1(t),\ldots,\breve{\mathfrak{x}}_m(t)\right)=-\nabla\mathcal{U}\left(\breve{\mathfrak{x}}_i(t)\right)-\sum_{j=1}^{m}k^{\text{poten}}\left(\breve{\mathfrak{x}}_i(t),\breve{\mathfrak{x}}_j(t)\right)r\left(\breve{\mathfrak{x}}_i(t),\breve{\mathfrak{x}}_j(t)\right). \tag{5.31}$$

Notice that for the constant relativeness $k^{\text{poten}}\left(\breve{\mathfrak{x}}_i(t),\breve{\mathfrak{x}}_j(t)\right)=\dfrac{\alpha}{m}=\text{const}$, the second term in Equation 5.31 is read as

$$\sum_{j=1}^{m}k^{\text{poten}}\left(\breve{\mathfrak{x}}_i(t),\breve{\mathfrak{x}}_j(t)\right)r\left(\breve{\mathfrak{x}}_i(t),\breve{\mathfrak{x}}_j(t)\right)=\frac{\alpha}{m}\sum_{j=1}^{m}r\left(\breve{\mathfrak{x}}_i(t),\breve{\mathfrak{x}}_j(t)\right)=\alpha r\left(\breve{\mathfrak{x}}_i(t),\breve{\mathfrak{x}}(t)\right).$$

In other words, the force F_i^{poten} influencing on the ith agent depends on the distance between the agent's location $\breve{\mathfrak{x}}_i(t)$ and the location $\left\langle\breve{\mathfrak{x}}(t)\right\rangle$ of the center mass of the swarm. Since during their motion in the constant external potential field $\mathcal{U}$ such swarms save the form of a circle, they are usually referred as harmonic swarms (Schweitzer, 2003). However, notice again that since harmonic swarm requires the knowledge of the location of the swarm's center mass, it can be implemented only for the agents that act in common maps and at each time share their coordinates to the members of the swarm.

Finally, let us consider the separation rule, which provides collision avoidance (the Reynolds rule 1 in Section 5.1). In the presented model, this rule can be defined by two ways:

1. By the changes of the external field $\mathcal{U}$ by the agents such that for all points $\mathfrak{x}\in\mathfrak{n}\left(\breve{\mathfrak{x}}_j(t)\right)$ in the close neighborhood $\mathfrak{n}\left(\breve{\mathfrak{x}}_j(t)\right)\subset\mathfrak{X}$ of each jth agent, $j = 1, 2,\ldots, m$, the gradient of the field $\mathcal{U}$ is $\nabla\mathcal{U}\left(\mathfrak{x}\right)<0$.
2. By the use of the function k^{poten} such that for all neighboring points $\mathfrak{x}\in\mathfrak{n}\left(\breve{\mathfrak{x}}_j(t)\right)$ its value is $k^{\text{poten}}\left(\breve{\mathfrak{x}}_j(t),\mathfrak{x}\right)<0$.

In the first case, the collision avoidance techniques follow the direction of obstacle avoidance (see Section 4.4.3), and in the second case, local repulsion of the agents is provided by the pairwise interactions. Certainly, these two methods can be applied simultaneously, and this possibility is utilized in the tasks of swarm foraging.

Notice that in the case of distance-based interactions, the force F_i^{poten} is often defined using direct specification of the attraction and repulsion of the agents. In such a case, instead of pairwise interaction field, the generalized Morse potential is applied. Namely, instead of the field $\mathcal{V}$ defined by Equation 5.30, it is assumed that (Romanczuk et al., 2012)

$$\mathcal{V}_i\left(\breve{\mathfrak{x}}_1(t),\ldots,\breve{\mathfrak{x}}_m(t)\right)=\sum_{j=1,j\neq i}^{m}\left(\alpha_{\text{r}}\exp\left[-\frac{1}{\beta_{\text{r}}}r\left(\breve{\mathfrak{x}}_i(t),\breve{\mathfrak{x}}_j(t)\right)\right]-\alpha_{\text{a}}\exp\left[-\frac{1}{\beta_{\text{a}}}r\left(\breve{\mathfrak{x}}_i(t),\breve{\mathfrak{x}}_j(t)\right)\right]\right), \quad (5.32)$$

where

$\alpha_{\text{r}} > 0$ and $\beta_{\text{r}} > 0$ are the amplitude and the range of repulsion, respectively
$\alpha_{\text{a}} > 0$ and $\beta_{\text{a}} > 0$ are the amplitude and the range of attraction

Then, instead of Equation 5.29, the force F_i^{poten}, $i = 1, 2,\ldots, m$, is defined as follows:

$$F_i^{\text{poten}}\left(\breve{\mathfrak{x}}_1(t),\ldots,\breve{\mathfrak{x}}_m(t)\right)=-\nabla\mathcal{U}\left(\breve{\mathfrak{x}}_i(t)\right)-\nabla\mathcal{V}_i\left(\breve{\mathfrak{x}}_1(t),\ldots,\breve{\mathfrak{x}}_m(t)\right). \quad (5.33)$$

Such definitions of the field $\mathcal{V}_i$ by Equation 5.32 and of the force F_i^{poten} by Equation 5.33 are inspired by the properties of the animal foraging, and, in contrast to the definitions based on the center mass of the swarm, does not require the shared maps or any other common knowledge of the agents. In the last time, similar definitions of attraction and repulsion of the agents are implemented in the swarm robotics systems (Hamann, 2010) and are used in general considerations of swarm dynamics (Gazi & Passino, 2011). The activity of the swarm with distance-based interactions according to the presented definitions of the forces F_i^{diss} and F_i^{poten}, $i = 1, 2, \ldots, m$, is illustrated by the following example; for additional considerations and examples, especially of harmonic swarms and the swarms with velocity alignment, see, for example, books by Schweitzer (2003) and by Romanczuk et al. (2012).

Example 5.7

Similar to previous examples, assume that the agents act in the square domain $\mathfrak{X}$ of the size $n = 100 \times 100$ with the impermeable bounds with the earlier defined potential field $\mathcal{U}$ (see Figure 5.16a). The motion of each agent i, $i = 1, 2,\ldots, m$, is governed by the Langevin equation (5.24) with the dissipative force F_i^{diss} defined by Equation 5.25 and velocity-dependent friction $\gamma(\breve{v}_i)$ defined by Equation 5.23 as it is prescribed by the energy depot model.

Following the distance-based interactions model, the relativeness $k^{\text{diss}}\left(\breve{\mathfrak{x}}_i(t),\breve{\mathfrak{x}}_j(t)\right)$ between the agents i and j is defined with respect to the distance $r\left(\breve{\mathfrak{x}}_i,\breve{\mathfrak{x}}_j\right)$ between the agents. In this example, it is assumed that the relativeness exponentially decreases with the distance, that is,

$$k^{\text{diss}}\left(\breve{\mathfrak{x}}_i(t),\breve{\mathfrak{x}}_j(t)\right)=k_0^{\text{diss}}\exp\left(-r\left(\breve{\mathfrak{x}}_i,\breve{\mathfrak{x}}_j\right)\right), \quad i\neq j, \quad i,j=1,2,\ldots,m.$$

Such definition follows general assumptions regarding animals' behavior; for simulations results on the movements of the modeled ants (see Section 5.2.1, especially—Example 5.2) with similar relativeness, see the work by Kagan et al. (2014).

In the example, the force F_i^{poten} is defined in two different forms with respect to the interactions field $\mathcal{V}$. The first follows the definition of the field $\mathcal{V}$ by Equation 5.30 and uses constant relativeness $k^{\text{poten}}\left(\breve{\mathfrak{x}}_i(t),\breve{\mathfrak{x}}_j(t)\right)=\alpha/m$, and the second implements the field specified by Equation 5.32.

The final Langevin equations for the velocities

$$\frac{d}{dt}\breve{\mathfrak{x}}_i(t)=\breve{v}_i(t), \quad i=1,2,\ldots,m,$$

are written as follows:
Case 1:

$$\frac{d}{dt}\breve{v}_i(t) = -\left(\gamma_0 - \frac{r_0 c_0}{\gamma_e + c_0 |\breve{v}_i|^2}\right)\breve{v}_i(t) - k_0^{\text{diss}} \sum_{j=1, j\neq i}^{m} \exp\left(-r\left(\breve{\mathfrak{x}}_i, \breve{\mathfrak{x}}_j\right)\right)\left(\breve{v}_i(t) - \breve{v}_j(t)\right)x$$

$$-\nabla\mathcal{U}\left(\breve{\mathfrak{x}}_i(t)\right) - \frac{\alpha}{m}\sum_{j=1}^{m} r\left(\breve{\mathfrak{x}}_i(t), \breve{\mathfrak{x}}_j(t)\right) + \sqrt{2\sigma^2}\mathcal{G}_{it}. \tag{5.34}$$

Case 2:

$$\frac{d}{dt}\breve{v}_i(t) = -\left(\gamma_0 - \frac{r_0 c_0}{\gamma_e + c_0 |\breve{v}_i|^2}\right)\breve{v}_i(t) - k_0^{\text{diss}} \sum_{j=1, j\neq i}^{m} \exp\left(-r\left(\breve{\mathfrak{x}}_i, \breve{\mathfrak{x}}_j\right)\right)\left(\breve{v}_i(t) - \breve{v}_j(t)\right)$$

$$-\nabla\mathcal{U}\left(\breve{\mathfrak{x}}_i(t)\right) - \nabla\sum_{j=1, j\neq i}^{m}\left(\alpha_r \exp\left[-\frac{1}{\beta_r} r\left(\breve{\mathfrak{x}}_i(t), \breve{\mathfrak{x}}_j(t)\right)\right] - \alpha_a \exp\left[-\frac{1}{\beta_a} r\left(\breve{\mathfrak{x}}_i(t), \breve{\mathfrak{x}}_j(t)\right)\right]\right) + \sqrt{2\sigma^2}\mathcal{G}_{it}. \tag{5.35}$$

The constants used in the equations in both cases are the following. Similar to the previous illustration (see Figure 5.17) the parameters of the energy depot are $\gamma_0 = 0.5$, $r_0 = 1$, $c_0 = 1$, and $\gamma_e = 1$. The dissipation coefficient is $k_0^{\text{diss}} = 1$. The parameters of interactions potential are $\alpha = 0.05$, $\alpha_r = 1$, $\beta_r = 1$, $\alpha_a = 100$, and $\beta_a = 1000$. Finally, the strength of the noise is $\sigma^2 = 0.5$. The agents number $m = 25$; initial locations of the agents are shown in Figure 5.18a.

The movements of the agents were simulated for the external potential field $\mathcal{U}$ shown in Figure 5.16a, and for constant external potential field such that $\nabla\mathcal{U}(\mathfrak{x}) = 0$ for all $\mathfrak{x} \in \mathfrak{X}$. An example of the trajectories of the agents, which are governed by the Langevin equation (5.35) with nonconstant external potential field, is shown in Figure 5.18b. For constant external potential field and the Langevin equation (5.34), an example of the trajectories of the agents is shown in Figure 5.18c and for the Langevin equation (5.35) in Figure 5.18d. In all cases, the time is $t = 0, 1, \ldots, 100$.

It is seen that while the agents act in external potential field, they tend to follow the field values (cf. distribution shown in Figure 5.16d), and for given parameters, the interactions' potential has minor influence. If external potential field is constant, then the trajectories of the agents mostly depend on the form of interactions potential. In Case 1 without repulsion, the agents are attracted by each other and tend to concentrate in certain region, while in Case 2 with both repulsion and attraction, the agents tend to move by small groups. Notice that in both cases, the interactions' potential is relatively small, and most of the time, the agents demonstrate simple Brownian motion before reaching the regions of repulsion or attraction. ■

Finally, let us consider the dynamics of the swarm using canonical Hamiltonian formalism (Glück, Hüffel, & Ilijić, 2009; Schweitzer, 2003; Schweitzer et al., 2001). In general, the dynamics of a mechanical system of m particles $j = 1, 2, \ldots, m$ is defined as follows:

$$\frac{d}{dt}\mathfrak{x}_j(t) = \frac{\partial}{\partial p_j} H\left(\mathfrak{x}_1, \mathfrak{x}_2, \ldots, \mathfrak{x}_m, p_1, p_2, \ldots, p_m, t\right),$$

$$\frac{d}{dt}p_j(t) = -\frac{\partial}{\partial \mathfrak{x}_j} H\left(\mathfrak{x}_1, \mathfrak{x}_2, \ldots, \mathfrak{x}_m, p_1, p_2, \ldots, p_m, t\right),$$

where

$\mathfrak{x}_j$ is a position of the jth particle
$p_j = m_j v_j$ is its impulse
H is the Hamilton function, which specifies the total energy of the system

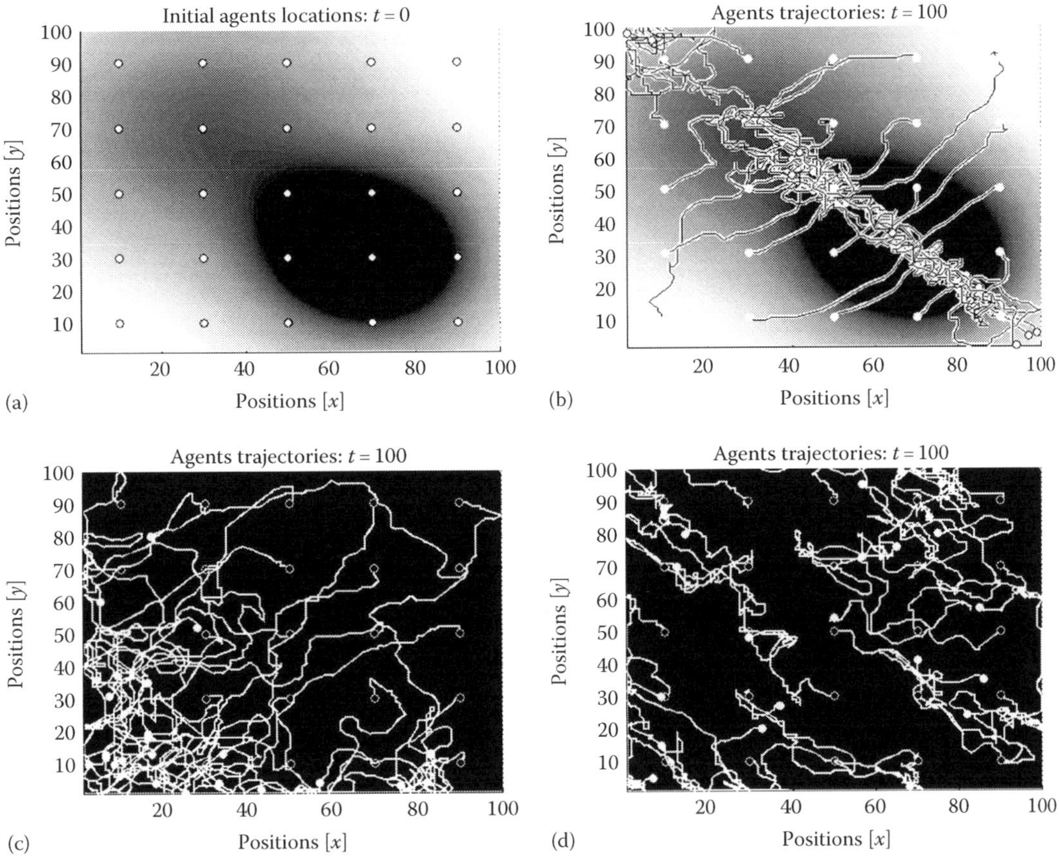

FIGURE 5.18 Examples of the trajectories of 25 active Brownian agents starting from the initial locations shown in Figure (a). figure (b) depicts the trajectories of the agents acting in external potential field shown in Figure 5.16a and governed by the Langevin equation (5.35). Trajectories of the agents, which act in constant external potential and are governed by the Langevin equation (5.34), are shown in figure (c) and trajectories of the agents, which are governed by the Langevin equation (5.35), are shown in figure (d). Parameters of the equations are $\gamma_0 = 0.5$, $r_0 = 1$, $c_0 = 1$, $\gamma_e = 1$, $k_0^{\text{diss}}=1$, $\alpha = 0.05$, $\alpha_r = 1$, $\beta_r = 1$, $\alpha_a = 100$, and $\beta_a = 1000$, and the time is $t = 0, 1, \ldots, 100$.

In the case of the system, in which the total energy is not constant, the second equation is written as follows (Schweitzer et al., 2001):

$$\frac{d}{dt} p_j(t) = -\frac{\partial H}{\partial \mathfrak{x}_j} - g\left(H\left(\mathfrak{x}_1, \mathfrak{x}_2, \ldots, \mathfrak{x}_m, p_1, p_2, \ldots, p_m, t\right)\right)\frac{\partial H}{\partial p_j},$$

where $g : \mathbb{R} \to \mathbb{R}$ is a function, which defines the changes of the total energy. Notice that in the dissipative systems, the function g is nondecreasing that defines the energy dropping off.

In the considered case of the multiagent system such that $m_j = 1$ for each j, it holds true that $p_j = m_j v_j = \breve{v}_j$, $j = 1, 2, \ldots, m$, and the Hamilton system obtains the following form:

$$\frac{d}{dt}\breve{\mathfrak{x}}_j(t) = \frac{\partial H}{\partial \breve{v}_j}, \quad \frac{d}{dt}\breve{v}_j(t) = -\frac{\partial H}{\partial \breve{\mathfrak{x}}_j} - g\left(H\left(\breve{\mathfrak{x}}_1, \breve{\mathfrak{x}}_2, \ldots, \breve{\mathfrak{x}}_m, \breve{v}_1, \breve{v}_2, \ldots, \breve{v}_m, t\right)\right)\frac{\partial H}{\partial \breve{v}_j}.$$

If, in addition, the motion of each agent is influenced by Gaussian white noise, then the system includes a stochastic term, and finally, it is written in the form of the Langevin equation (Schweitzer, 2003):

$$\frac{d}{dt}\breve{x}_j(t)=\frac{\partial H}{\partial \breve{v}_j},\quad \frac{d}{dt}\breve{v}_j(t)=-\frac{\partial H}{\partial \breve{x}_j}-g\left(H\left(\breve{x}_1,\breve{x}_2,\ldots,\breve{x}_m,\breve{v}_1,\breve{v}_2,\ldots,\breve{v}_m,t\right)\right)\frac{\partial H}{\partial \breve{v}_j}+\sqrt{2\sigma^2(H)}\mathcal{G}_{jt}. \tag{5.36}$$

Similar to the Langevin equations (5.10 and 5.22), which define the motion of a single agent, and Equations 5.24, 5.25, and 5.29, which define the dynamics of the group of m agents, in Equation 5.36 the term $F_j^{\text{poten}}=-\frac{\partial H}{\partial \breve{x}_j}$ determines the influence of the potential field, and the term $F_j^{\text{diss}}(t)=-g\left(H\left(\breve{x}_1,\breve{x}_2,\ldots,\breve{x}_m,\breve{v}_1,\breve{v}_2,\ldots,\breve{v}_m,t\right)\right)\frac{\partial H}{\partial \breve{v}_j}$ specifies the dissipative force and the term $F_j^{\text{stoch}}(t)=\sqrt{2\sigma^2(H)}\mathcal{G}_{jt}$.

In the terms of the Hamilton function H, the Fokker–Planck equation for the probability distribution of the swarm is defined as follows. Denote by $v\left(\breve{x}_{1\ldots m},\breve{v}_{1\ldots m},t\right)=v\left(\breve{x}_1,\breve{x}_2,\ldots,\breve{x}_m,\breve{v}_1,\breve{v}_2,\ldots,\breve{v}_m,t\right)$ the probability distribution function of m agents. The dynamics of this distribution with respect to the Langevin equation (5.36) is defined as follows (Schweitzer, 2003; Schweitzer et al., 2001):

$$\frac{\partial}{\partial t}v\left(\breve{x}_{1\ldots m},\breve{v}_{1\ldots m},t\right)=\sum_{j=1}^{m}\frac{\partial}{\partial \breve{v}_j}\left(\sigma^2(H)\frac{\partial}{\partial \breve{v}_j}v\left(\breve{x}_{1\ldots m},\breve{v}_{1\ldots m},t\right)+g(H)\frac{\partial H}{\partial \breve{v}_j}v\left(\breve{x}_{1\ldots m},\breve{v}_{1\ldots m},t\right)\right)$$
$$-\sum_{j=1}^{m}\breve{v}_j\frac{\partial}{\partial \breve{x}_j}v\left(\breve{x}_{1\ldots m},\breve{v}_{1\ldots m},t\right)-+\sum_{j=1}^{m}\frac{\partial H}{\partial \breve{v}_j}\frac{\partial}{\partial \breve{v}_j}v\left(\breve{x}_{1\ldots m},\breve{v}_{1\ldots m},t\right). \tag{5.37}$$

It is clear that for a single agent, this equation with the Hamiltonian $H\left(\breve{x},\breve{v},t\right)$ defined with respect to the nonlinear friction $\gamma\left(\breve{x},\breve{v}\right)$ is equivalent to the Fokker–Planck equation (5.17). In addition, notice that for the group of noninteracting agents, the Hamiltonian $H\left(\breve{x}_1,\breve{x}_2,\ldots,\breve{x}_m,\breve{v}_1,\breve{v}_2,\ldots,\breve{v}_m,t\right)$ is

$$H\left(\breve{x}_1,\breve{x}_2,\ldots,\breve{x}_m,\breve{v}_1,\breve{v}_2,\ldots,\breve{v}_m,t\right)=\sum_{j=1}^{m}H\left(\breve{x}_j,\breve{v}_j,t\right)=\frac{1}{2}\sum_{j=1}^{m}\breve{v}_j^2(t),$$

and the probability distribution $v\left(\breve{x}_{1\ldots m},\breve{v}_{1\ldots m},t\right)$ is reduced to the product of the distributions specifying the location probabilities of each agent in the group, that is,

$$v\left(\breve{x}_{1\ldots m},\breve{v}_{1\ldots m},t\right)=\prod_{j=1}^{m}v\left(\breve{x}_j,\breve{v}_j,t\right).$$

Certainly, the dynamics of the swarm defined according to the Hamiltonian approach is equivalent to the earlier considered dynamics based on the direct application of the Langevin equations. However, the application of the Hamilton function stresses the connection of swarm dynamics with general approach considered by the dynamical systems theory that can provide additional results and insights regarding self-organization stability of the swarms. For additional information and examples, see the book by Schweitzer (2003), and for different implementations and considerations, see the earlier indicated literature on swarm intelligence and collective behavior.

5.3.3 Distributions of Turning Angles for Active Brownian Agents

In the algorithms of individual foraging presented in Chapter 4, the movements of the agents were defined according to different types of Brownian motion and of the Lévy flights that implied the exact definition of the turn angles of the agents at each time t. In the case of simple Brownian walks and Lévy flights, such angles were drown randomly from the interval $[-\pi, \pi)$, and for correlated and fractional motion, the angles were defined with respect to the previous angles: at short time for the correlated Brownian motion and at long time for the fractional Brownian motion. In case of search and foraging in the heterogeneous environment, for example, with the given target location probabilities, the turn angles were defined with respect to these probabilities. In the active Brownian motion, an additional factor, which influences to the specification of the turning angles, is an energy depot model and the energy pumping and dissipation. The succeeding discourse addresses this issue following the paper by Romanczuk et al. (2012).

At first, let us consider the motion of the agent in 1D domain $\mathfrak{X} \subset \mathbb{R}$, and assume that the Langevin equation, which governs its motion, is (cf. Equation 5.16)

$$\frac{d}{dt}\breve{x}(t) = \breve{v}(t), \quad \frac{d}{dt}\breve{v}(t) = -\gamma(\breve{v})\breve{v}(t) + \sqrt{2\sigma_v^2}\mathcal{G}_t^v, \tag{5.38}$$

where

$\mathcal{G}_t^v$ stands for the Gaussian white noise governing the velocity
σ_v^2 is a constant specifying its strength

For this motion, consider the Fokker–Planck equation in the form (cf. Equation 5.17)

$$\frac{\partial}{\partial t}\nu(\breve{x},\breve{v},t) = \sigma^2(\breve{v})\frac{\partial^2}{\partial \breve{v}^2}\nu(\breve{x},\breve{v},t) + \frac{\partial}{\partial \breve{v}}\gamma(\breve{v})\breve{v}(t)\nu(\breve{x},\breve{v},t) - \breve{v}(t)\frac{\partial}{\partial \breve{x}}\nu(\breve{x},\breve{v},t). \tag{5.39}$$

The goal is to define the diffusion rate $\sigma^2(\breve{v})$ with respect to the nonlinear friction $\gamma(\breve{v})$.

Denote by $\Phi(\breve{v})$ the velocity potential over the domain $\mathfrak{X}$. In particular, for the Rayleigh–Helmholtz friction defined by Equation 5.18, the velocity potential is (cf. Equation 5.19)

$$\Phi(\breve{v}) = \frac{1}{4}\beta\breve{v}^4 - \frac{1}{2}\alpha\breve{v}^2,$$

and for the friction defined by Equation 5.23, which is used in the energy depot model, the velocity potential is

$$\Phi(\breve{v}) = \frac{\gamma_0}{2}\left(\breve{v}^2 - \frac{r_0}{\gamma_0}\ln\left(1 + \breve{v}^2\frac{c_0}{\gamma_e}\right)\right),$$

where, as mentioned earlier, $\gamma_0 > 0$ is the constant coefficient of the Stokes friction, $r_0 > 0$ is the energy intake rate, $c_0 > 0$ is the energy conversion rate, and $\gamma_e > 0$ is the energy dissipation rate.

Using the velocity potential $\Phi(\breve{v})$, the diffusion rate $\sigma^2(\breve{v})$ is specified as follows (Romanczuk et al., 2012):

$$\sigma^2(\breve{v}) = \frac{\left(\sigma_v^2\right)^{-2}\int_0^\infty e^{\Phi(\breve{v}_1)/\sigma_v^2}\left(\int_{\breve{v}_1}^\infty \breve{v}_1 e^{\Phi(\breve{v}_2)/\sigma_v^2}d\breve{v}_2\right)^2 d\breve{v}_1}{\int_0^\infty e^{\Phi(\breve{v})/\sigma_v^2}d\breve{v}}. \tag{5.40}$$

In particular, if friction $\gamma(\breve{v})$ is defined by the power law

$$\gamma(\breve{v}) = \gamma_0 \breve{v}^{\alpha 2}, \quad \alpha > 0,$$

then the diffusion rate is also scaled by the power law

$$\sigma^2(\breve{v}) \sim \gamma^{-2/(1+\alpha)} \left(\sigma_v^2\right)^{(1-\alpha)/1+\alpha},$$

which defines the larger diffusion for slow motion and, consequently, smaller diffusion for fast motion of the agent.

Now, assume that $\mathfrak{X} \subset \mathbb{R}^2$ be a 2D domain with the points $\mathfrak{x} = (x, y)$, and consider the over-damped motion of the active Brownian agents governed by Equations 5.13 and 5.14. Recall that in such motion it is assumed that the friction is relatively high and, consequently, inertia of the particle is neglected. Similar to Section 4.2.2, denote by $\alpha(t)$ the turning angle of the agent at time $t \in [0, T]$ and by δ the length of the agent's step in the direction defined by $\alpha(t)$ (see Figure 4.7).

Assume that the time is discrete $t = 0, 1, 2, \ldots, T$ and consider the distance $r(t) = \left\|\breve{\mathfrak{x}}(t) - \breve{\mathfrak{x}}(0)\right\|$ between the agent's location $\breve{\mathfrak{x}}(t)$ at time t and its starting point $\breve{\mathfrak{x}}(0)$. Denote by $\beta(t)$ the angle between the x-axis and the agent's heading at time t, that is, $\beta(t) = \beta(t-1) + \alpha(t)$. Then, the displacement $\left|r(t) - r(t-1)\right|$ at time t is

$$\left|r(t) - r(t-1)\right| = \delta\left(e_x \cos\left(\beta(t)\right) + e_y \sin\left(\beta(t)\right)\right), \tag{5.41}$$

where e_x and e_y are unit vectors at the x and y axes, respectively. Then, the squared distance $r^2(t)$ is specified as follows:

$$r^2(t) = \left(\sum_{\tau_1=1}^{t} \left|r(\tau_1) - r(\tau_1 - 1)\right|\right) \times \left(\sum_{\tau_2=1}^{t} \left|r(\tau_2) - r(\tau_2 - 1)\right|\right).$$

That yields that the average $r^2(t)$ of the squared distance is written as (Ibe, 2013; Romanczuk et al., 2012)

$$\begin{aligned} r^2(t) &= \sum_{\tau=1}^{t} \left|r(\tau) - r(\tau-1)\right|^2 + 2\sum_{\tau_1=1}^{t-1} \sum_{\tau_2>\tau_1}^{t} \left|r(\tau_1) - r(\tau_1 - 1)\right| \times \left|r(\tau_2) - r(\tau_2 - 1)\right| \\ &= t\delta^2 + 2\delta^2 \sum_{\tau_1=1}^{t-1} \sum_{\tau_2>\tau_1}^{t} \cos\left(\beta(\tau_1) - \beta(\tau_2)\right). \end{aligned} \tag{5.42}$$

Denote by $\mathcal{P}_\alpha$ the probability distribution of the turn angles. Then, the average of the turning angles' cosine is

$$\cos(\alpha) = \int_{-\pi}^{\pi} \cos(\alpha) \mathcal{P}_\alpha d\alpha,$$

and, in general (Romanczuk et al., 2012),

$$\cos\left(\beta(t)-\beta(t+2)\right)=\cos(\alpha)^2,\ldots,\cos\left(\beta(t)-\beta(t+\tau)\right)=\cos(\alpha)^k,$$

for any integer $k > 2$. Using these dependences, for Equation 5.42, one obtains (Romanczuk et al., 2012)

$$r^2(t)=\delta^2\left(t\frac{1+\cos(\alpha)}{1-\cos(\alpha)}-2\cos(\alpha)\frac{1-\cos(\alpha)^t}{\left(1-\cos(\alpha)\right)^2}\right)\underset{t\to\infty}{\rightarrow}\delta^2\left(t\frac{1+\cos(\alpha)}{1-\cos(\alpha)}-2\cos(\alpha)\frac{1}{\left(1-\cos(\alpha)\right)^2}\right),$$

and diffusion rate σ^2, which provides such average distance, is

$$\sigma^2(\breve{v})=\frac{\delta^2\nu}{4}\frac{1+\cos(\alpha)}{1-\cos(\alpha)}=\frac{\delta^2}{4\tau}\frac{1+\cos(\alpha)}{1-\cos(\alpha)}, \quad (5.43)$$

where, similarly to Section 4.2.1, ν and τ are characteristic speed and time of the agent, respectively.

It is clear that if turn angles are distributed equally for each t, then $\cos(\alpha) = 0$ and diffusion rate becomes $\sigma^2 = \delta^2\nu/4 = \delta^2/(4\tau)$ that corresponds to the simple Brownian motion (cf. Section 4.2.1).

Finally, let us consider the active Brownian agent moving in the medium with velocity-dependent friction and constant potential field (see the Langevin equation (5.16)). In addition, assume that the turning angles do not depend on the velocity of the agent and also are governed by the Gaussian white noise. Namely, the equations governing the dynamics of the agent are the following (Romanczuk et al., 2012):

$$\frac{d}{dt}\breve{v}(t)=-\gamma(\breve{v})\breve{v}(t)+\sqrt{2\sigma_v^2}\mathcal{G}_t^v, \quad (5.44)$$

$$\frac{d}{dt}\alpha(t)=\sqrt{2\sigma_\alpha^2}\mathcal{G}_t^\alpha \quad (5.45)$$

where

$\mathcal{G}_t^v$ and $\mathcal{G}_t^\alpha$ stand for the Gaussian white noise governing the velocity and the turning angles of the agent, respectively

σ_v^2 and σ_α^2 are constants.

Assume that both the decays of the velocities and of the turning angles are exponential, that is,

$$\breve{v}(0)\breve{v}(t)=\left|\breve{v}\right|^2 e^{-\frac{1}{\tau_v}t}+\left|\breve{v}\right|^2\left(1-e^{-\frac{1}{\tau_v}t}\right),\quad e_v(0)e_v(t)=e^{-\frac{1}{\tau_\alpha}t},$$

where

$e_v(t)$ is a unit vector such that the $\breve{v}(t)=\left|\breve{v}(t)\right|e_v(t)$

$1/\tau_v$ is a relaxation rate (with relaxation time τ_v)

$1/\tau_\alpha$ is a correlation decay rate

Then, the average squared distance $r^2(t)$ between the agent's starting point $\breve{\mathfrak{x}}(0)$ and its location $\breve{\mathfrak{x}}(t)$ at time t is (Romanczuk et al., 2012):

$$\left\langle r^2(t)\right\rangle = 2\tau_\alpha \left\langle |\breve{v}|\right\rangle^2 \left(t + \tau_\alpha \left(e^{-\frac{1}{\tau_\alpha}t} - 1 \right)\right) + 2\frac{\left\langle |\breve{v}|^2\right\rangle - \left\langle |\breve{v}|\right\rangle^2}{1/\tau_v + 1/\tau_\alpha}\left(t + \frac{1}{1/\tau_v + 1/\tau_\alpha}\left(e^{-\left(\frac{1}{\tau_v}+\frac{1}{\tau_\alpha}\right)t} - 1\right)\right), \quad (5.46)$$

In this equation, the first term defines the displacement by the self-propelled motion, and the second term defines the influence of the velocity fluctuations. For additional considerations of the mean squared displacements in active Brownian motion, see the paper by Ebeling (2004).

If the agent moves with the constant speed $\left\langle \breve{v}\right\rangle = \breve{v}_0 = \text{const}$, then there are two possible regimes of the agent's motion: ballistic motion, in which the average distance $\left\langle r^2(t)\right\rangle$ is proportional to the squared time, that is, $\left\langle r^2(t)\right\rangle \sim t^2$, and diffusive motion, in which the average distance $\left\langle r^2(t)\right\rangle$ is scaled by the time t, that is, $\left\langle r^2(t)\right\rangle \sim t^2$ while $t \to \infty$. In the second regime, the diffusion rate is

$$\sigma^2(\breve{v}) = \frac{1}{2}\left(\tau_\alpha \left\langle |\breve{v}|\right\rangle^2 + \frac{\left\langle |\breve{v}|^2\right\rangle - \left\langle |\breve{v}|\right\rangle^2}{1/\tau_v + 1/\tau_\alpha}\right). \quad (5.47)$$

It is clear that these two regimes of the movements of active Brownian agent are the same as the phase 1 movements of the searcher with intermittent strategies considered in Section 4.3.3. The motion with $\left\langle r^2(t)\right\rangle \sim t^2$ corresponds to the ballistic mode of the searcher, in which it moves by a straight line, and the motion with $\left\langle r^2(t)\right\rangle \sim t$ corresponds to the diffusive mode, in which the searcher moves according to Brownian motion with the diffusion rate given by Equation 5.47.

Such a relation between active Brownian motion and search strategies provides a basis for modeling the search and foraging activities using the methods of active Brownian motion that will be utilized in the next section. For further information regarding the relation between diffusion rate and nonlinear friction and physical interpretation of this relation, see the paper by Romanczuk et al. (2012).

5.4 TURING SYSTEM FOR THE SWARM FORAGING

Assume that at time $t = 0$ the domain $\mathfrak{X} \subset \mathbb{R}^2$ contains a finite quantity of prey distributed over the points of $\mathfrak{X}$ with respect to the density function $f : \mathfrak{X} \times [0,T] \to \mathbb{R}^+$. Then, initial distribution of the prey is $f(\mathfrak{x},0)$, $\mathfrak{x} \in \mathfrak{X}$, and its initial quantity is $f_0(\mathfrak{X}) = \int_{\mathfrak{X}} f(\mathfrak{x},0)\,d\mathfrak{x}$. In addition, let $g : \mathfrak{X} \times [0,T] \to \mathbb{R}^+$ be a density function, which specifies the distribution of the agents over the domain $\mathfrak{X}$ at times $t \in [0, T]$. Similar to the prey, initial distribution of the agents is $g(\mathfrak{x},0)$, $\mathfrak{x} \in \mathfrak{X}$, but in contrast to the prey, the number of the agents is constant, so $g_t(\mathfrak{X}) = \int_{\mathfrak{X}} g(\mathfrak{x},t)\,d\mathfrak{x} = \int_{\mathfrak{X}} g(\mathfrak{x},0)\,d\mathfrak{x}$ for any time $t \in [0, T]$. Usually, the values $f(\mathfrak{x},t)$ and $g(\mathfrak{x},t)$ are, respectively, considered as concentrations of the prey and of the predators or of the targets and the searchers in given point $\mathfrak{x}$ at time t.

This section addresses the mutual dynamics of the densities f and g or, in the terms of Section 3.2 (see Equation 3.7), the dynamics of the probability densities u and v, which define the probabilities of finding the prey and the agents in the points $\mathfrak{x} \in \mathfrak{X}$ of the domain at the times $t \in [0, T]$.

To clarify the considered dynamics, let us start with the simple example of foraging with constant eat rate or energy intake rate r_e = const (cf. Section 3.3). Consider the gain function $\xi : \mathfrak{X} \times [0,T] \to \mathbb{R}^+$ such that $\xi(\mathfrak{x},t)$ is a quantity of prey eaten by the agents in the point $\mathfrak{x} \in \mathfrak{X}$ up to time t (cf. Section 4.3.2). For destructive and nondestructive search scenarios (see Section 4.3.2), the gain ξ is specified as follows.

In the nondestructive search, the gain depends on the initial quantity $f(\mathfrak{x},0)$ of the prey in the point $\mathfrak{x}$ and on the distribution $g(\mathfrak{x},t)$ of the agents at time t and is governed by the following equation:

$$\frac{\partial}{\partial t}\xi(\mathfrak{x},t) = -r_c g(\mathfrak{x},t) f(\mathfrak{x},0)\xi(\mathfrak{x},t). \tag{5.48}$$

If $f(\mathfrak{x},0) = \text{const}$ are equal for all $\mathfrak{x} \in \mathfrak{X}$, then this equation obtains a simple form (Romanczuk et al., 2012)

$$\frac{\partial}{\partial t}\xi(\mathfrak{x},t) = -r_c g(\mathfrak{x},t)\xi(\mathfrak{x},t).$$

In the destructive search, in contrast, the gain depends both on the agent's distribution and on the distribution $f(\mathfrak{x},t)$ of the prey, that is,

$$\frac{\partial}{\partial t}\xi(\mathfrak{x},t) = -r_c g(\mathfrak{x},t) f(\mathfrak{x},t)\xi(\mathfrak{x},t). \tag{5.49}$$

Notice that Equations 5.48 and 5.49 have the same meaning as the energy balance equations considered in the depot model of active Brownian particle.

Recall that $f(\mathfrak{x},t)$ is a quantity of the prey remained in the point $\mathfrak{x}$ at time t and $\xi(\mathfrak{x},t)$ is a quantity of prey eaten in the point $\mathfrak{x}$ up to time t. Thus for each point $\mathfrak{x}$ and each time t it holds true that $f(\mathfrak{x},0) = f(\mathfrak{x},t) + \xi(\mathfrak{x},t)$, and Equations 5.48 and 5.49 can be used for definition of the prey density f in nondestructive and destructive search, respectively,

$$\frac{\partial}{\partial t} f(\mathfrak{x},t) = r_c g(\mathfrak{x},t) f(\mathfrak{x},0)\big(f(\mathfrak{x},0) - f(\mathfrak{x},t)\big), \tag{5.50}$$

$$\frac{\partial}{\partial t} f(\mathfrak{x},t) = r_c g(\mathfrak{x},t)\big(f(\mathfrak{x},0) - f(\mathfrak{x},t)\big) f(\mathfrak{x},t), \tag{5.51}$$

where, in terms of Section 3.2, the first equation defines the Malthusian dynamics of the prey in the point $\mathfrak{x} \in \mathfrak{X}$ and the second its logistic dynamics, both with respect to the density $g(\mathfrak{x},t)$ of the agents in this point.

The evolution of the density g depends on the agents' sensing and motion abilities. In particular, for the simple Brownian agents, the evolution of the agent's density g is governed by the earlier considered diffusion equation

$$\frac{\partial}{\partial t} g(\mathfrak{x},t) = \sigma_g^2 \Delta g(\mathfrak{x},t). \tag{5.52}$$

Then, the system of Equations 5.50 and 5.52 (for Malthusian prey) or of Equations 5.51 and 5.52 (for logistic prey) defines the required evolution of the densities of the prey and the agents or, in the terms of Section 3.2, the dynamics of the system "target–searcher" or "prey–predator". The dynamics of these systems are illustrated in Figure 5.19. In both cases, $\sigma_g^2 = 0.5$ for both axes and the initial distribution of the agents is defined by the binormal distribution with the central point $\breve{\mathfrak{x}}(0) = (50,50)$ and variances over the axes $\sigma_1 = \sigma_2 = 10$ (see, e.g., Figure 5.17c), and at $t = 0$ each point $\mathfrak{x} \in \mathfrak{X}$ of the domain contains the same quantity of the prey $f(\mathfrak{x},0) = \text{const}$.

It is seen that for Malthusian dynamics (see Figure 5.19a and b), the initial distribution of the prey such that $f(\mathfrak{x},0) = \text{const}$ are equal for all points $\mathfrak{x} \in \mathfrak{X}$ is immediately ordered with respect to

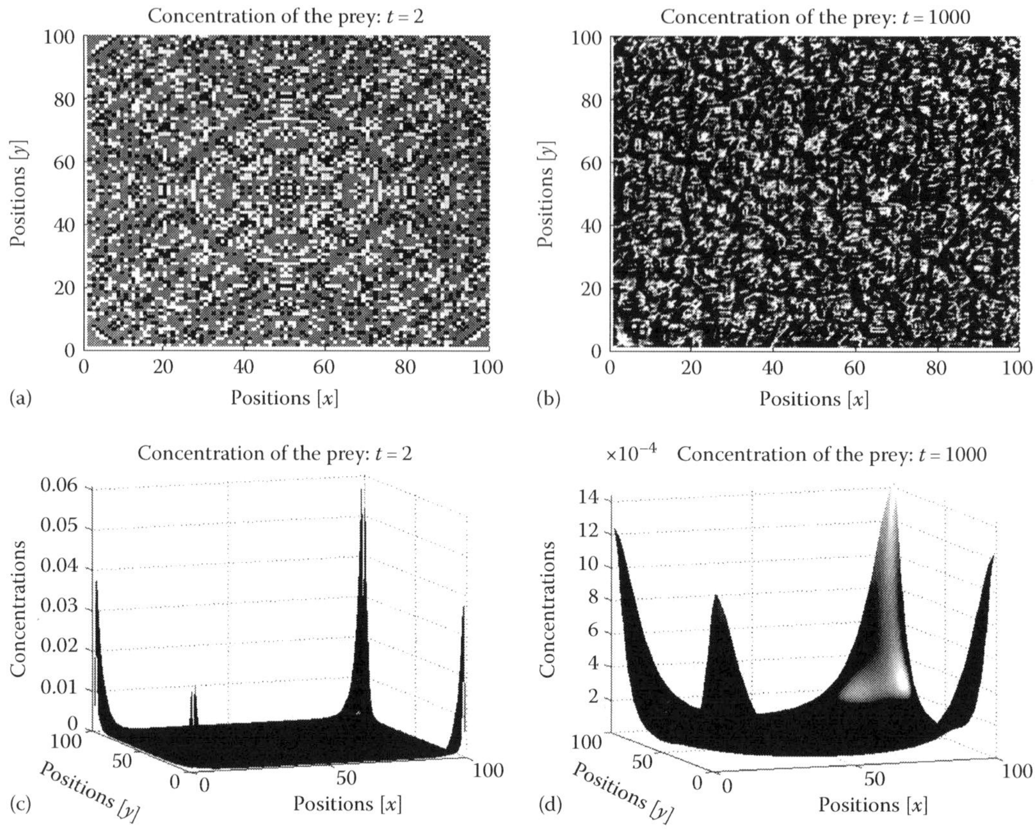

FIGURE 5.19 Concentrations of the prey for Malthusian and logistic dynamics. Figures (a) and (b) depict the distributions of the prey governed by Equation 5.50 immediately after the beginning of the process and at time $t = 1000$, and figures (c) and (d) show the distribution of the prey governed by Equation 5.51 at these times. In both cases, it is assumed that the agents are Brownian and their distribution is governed by Equation 5.52 with $\sigma_g^2 = 0.5$ for both axes.

the distribution of the agents, while with time, the concentrations of the prey become random with small deviation around initial values. In contrast, for logistic dynamics (see Figure 5.19c and d), the distribution of the prey follows the distribution of the agents and with time the most of the prey is concentrated at the corners and the borders of the domain.

In the presented example, it was assumed that the prey is immobile and the agents' dynamics are defined by the simple Brownian motion, while the concentration of prey depends only on the concentration of the agents in the points in the domain with respect to definite gain function.

Notice that for Malthusian dynamics, using the solution $g(\mathfrak{x},t) = \frac{1}{4\pi\sigma_g t}\exp\left(-\frac{\mathfrak{x}^2(t)}{4\sigma_g^2 t}\right)$ of Equation 5.52, the gain function for the amount of prey, which is constant $f(\mathfrak{x},0) = \text{const}$ and equal for all points $\mathfrak{x} \in \mathfrak{X}$, the gain function ξ is specified as follows (cf. gain functions shown in Figure 3.4):

$$\xi(\mathfrak{x},t) = \xi_0(\mathfrak{x})\exp\left(-r_e\int_0^t g(\mathfrak{x},\tau)\,d\tau\right) = \xi_0(\mathfrak{x})\exp\left(-\frac{r_e}{4\pi\sigma_g}\int_0^t t^{-1}\exp\left(-\frac{\mathfrak{x}^2(t)}{4\sigma_g^2 t}\right)d\tau\right),$$

where $\xi_0(\mathfrak{x})$ is an initial gain in the points $\mathfrak{x} \in \mathfrak{X}$.

In general, both the prey and the agents are allowed to move over a domain and their concentrations are governed by the system (cf. Equation 3.7)

$$\begin{cases} \frac{\partial}{\partial t} f(\mathfrak{x},t) = s_f(\mathfrak{x},t)\Delta f(\mathfrak{x},t) + \mathbb{F}(f,g), \\ \frac{\partial}{\partial t} g(\mathfrak{x},t) = s_g(\mathfrak{x},t)\Delta g(\mathfrak{x},t) + \mathbb{G}(f,g), \end{cases} \tag{5.53}$$

where the function $\mathbb{F}$ specifies the dependence of prey concentration on the concentration of the predators and the function $\mathbb{G}$ specifies the predators' concentration with respect to the concentration of the prey. As in Equation 3.7, the values $s_f(\mathfrak{x},t)$ and $s_g(\mathfrak{x},t)$ specify the diffusion rates of the prey and of the agents.

The system defined by Equation 5.53, which is often called the Turing system and also is well-known as reaction–diffusion system or activator–inhibitor system, was suggested in 1952 by Turing (1952) as a basic model of dynamics and pattern formation in distributed two-component chemical reactions and in the swarms of living organisms. Later, the Turing system was used as a basis for the approaches suggested by Haken (1977) and by Prigogine and his colleagues (1977), and currently, its different versions are widely used for modeling and analysis of different systems, which allow formation of spatial and temporal patterns (see, e.g., Rabinovich, Ezersky, & Weidman, 2000; Romanovsky, Stepanova, & Chernavsky, 1984; Vasiliev, Romanovsky, Chernavskii, & Yakhno, 1987). For numerical simulations of different versions of the Turing system and examples of the patterns in two and three dimensions, see the works by Leppänen (2004, 2005) and by Shoji, Yamada, Ueyama, and Ohta (2007).

It is clear that each equation in the Turing system has similar form as the Fokker–Planck equations specifying the distributions of the active Brownian particles in Section 5.3.1, while the system itself is the same as the system "target–searcher" or "prey–predator" defined by Equation 3.7 (see also previous considerations of spatial dynamics of populations in Section 3.2). The succeeding discourse implements this similarity and applies the methods and results achieved in the studies of reaction–diffusion for analysis of search and foraging by swarms of mobile agents.

5.4.1 Turing System for the Target–Searcher Dynamics

The Turing system defined by Equation 5.53 provides general framework for the description of a wide range of complex systems such that the particular dynamics are defined by the dependence between the concentrations of activator and inhibitor. In the case of target–searcher system, the system deals with the location probabilities of the target and the searcher, and the relation between these probabilities is defined by the detection function φ, which defines the probability of detection the target in the point $\mathfrak{x} \in \mathfrak{X}$ given that the target is located in this point and the searcher observes this point (see Chapter 2).

5.4.1.1 Remarks on the Dynamics of Turing System

Let us start with general remarks on the dynamics of the Turing system; this brief overview follows the paper by Kagan (2010). Let $\mathfrak{X} \subset \mathbb{R}^2$ be a 2D domain with the points $\mathfrak{x} = (x,y)$ and consider the Turing system defined over a $\mathfrak{X} \subset \mathbb{R}^2$ with constant diffusion rates $s_f(\mathfrak{x},t) = \sigma_f^2 = \text{const}$ and $s_g(\mathfrak{x},t) = \sigma_g^2 = \text{const}$. Assume that the dependences between the concentrations of activator and inhibitor are characterized by the parameter $\alpha \in \mathbb{R}$, for example, by friction coefficient γ or energy intake rate r_e. Accordingly, the functions $\mathbb{F}$ and $\mathbb{G}$ appearing in Equation 5.53 are equipped by the index α, that is, $\mathbb{F}_\alpha$ and $\mathbb{G}_\alpha$. The Turing system in this case is written as follows:

$$\begin{cases} \frac{\partial}{\partial t} f(\mathfrak{x},t) = \sigma_f^2\left(\frac{\partial^2}{\partial x^2} + \frac{\partial^2}{\partial y^2}\right) f(\mathfrak{x},t) + \mathbb{F}_\alpha(f,g), \\ \frac{\partial}{\partial t} g(\mathfrak{x},t) = \sigma_g^2\left(\frac{\partial^2}{\partial x^2} + \frac{\partial^2}{\partial y^2}\right) g(\mathfrak{x},t) + \mathbb{G}_\alpha(f,g). \end{cases} \tag{5.54}$$

If in this system the functions $\mathbb{F}_\alpha$ and $\mathbb{G}_\alpha$ are linear, then the behavior of the system described by Equation 5.5.4 is relatively simple. For example, if $\mathbb{F}_\alpha(f,g) \equiv g$ and $\mathbb{G}_\alpha(f,g) \equiv -\omega^2 f$, then instead of Equation 5.54, one obtains the system

$$\begin{cases} \dfrac{\partial}{\partial t} f(\mathfrak{x},t) = \sigma_f^2 \left(\dfrac{\partial^2}{\partial x^2} + \dfrac{\partial^2}{\partial y^2} \right) f(\mathfrak{x},t) + g(\mathfrak{x},t), \\ \dfrac{\partial}{\partial t} g(\mathfrak{x},t) = \sigma_g^2 \left(\dfrac{\partial^2}{\partial x^2} + \dfrac{\partial^2}{\partial y^2} \right) g(\mathfrak{x},t) - \omega^2 f(\mathfrak{x},t), \end{cases} \tag{5.55}$$

which describes the media that consists of the connected oscillators, where ω is a frequency of oscillations (Landa, 1996). The behavior of such system is illustrated in Figure 5.20. As mentioned earlier, initial concentrations of the reagents are defined by binormal distributions around $\mathfrak{x} = (50,50)$ with the variances $\sigma_1 = \sigma_2 = 10$ (see Figure 5.17c). Parameters of the system are $\sigma_f^2 = \sigma_g^2 = 0.5$ and $\omega^2 = 1$.

It is seen that at the beginning of the process, the reagents are distributed regularly around the central point, and with time, because of the interactions between activator and inhibitor, the regularity is broken. However, the concentration patterns of both reagents are preserved similarly. Certainly, if in this system both diffusion rates are $\sigma_f^2 = \sigma_g^2 = 0$, then it is reduced to the equation of linear oscillator without dissipation.

In contrast, if the functions $\mathbb{F}_\alpha$ and $\mathbb{G}_\alpha$ are nonlinear and such that both $\frac{\partial}{\partial f}\mathbb{F}_\alpha(f,g) > 0$ and $\frac{\partial}{\partial g}\mathbb{G}_\alpha(f,g) > 0$, then the system can demonstrate different types of behavior, which are defined by the value of the parameter α. In most cases, there exists a value α_0 such that if $\alpha = \alpha_0$, then the initially homogeneous state of the media described by the system (5.54) is unstable, and the system behaves as follows (see, e.g., Kerner & Osipov, 1994):

- If $\alpha < \alpha_0$, then a short external influence on the system gives a rise to static or pulse autowaves.
- If $\alpha > \alpha_0$, then autowaves and wave structures appear without external influence.

In addition, if in the two-component system (5.54), it holds true that $\sigma_f^2 \neq \sigma_g^2$, then the system allows an existence of dissipative structures (Vasiliev et al., 1987); these structures are often called Turing structures or Turing patterns. In special cases, the system (5.54) can also have the soliton solutions, which represent the solitary running or static pulse waves.

To clarify the introduced concepts, let us consider the concentrated Turing system (i.e., such that $\sigma_f^2 = \sigma_g^2 = 0$) that corresponds to the single van-der-Pol oscillator (Klimontovich, 1989, 1991). The equation of such oscillator is the following:

$$\frac{d^2}{dt^2} s(t) + \left(-\alpha + \beta s^2(t)\right) \frac{d}{dt} s(t) + \omega^2 s(t) = 0, \tag{5.56}$$

where

$s(t)$ is a state of the oscillator
α is a friction coefficient
β is a degree of the feedback
ω is a frequency of the oscillations.

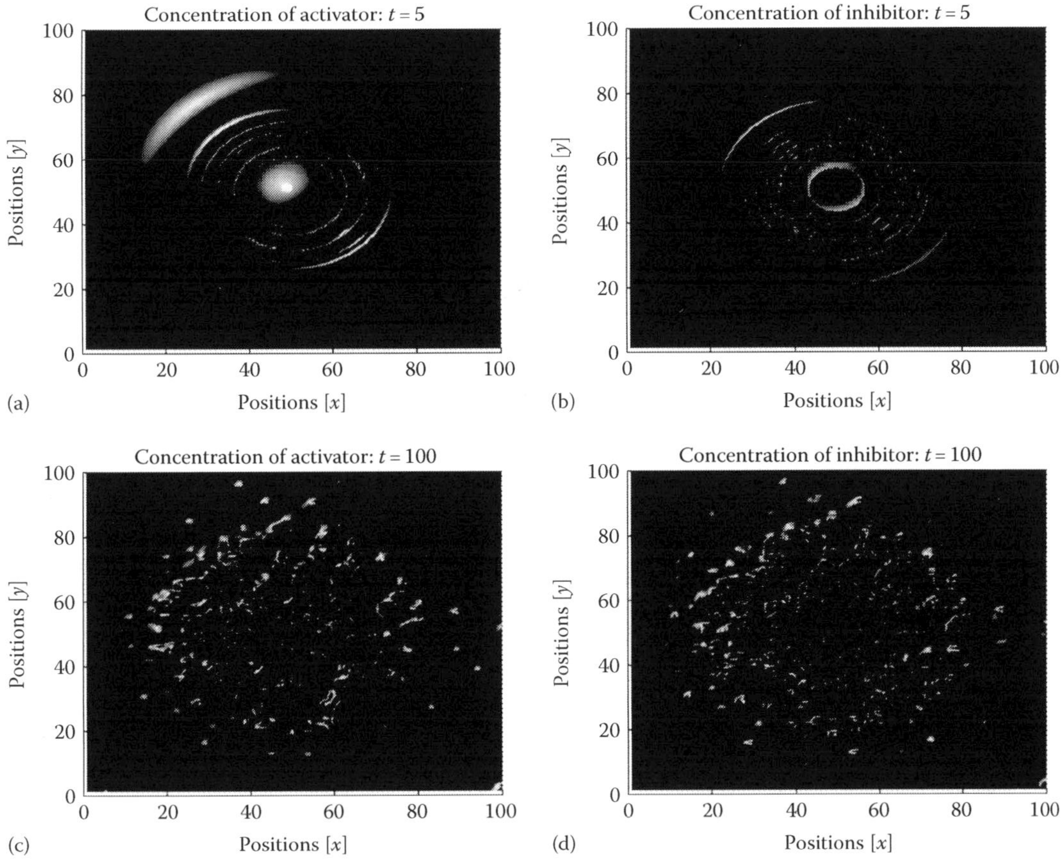

FIGURE 5.20 Concentrations of activator and inhibitor defined by the system (5.55) of connected van-der Pol oscillators. The process starts with equal distributions of the reagents defined by the binormal distributions with central point $\mathfrak{x} = (50, 50)$ and variances $\sigma_1 = \sigma_2 = 10$. Figures (a) and (b) depict the concentrations of the reagents at the beginning of the process at time $t = 5$, and figures (c) and (d) show the concentrations at time $t = 100$. Parameters of the system are $\sigma_f^2 = \sigma_g^2 = 0.5$ and $\omega^2 = 1$.

In the form of the concentrated Turing system, in which both f and g do not depend on the coordinate $\mathfrak{x}$ and are functions of time only, the oscillator defined by Equation 5.56 is specified as follows (Klimontovich, 1991):

$$\begin{cases} \frac{d}{dt} f(t) = \frac{1}{2}\left(\alpha - \beta E(t)\right) f(t) + g(t), \\ \frac{d}{dt} g(t) = \frac{1}{2}\left(\alpha - \beta E(t)\right) g(t) - \omega^2 f(t), \end{cases} \tag{5.57}$$

where $E(t) = \frac{1}{2}\left(g^2(t) + \omega^2 f^2(t)\right)$ stands for the energy of oscillations with frequency ω.

From this equation, it follows that that the energy of the oscillations is governed by the following equation (cf. the energy of active Brownian motion considered in Section 5.3.1):

$$\frac{d}{dt} E(t) = \left(\alpha - \beta E(t)\right) E(t),$$

and the exact solution of the system (5.57) is

$$\begin{cases} f(t) = \frac{1}{\omega}\sqrt{2E(t)}\cos(\omega t + \xi_0), \\ g(t) = -\sqrt{2E(t)}\sin(\omega t + \xi_0), \end{cases}$$

where $E(t) = \frac{\alpha}{\beta}E(0)\left[E(0) - \left(E(0) - \frac{\alpha}{\beta}\right)\exp(-\alpha t)\right]^{-1}$, $E(0)$ stands for the initial energy of the system, and ξ_0 is a phase.

Now, let us consider the distribution $p(E, t)$ of the energy at time. The Fokker–Planck equation of the energy of the system (5.57) is defined as follows (Klimontovich, 1991):

$$\frac{\partial}{\partial t}p(E,t) = S\frac{\partial}{\partial E}\left(E(t)\frac{\partial}{\partial E}p(E,t)\right) + \frac{\partial}{\partial E}\big(p(E,t)(-\alpha + \beta E(t))E(t)\big), \tag{5.58}$$

where S is the intensity of the external sources of fluctuations. Notice that if the fluctuations depend on the energy and do not depend on the external sources, then $S(E) = (-\alpha + \beta E(t))k_B T$ and

$$\frac{\partial}{\partial t}p(E,t) = k_B T\frac{\partial}{\partial E}\left((-\alpha + \beta E(t))E(t)\frac{\partial}{\partial E}p(E,t)\right) + \frac{\partial}{\partial E}\big(p(E,t)(-\alpha + \beta E(t))E(t)\big),$$

where

k_B is the Boltzmann constant
T is a temperature.

The stationary solution $p(E)$ of the Fokker–Planck equation (5.58) is given by the Gibbs distribution (Klimontovich, 1987, 1991):

$$p(E) = \exp\left[\frac{1}{S}(E_0 - H(E))\right], \quad \int_0^\infty p(E)dE = 1, \tag{5.59}$$

where

$H(E) = -\alpha E + \frac{1}{2}\beta E^2$ is the Hamiltonian of the system
E_0 is a free energy

and the Shannon entropy of the distribution $p(E)$ is

$$\mathbb{S}(p(E)) = -\int_0^\infty p(E)\ln(p(E))dE. \tag{5.60}$$

Assume that the friction coefficient α is defined as $\alpha = \gamma - \gamma_0$, where γ_0 is a coefficient of linear friction and γ is an energy intake (cf. definition of the dissipation force F^{diss} in Section 5.3.1). Then, with respect to the value of the parameter α, the types of the behavior of the system (5.57), are the following:

1. If energy intake is $\gamma = 0$, that is, $\alpha = -\gamma_0$, and $S\frac{\beta}{\gamma_0^2} \ll 1$, then the system demonstrates chaotic behavior with

$$p_1(E) = \frac{1}{S}\gamma_0\exp\left(-\frac{1}{S}\gamma_0 E\right), \quad \mathbb{S}(p_1(E)) = \ln\left(S\frac{1}{\gamma_0}\right) + 1, \quad E_1 = S\frac{1}{\gamma_0}.$$

2. If energy intake γ is equal to the dissipation γ_0 by linear friction, $\gamma = \gamma_0$, that is $\alpha = 0$, then for this threshold state, it holds true that

$$p_2(E)=\sqrt{\frac{2\beta}{\pi S}}\exp\left(-\frac{1}{2S}\beta E^2\right),\quad \mathbb{S}(p_2(E))=\ln\left(\sqrt{\frac{\pi S}{2\beta}}\right)+\frac{1}{2},\quad E_2=\sqrt{\frac{2S}{\pi\beta}}.$$

3. If energy intake γ is essentially greater than the dissipation γ_0, $\gamma \gg \gamma_0$, that is, $\alpha \approx \gamma$ and $S\frac{\beta}{\gamma_0^2} \ll 1$, then the system is in generating regime with

$$p_3(E)=\sqrt{\frac{\beta}{2\pi S}}\exp\left(-\frac{(E-\alpha/\beta)^2}{2S/\beta}\right),\quad \mathbb{S}(p_3(E))=\ln\left(\sqrt{\frac{2\pi S}{\beta}}\right)+\frac{1}{2},\quad E_3=\frac{|\alpha|}{\beta}.$$

These types of the system behavior are illustrated in Figure 5.21. In all cases, $\omega = 0.5$ and $\gamma_0 = 1$. Notice, that for these three types of the behavior, it holds true that

$$\mathbb{S}(p_1(E)) < S(p_2(E)) < S(p_3(E)) \quad \text{and} \quad E_1 < E_2 < E_3.$$

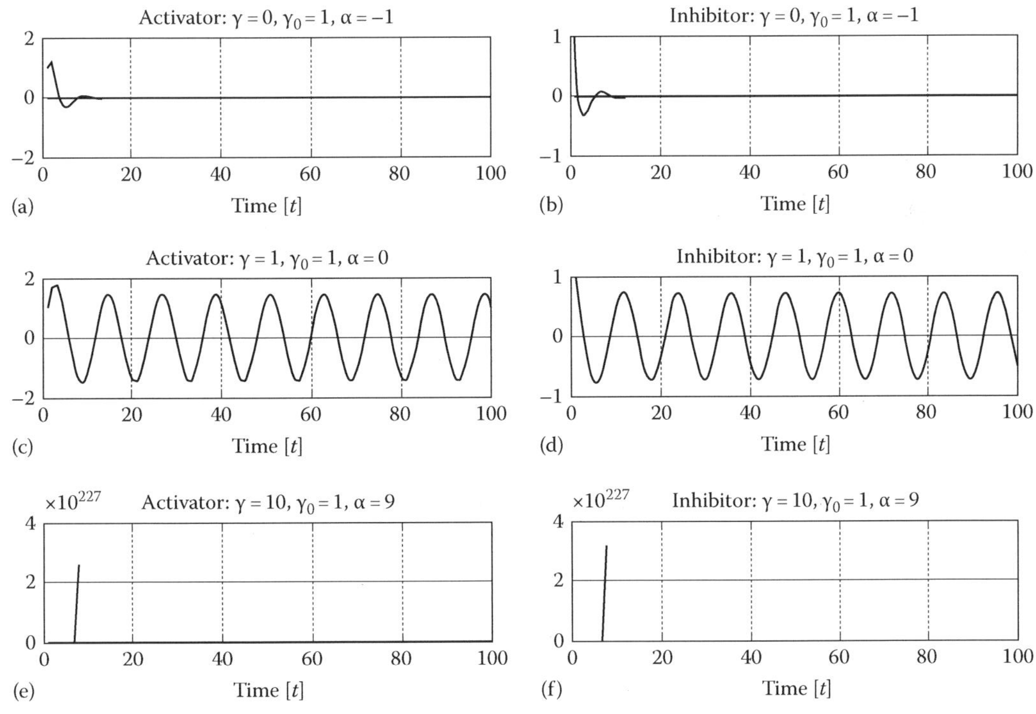

FIGURE 5.21 Concentrations of activator and inhibitor defined by the concentrated Turing system, which correspond to the van der Pol oscillator with the frequency $\omega = 0.5$. Figures (a) and (b) show the concentrations of the reagents for energy intake $\gamma = 0$ and $\alpha = -\gamma_0 = -1$, figures (c) and (d) the concentrations for energy intake $\gamma = \gamma_0 = 1$ and $\alpha = \gamma - \gamma_0 = 0$, and figures (e) and (f) the concentrations for energy intake $\gamma = 10$ and $\alpha = \gamma - \gamma_0 = 9$. In all cases, time is $t = 1,\ldots,100$.

After renormalization of the entropies using the corresponding energies

$$\tilde{\mathbb{S}}\left(p_1(E)\right) = \frac{\mathbb{S}\left(p_1(E)\right)}{E_1}, \quad \tilde{\mathbb{S}}\left(p_2(E)\right) = \frac{\mathbb{S}\left(p_2(E)\right)}{E_2}, \quad \tilde{\mathbb{S}}\left(p_3(E)\right) = \frac{\mathbb{S}\left(p_3(E)\right)}{E_3},$$

one obtains (Klimontovich, 1987)

$$\tilde{\mathbb{S}}\left(p_1(E)\right) > \tilde{\mathbb{S}}\left(p_2(E)\right) > \tilde{\mathbb{S}}\left(p_3(E)\right). \tag{5.61}$$

The inequality (5.61) is known as the Klimontovich $\mathbb{S}$-theorem, which specifies that while the control parameter α increases and the system moves from the chaotic behavior (type 1), to the threshold state (type 2) to the generating regime (type 3), the entropy of the system decreases. In the other words, the system demonstrates the self-organization behavior controlled by the parameter α.

Let us extend these considerations to the distributed system of van-der-Pol oscillators. In this case, the energy distribution $p(E,\mathfrak{x},t)$ depends both on spatial coordinates $\mathfrak{x} \in \mathfrak{X}$ and on time t, and the Fokker–Planck equation (5.58) includes additional term that corresponds to the spatial diffusion of the energy. For noninteracting oscillators, this equation is written as follows (Klimontovich, 1989, 1991):

$$\frac{\partial}{\partial t} p(E,\mathfrak{x},t) = \sigma^2 \Delta p(E,\mathfrak{x},t) + S\frac{\partial}{\partial E}\left(E(t)\frac{\partial}{\partial E} p(E,\mathfrak{x},t)\right) + \frac{\partial}{\partial E}\left(p(E,\mathfrak{x},t)\left(-\alpha + \beta E(t)\right)E(t)\right), \tag{5.62}$$

where
σ^2 is a diffusion rate

$$\int_0^\infty dE \int_{\mathfrak{X}} p(E,\mathfrak{x},t)\,d\mathfrak{x} = 1.$$

The Shannon entropy of the distribution $p(E,\mathfrak{x},t)$ is defined as follows:

$$\mathbb{S}\left(p(E,\mathfrak{x},t)\right) = -\int_0^\infty dE \int_{\mathfrak{X}} p(E,\mathfrak{x},t)\ln\left(p(E,\mathfrak{x},t)\right)d\mathfrak{x}. \tag{5.63}$$

Consider the dynamics of the entropy $\mathbb{S}\left(p(E,\mathfrak{x},t)\right)$ in time, that is,

$$\frac{d}{dt}\mathbb{S}\left(p(E,\mathfrak{x},t)\right) = -\int_0^\infty dE \int_{\mathfrak{X}} \frac{\partial}{\partial t} p(E,\mathfrak{x},t)\ln\left(p(E,\mathfrak{x},t)\right)d\mathfrak{x} - \int_0^\infty dE \int_{\mathfrak{X}} \frac{\partial}{\partial t} p(E,\mathfrak{x},t)\,d\mathfrak{x}. \tag{5.64}$$

If $\frac{\partial}{\partial t} p(E,\mathfrak{x},t) = 0$, then $\frac{d}{dt}\mathbb{S}\left(p(E,\mathfrak{x},t)\right) = 0$ and the entropy is constant. With the changes of distribution, the entropy also changes and specifies different patterns of system dynamics. For example, for harmonic distribution $p(E,\mathfrak{x},t) \sim \cos(x)\cos(y)\cos^2(t)$, the entropy changes periodically such that $\frac{d}{dt}\mathbb{S}\left(p(E,\mathfrak{x},t)\right) \sim \cos(t)\sin(t)$ and corresponds to the running waves.

Notice that in the case of interacting oscillators, Equation 5.62 includes additional dependences of the parameters and results in complex dynamics with transitions of the dynamics patterns.

Certainly, the earlier presented remarks do not exhaust the endless variety of the Turing system dynamics and demonstrate the basic ideas only. In particular cases, the description of the system includes additional equations, which govern the velocities of the activator and inhibitor generation (Balckarey, 1987; Kagan, 1995, 1998) or parameters of the system (Rabinovich et al., 2000; Tsimring et al., 1995).

5.4.1.2 Target–Searcher System

Let us implement the presented approaches for definition of the probabilistic dynamics of the target and the searchers. Recall that in Section 3.2, the activities of the prey and predator were considered based on the population dynamics, and the prey–predator system (3.7) was specified. In particular, in the equation of the target's location probabilities, it was taken into account that the probabilities of finding the target depend on the searcher's activity and observation results and was represented using the detection function (see Section 2.3). In Section 5.3, following the model of active Brownian motion, the dynamics of the searcher's location probabilities with respect to the external field were defined. In the succeeding text, the presented models are applied for the specification of the Turing system, which defines the location probabilities of the target and the searchers and, as in Section 3.2, is called the target–searcher system.

Similar to the previous sections, let $\mathfrak{X} \subset \mathbb{R}^2$ be a 2D domain with the points $\mathfrak{x} = (x, y)$ defined by their Cartesian coordinates x and y. The searcher's and the target's trajectories over the domain $\mathfrak{X}$ are defined by the functions $\breve{\mathfrak{x}} : \mathfrak{X} \times [0, T] \to \mathfrak{X}$ and $\hat{\mathfrak{x}} : \mathfrak{X} \times [0, T] \to \mathfrak{X}$, correspondingly; thus, $\breve{\mathfrak{x}}(t) \in \mathfrak{X}$ is the searcher's coordinate at time t and $\hat{\mathfrak{x}}(t) \in \mathfrak{X}$ is the target's coordinate at this time. The target's motion is defined by the probability density function $u : \mathfrak{X} \times [0, T] \to \mathbb{R}^+$, which specifies conditional target's location probabilities $u(\mathfrak{x}, t)$ in the points $\mathfrak{x} \in \mathfrak{X}$ given that the search up to time $t \in [0, T]$ was unsuccessful. Consequently, the searcher's motion is defined by the probability density function $v : \mathfrak{X} \times [0, T] \to \mathbb{R}^+$.

Recall (see Section 2.3.3) that the observation process is defined based on the function $\psi : \mathfrak{X} \times \mathfrak{X} \to \mathbb{R}^+$, which specifies the probability $\psi(\hat{\mathfrak{x}}(t), \breve{\mathfrak{x}}(t))dt$ of detecting the target during the period $[t, t + dt)$, while the target's location is $\hat{\mathfrak{x}}(t)$ and the searcher's location is $\breve{\mathfrak{x}}(t)$. Then, the value according to the value $\int_{\mathfrak{X}} \psi(\mathfrak{x}, \breve{\mathfrak{x}}(t)) u(\mathfrak{x}, t) d\mathfrak{x}$ is the probability of detecting the target in the domain at time t, while the searcher is in the point $\breve{\mathfrak{x}}(t)$. More generally, denote by $\psi : \mathbb{R}^2 \to \mathbb{R}^+$ the function such that the value $\psi(u,v)$ specifies the possibility of detecting the target, while the probability of target's location in the point $\mathfrak{x}$ at time t is $u(\mathfrak{x}, t)$ and the probability of searcher's location in the point $\mathfrak{x}$ at time t is $v(\mathfrak{x}, t)$.

Consider the target, which without search is moving according to the Brownian process with respect to the environment, and the searcher, which follows the active Brownian motion in external potential field $\mathcal{U} : \mathfrak{X} \to \mathbb{R}$. The diffusion rates of the target and the searcher in the point $\mathfrak{x}$ at time t are denoted by $\hat{s}(\mathfrak{x}, t, u, v)$ and $\breve{s}(\mathfrak{x}, t, u, v)$, respectively. Accordingly, the drift coefficients of the target and the searcher are denoted by $\hat{m}(\mathfrak{x}, t, u, v)$ and $\breve{m}(\mathfrak{x}, t, u, v)$. To stress the axes, for which the rates ad drifts are defined, the corresponding indices are used if needed.

According to the algorithms of probabilistic search (see Section 4.4), assume that the searcher is attracted by the regions with maximal probabilities of finding the target. Then, at each time t, the potential $\mathcal{U}(\mathfrak{x})$ in the point $\mathfrak{x} \in \mathfrak{X}$ is $\mathcal{U}(\mathfrak{x}) = -\mathcal{U}_0 u(\mathfrak{x}, t)$, where $\mathcal{U}_0 \geq 0$ is a normalization coefficient.

Following the prey–predator system (3.7) and the model of active Brownian motion in external field (see, e.g., the Fokker–Planck equation [5.17]), the Turing system that governs the probabilities of the target and the searcher is defined as follows:

$$\begin{cases} \dfrac{\partial}{\partial t} u(\mathfrak{x}, t) = \hat{s}(\mathfrak{x}, t, u, v) \Delta u(\mathfrak{x}, t) - \hat{m}(\mathfrak{x}, t, u, v) \nabla u(\mathfrak{x}, t) - \mathcal{V}_0 \psi(u, v) u(\mathfrak{x}, t), \\ \dfrac{\partial}{\partial t} v(\mathfrak{x}, t) = \breve{s}(\mathfrak{x}, t, u, v) \Delta v(\mathfrak{x}, t) - \breve{m}(\mathfrak{x}, t, u, v) \nabla v(\mathfrak{x}, t) + \mathcal{U}_0 \nabla u(\mathfrak{x}, t) \nabla v(\mathfrak{x}, t), \end{cases} \tag{5.65}$$

where $\mathcal{V}_0 \geq 0$ is a normalization coefficient, and for both diffusion rates $\hat{s}(\mathfrak{x},t,u,v)$ and $\breve{s}(\mathfrak{x},t,u,v)$ and both drift coefficients $\hat{m}(\mathfrak{x},t,u,v)$ and $\breve{m}(\mathfrak{x},t,u,v)$, it is denoted

$$s(\mathfrak{x},t,u,v)\Delta p(\mathfrak{x},t) = \left(s_x(\mathfrak{x},t,u,v)\frac{\partial^2}{\partial x^2} + s_y(\mathfrak{x},t,u,v)\frac{\partial^2}{\partial y^2} \right) p(\mathfrak{x},t),$$

$$m(\mathfrak{x},t,u,v)\nabla p(\mathfrak{x},t) = \left(m_x(\mathfrak{x},t,u,v)\frac{\partial}{\partial x} + m_y(\mathfrak{x},t,u,v)\frac{\partial}{\partial y} \right) p(\mathfrak{x},t),$$

where indices x and y indicated the diffusion rates and drifts over the axes x and y, respectively.

Certainly, without searching both $\mathcal{V}_0 = 0$ and $\mathcal{U}_0 = 0$, the system (5.65) defines the probabilities of two agents moving with respect to the location probabilities of each other according to the diffusion rates and drifts. If also the rates and drifts do not depend on the probabilities $u(\mathfrak{x},t)$ and $v(\mathfrak{x},t)$, then the system is divided into two independent diffusion equations, which govern the evolution of the target's and the searcher's location probabilities over the domain. The dynamics of the target's and the searcher's location probabilities defined by the target–searcher system (5.65) is illustrated by the following example.

Example 5.8

Consider the target and the searcher moving according to the target–searcher system (5.65) in the square domain $\mathcal{X}$ of the size $n = 100 \times 100$ with the impermeable bounds.

To stress the influence of the observation process and attraction by the gradient field, it is assumed that the diffusion rates of the target and the searcher are constant and equal for both axes, that is, $\hat{s}(\mathfrak{x},t,u,v) = \hat{\sigma}^2 = 1$ and $\breve{s}(\mathfrak{x},t,u,v) = \breve{\sigma}^2 = 1$, and the drift coefficients are $\hat{m}(\mathfrak{x},t,u,v) = \breve{m}(\mathfrak{x},t,u,v) \equiv 0$. In addition, it is assumed that the values $\psi(u, v)$ in the points $\mathfrak{x} \in \mathcal{X}$ are defined as $\psi(u,v) = u(\mathfrak{x},t)v(\mathfrak{x},t)$, that is, the probability of detecting the target in the point $\mathfrak{x}$ at time t given that the probability of the searcher's location in this point is $u(\mathfrak{x},t)$ and the probability of the target's location in this point is $v(\mathfrak{x},t)$.

Initial location probabilities of both the target and the searcher at time $t = 0$ are defined by the binormal distribution with the central point $\mathfrak{x}(0) = (50,50)$ and variances over the axes $\sigma_1 = \sigma_2 = 10$. The normalization coefficients used in the system are $\mathcal{V}_0 = 10^5$ and $\mathcal{U}_0 = 10^4$.

The probability distributions of the target's and the searcher's locations in the points of the domain $\mathcal{X}$ for times $t = 0$, $t = 100$, $t = 500$, and $t = 1000$ are shown in Figure 5.22.

As it is expected, the target's distribution follows the diffusion with respect to the observations conducted by the searcher, while the searcher follows maximum gradient of the target's distribution. Notice that the same distributions of the searcher's location probabilities were obtained by the algorithms of probabilistic search presented in Section 4.4.1; for example of such distributions, see the paper by Kagan, Goren, and Ben-Gal (2012). ■

Given the target–searcher system (5.65), the trajectories of the target and the searcher are governed by the corresponding Langevin equations. In the simplest case, such equations define the motion of the Brownian particles in the external field provided by the probabilities $u(\mathfrak{x},t)$ and $v(\mathfrak{x},t)$ of, respectively, the target's and the searcher's location in the points $\mathfrak{x} \in \mathcal{X}$ at time t. The system of these equations is written as follows:

$$\begin{cases} \dfrac{d}{dt}\hat{\mathfrak{x}}(t) = \hat{v}(t), & \dfrac{d}{dt}\hat{v}(t) = -\hat{\gamma}_0\hat{v}(t) + \nabla U(\mathfrak{x},t) + \sqrt{2\hat{S}}\hat{\mathcal{G}}_t, \\ \dfrac{d}{dt}\breve{\mathfrak{x}}(t) = \breve{v}(t), & \dfrac{d}{dt}\breve{v}(t) = -\breve{\gamma}_0\breve{v}(t) + \nabla V(\mathfrak{x},t) + \sqrt{2\breve{S}}\breve{\mathcal{G}}_t, \end{cases} \quad (5.66)$$

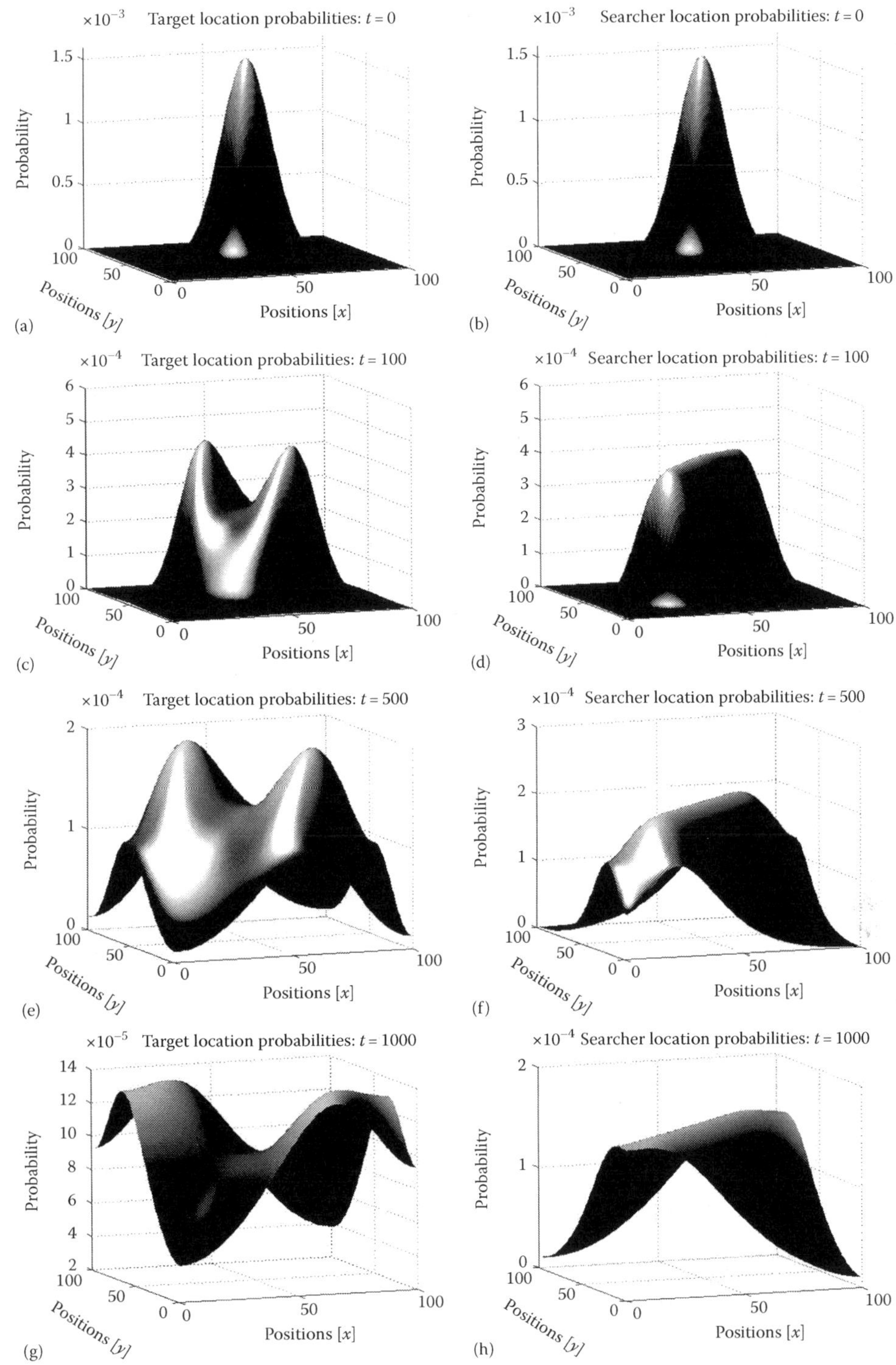

FIGURE 5.22 Evolution of the target's and the searcher's location probabilities defined by the target–searcher system (5.65). Figures (a) and (b) show the initial equivalent distributions of the target and the searcher, and figures (c) and (d) show the distributions at time $t = 100$, figures (e) and (f) at time $t = 500$, and figures (g) and (h) at time $t = 1000$. The diffusion rates of both the target and the searcher are $\hat{s}(\mathfrak{x},t,u,v)=\breve{s}(\mathfrak{x},t,u,v)\equiv 1$, drift coefficients are $\hat{m}(\mathfrak{x},t,u,v)=\breve{m}(\mathfrak{x},t,u,v)\equiv 0$, and the normalization coefficients are $\mathcal{V}_0 = 10^5$ and $\mathcal{U}_0 = 10^4$.

where $\mathcal{U}$ and $\mathcal{V}$ are potential fields defined by the target's and the searcher's location probabilities, that is, $\mathcal{U}(\mathfrak{x},t) \sim u(\mathfrak{x},t)$ and $\mathcal{V}(\mathfrak{x},t) \sim v(\mathfrak{x},t)$ for all $\mathfrak{x} \in \mathfrak{X}$ and $t \in [0, T]$; $\hat{\mathfrak{x}}(t)$ and $\check{\mathfrak{x}}(t)$ stand for the locations of the target and the searcher at time t and $\hat{v}(t)$ and $\check{v}(t)$ for their velocities at this time; $\hat{\gamma}_0$ and $\check{\gamma}_0$ denote the friction coefficients for the target and the searcher; $\hat{\mathcal{G}}_t$ and $\check{\mathcal{G}}_t$ are independent Gaussian white noise processes for the target and the searcher; and $\hat{S}$ and $\check{S}$ are intensities of the noise for each agent.

Notice that the system (5.66) is a general system of two independent Langevin equations, each of which defines the movements of the Brownian particle in the external field, while these fields are provided by the target–searcher system (5.65). The Langevin equations, which define the search process independently on the target–searcher system (5.65), involve specification of the nonlinear friction with respect to the target's and the searcher's locations and require additional studies.

The next example illustrates the trajectories of the target and the searcher defined by the systems (5.65) and (5.66).

Example 5.9

Let us follow the settings of the previous example and consider the trajectories, which follow the target's and the searcher's location probabilities defined by the target–searcher system (5.65) with the indicated parameters (see Figure 5.22). In the system (5.66), it is assumed that the friction coefficients are $\hat{\gamma}_0 = \check{\gamma}_0 = 0.5$ and the noise strengths are $\hat{S} = \check{S} = 0.5$. The values $\mathcal{U}(\mathfrak{x},t)$ and $\mathcal{V}(\mathfrak{x},t)$ of the potential fields $\mathcal{U}$ and $\mathcal{V}$ are specified by the normalized in time location probabilities $u(\mathfrak{x},t)$ and $v(\mathfrak{x},t)$, respectively.

Both the searcher and the target start in the center of the domain, that is, $\hat{\mathfrak{x}}(0) = (50,50)$ and $\check{\mathfrak{x}}(0) = (50,50)$, and the systems (5.65) and (5.66) are evaluated in parallel for times $t = 1,\ldots,1000$. The trajectories of the target and the searcher are shown in Figure 5.23.

It is seen that according to the Langevin equations appearing in the system (5.66), the agents follow maximum gradient of the corresponding probability distributions and tend to move in the regions with maximum probabilities. ■

Let us stress again that, similarly to the earlier considered search and foraging problem, the target's location probabilities defined by the target–searcher system (5.65) and its trajectory defined by the corresponding Langevin equation from the system (5.66) are the probabilities as they are

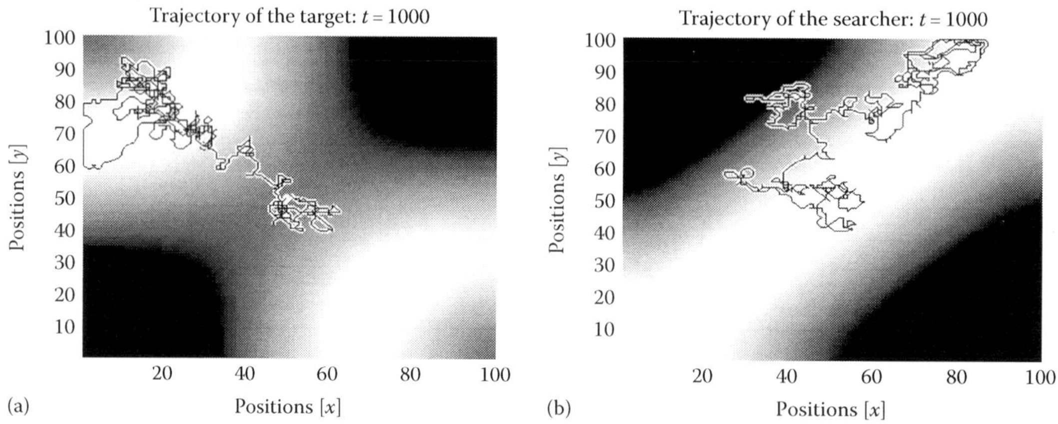

FIGURE 5.23 Trajectories of (a) the target and (b) the searcher defined by the target–searcher system (5.65) and the system (5.66) of Langevin equations for time $t = 1,\ldots,1000$. The settings and the parameters of the system (5.65) follow Example 5.8. In the system (5.66), the friction coefficients are $\hat{\gamma}_0 = \check{\gamma}_0 = 0.5$ and the noise strengths are $\hat{S} = \check{S} = 0.5$. The starting point of both agents is $\mathfrak{x} = (50,50)$.

considered by the searcher with respect to the observation results. In other words, the target–searcher system (5.65) provides the probabilities of finding the target with respect to the searcher's movements and detection function, and the system (5.66) defines the trajectory of the target as it is expected by the searcher with respect to the obtained observation results.

5.4.2 Ordering and Disordering Phenomena and Stability of the Patterns

As indicated earlier, the Turing systems allow different types of behavior and transitions between these types with respect to the values of control parameters. In the case of certain feedback, such transitions can hold without external influence; such process is usually considered as self-organization, that is, the appearance of ordered spatial or temporal patterns in initially disordered or homogeneous system.

5.4.2.1 Self-Organization and Stability of Turing Patterns

Let us return to general Turing system (5.53) with the parameter α as follows:

$$\begin{cases} \dfrac{\partial}{\partial t} f(\mathfrak{x},t) = s_f(\mathfrak{x},t)\Delta f(\mathfrak{x},t) + \mathbb{F}_\alpha(f,g), \\ \dfrac{\partial}{\partial t} g(\mathfrak{x},t) = s_g(\mathfrak{x},t)\Delta g(\mathfrak{x},t) + \mathbb{G}_\alpha(f,g), \end{cases} \tag{5.67}$$

where, as mentioned earlier, $\mathfrak{x} = (x,y) \in \mathfrak{X} \subset \mathbb{R}^2$ and $t \in \mathbb{R}^+$. Notice again that the existence of spatial and temporal patterns in the system is provided by the values $s_f(\mathfrak{x},t)$ and $s_g(\mathfrak{x},t)$ of the diffusion rates and the functions $\mathbb{F}_\alpha(f,g)$ and $\mathbb{G}_\alpha(f,g)$.

Assume that initially, the system is in homogeneous spatial state, that is, at time $t = 0$ there cannot be any distinguished regular patterns in the values $f(\mathfrak{x},0)$ and $g(\mathfrak{x},0)$ since these values are equal or random over the domain $\mathfrak{X}$. In addition, assume that the diffusion rates s_f and s_g and the functions $\mathbb{F}_\alpha$ and $\mathbb{G}_\alpha$ are such that the spatial patterns are allowed. The process of formation of such regular spatial patterns from initially homogeneous state is usually called self-organization, and the system, which support such process, is called self-organizing system.

Certainly, "self-organization" is more philosophical or methodological rather than mathematical or physical term, which allows consideration of different phenomena observed in nonequilibrium nonlinear systems from a unified point of view (Prigogine, 1980). Originally, in the cybernetics literature, this term indicated the principle "the order from noise," while in later physical considerations, it mainly stands for the principle "the order from fluctuations" (Nosonovsky & Mortazavi, 2014); the spatial patterns resulted from self-organization are called dissipative structures. For complete information on self-organization and synergetics, see the earlier indicated sources and the recent volume by Haken (2012).

To clarify the behavior of self-organizing system, let us consider the dissipative patterns demonstrated by the bacterial colonies (Ben-Jacob & Garik, 1990; Ben-Jacob & Levine, 2001) using the simplified version of the reaction–diffusion system, which describes the growth of lubricating bacteria (Ben-Jacob, Cohen, Golding, & Kozlovsky, 2001; Golding, Kozlovsky, Cohen, & Ben-Jacob, 1998; Tsimring et al., 1995). The considered system is the following:

$$\begin{cases} \dfrac{\partial}{\partial t} f(\mathfrak{x},t) = s_f(t)\Delta f(\mathfrak{x},t) + f(\mathfrak{x},t)g(\mathfrak{x},t) - \alpha f(\mathfrak{x},t), \\ \dfrac{\partial}{\partial t} g(\mathfrak{x},t) = s_g(t)\Delta g(\mathfrak{x},t) - f(\mathfrak{x},t)g(\mathfrak{x},t), \end{cases}$$

where $s_f(t) = s_g(t) = \sin(\omega t)$. In this system, activator f represents the concentration of bacteria, and inhibitor g is a concentration of nutrients. In contrast to the original one (Golding et al., 1998;

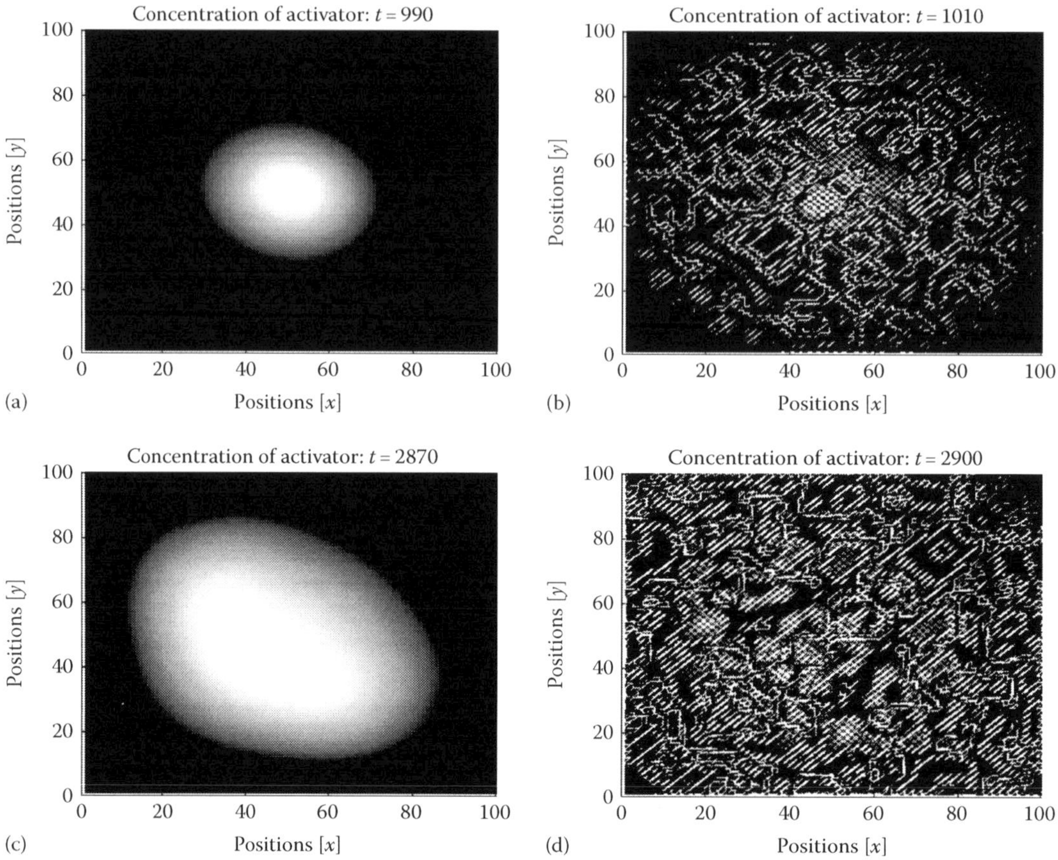

FIGURE 5.24 Concentrations of activator (concentration of bacteria) with harmony in time lubrication sin (0.01t). Initially, at time t = 0, the activator is concentrated in the central point $\mathfrak{x}$ = (50,50) of the domain and the inhibitor is distributed uniformly. Figure (a) depicts the concentration of activator at time t = 990, figure (b) at time t = 1010, figure (c) at time t = 2870, and figure (d) at time t = 2900.

Ben-Jacob et al., 2001), here it is assumed that the lubrication is harmonic and does not depend on the bacteria growth. The patterns appearing in the system are illustrated by Figure 5.24. Initially, the bacterial are concentrated in the central point of the domain, that is, $f(\mathfrak{x},0)=1$ in $\mathfrak{x}=(50,50)$ and $f(\mathfrak{x},0)=0$ in all other points $\mathfrak{x}\in\mathfrak{X}\setminus\{(50,50)\}$, and the inhibitor is distributed uniformly over a domain, that is, $g(\mathfrak{x},0)=1$ for all $\mathfrak{x}\in\mathfrak{X}$. The parameter is $\alpha = 0.5$ and $\omega = 0.01$.

It is seen that in the system, the disordered state is substituted by certain regular spatial pattern. Then, the system returns to the disordered state, which again is transformed to the regular pattern. The form of the pattern is defined by the functions $\mathbb{F}_\alpha(f,g)=f(\mathfrak{x},t)g(\mathfrak{x},t)-\alpha f(\mathfrak{x},t)$ and $\mathbb{G}_\alpha(f,g)=-f(\mathfrak{x},t)g(\mathfrak{x},t)$ and diffusion rates $s_f(t) = s_g(t) = \sin(\omega t)$.

Notice that if the diffusion rates depend on the lubrication or bacterial concentration by the power law (Ben-Jacob et al., 2001; Golding et al., 1998) (cf. Section 4.2.3), then the obtained patterns are the same as the patterns demonstrated by the multifractional Brownian motion (Havlin et al., 1993).

Finally, let us give some remarks regarding the stability of the spatial patterns. The succeeding consideration follows the extension of the Lyapunov method suggested by Sirazetdinov (1987). The idea of the method with respect to the considered systems is the following (Kagan, 2008).

Assume that there is an abstract process $\mathcal{P}:\mathbb{R}^+\to\mathfrak{S}$ with the states $\mathfrak{s}\in\mathfrak{S}$ and consider two functions $\mathcal{M}_0$ and $\mathcal{M}$ such that $\mathcal{M}_0(\mathcal{P},t)$ is the distance between the states of disturbed and nondisturbed processes and $\mathcal{M}(\mathcal{P},t)$ is the distance between the states of the process at times $(t-dt)$ and t.

The process $\mathcal{P}$ is called stable by the measures $\mathcal{M}_0$ and $\mathcal{M}$, if for every $\varepsilon > 0$ there exists $\delta(\varepsilon) > 0$ such that for every initial state of $\mathcal{P}(t_0)$ such that $\mathcal{M}_0(\mathcal{P}, t_0) < \delta(\varepsilon)$, it holds true that $\mathcal{M}(\mathcal{P}, t) < \varepsilon$, where $t, t_0 \in \mathbb{R}^+$ and $t \geq t_0$. In other words, the process is called stable if a bounded initial disturbance does not lead to the divergence of the process.

With respect to the Turing system (5.67), this definition leads to the following. Each equation in this system is a diffusion equation with the disturbance provided by the functions $\mathbb{F}_\alpha$ and $\mathbb{G}_\alpha$, and if these functions avoid the influence of the reagents on each other, then the system is divided into two independent diffusion processes. Following this observation, the measure $\mathcal{M}_0$ for the system (5.67) is defined as follows:

$$\mathcal{M}_0(t) = \int_{\mathfrak{X}} \left| \mathbb{F}_\alpha(f, g) - \mathbb{G}_\alpha(f, g) \right| d\mathfrak{x}. \tag{5.68}$$

For example, if the system specifies the dynamics of van der Pol equations, then following Equation 5.57, functions $\mathbb{F}_\alpha$ and $\mathbb{G}_\alpha$ are

$$\mathbb{F}_\alpha(f, g) = \frac{1}{2}\left(\alpha - \beta E(\mathfrak{x}, t)\right) f(\mathfrak{x}, t) + g(\mathfrak{x}, t),$$

$$\mathbb{G}_\alpha(f, g) = \frac{1}{2}\left(\alpha - \beta E(\mathfrak{x}, t)\right) g(\mathfrak{x}, t) - \omega^2 f(\mathfrak{x}, t),$$

Then, the measure $\mathcal{M}_0$ is defined as follows:

$$\mathcal{M}_0(t) = \int_{\mathfrak{X}} \left| \frac{1}{2}\alpha\left(f(\mathfrak{x}, t) - g(\mathfrak{x}, t)\right) - \frac{1}{2}\beta E(\mathfrak{x}, t)\left(f(\mathfrak{x}, t) - g(\mathfrak{x}, t)\right) + \left(\omega^2 f(\mathfrak{x}, t) + g(\mathfrak{x}, t)\right) \right| d\mathfrak{x}.$$

For equal concentrations $f(\mathfrak{x}, t) = g(\mathfrak{x}, t)$, $\mathfrak{x} \in \mathfrak{X}$, it gives

$$\mathcal{M}_0(t) = \int_{\mathfrak{X}} \left| \left(\omega^2 f(\mathfrak{x}, t) + g(\mathfrak{x}, t)\right) \right| d\mathfrak{x}.$$

Notice that if the concentrations f and g are constant in time, that is, $\frac{\partial}{\partial t} f(\mathfrak{x}, t) = 0$ and $\frac{\partial}{\partial t} g(\mathfrak{x}, t) = 0$, the measure $\mathcal{M}_0(t)$ is

$$\mathcal{M}_0(t) = \int_{\mathfrak{X}} \left| s_f(\mathfrak{x}, t)\Delta f(\mathfrak{x}, t) - s_g(\mathfrak{x}, t)\Delta g(\mathfrak{x}, t) \right| d\mathfrak{x},$$

and depends on the initial distribution of the reagents and diffusion rates. If, in addition, $f(\mathfrak{x}, t) = g(\mathfrak{x}, t)$ for all $\mathfrak{x} \in \mathfrak{X}$, then $\mathcal{M}_0(t) = C\int_{\mathfrak{X}} \left| s_f(\mathfrak{x}, t) - s_g(\mathfrak{x}, t) \right| d\mathfrak{x}$, $C = \text{const}$, in coincidence with the previous remark regarding the existence of the patterns in the case of different diffusion rates of activator and inhibitor. If $\mathbb{F}_\alpha(f, g) \equiv \mathbb{G}_\alpha(f, g)$, then $\mathcal{M}_0(t) = 0$ and, as indicated earlier, the system defines two independent processes. Certainly, each of these processes can be unstable with respect to its diffusion rate $s_f(\mathfrak{x}, t)$ or $s_g(\mathfrak{x}, t)$, respectively, and its stability can be considered independently.

Now, let us define the measure $\mathcal{M}$. In general, this measure can be defined as follows:

$$\mathcal{M}(t)=\int_{\mathfrak{X}}\left|\frac{\partial}{\partial t}\left(\nabla f(\mathfrak{x},t)-\nabla g(\mathfrak{x},t)\right)\right|d\mathfrak{x}, \tag{5.69}$$

which for the functions constant in space coordinates, as well as for the functions such that $f(\mathfrak{x},t)=g(\mathfrak{x},t)$ for all $\mathfrak{x}\in\mathfrak{X}$, immediately results in $\mathcal{M}(t)=0$. However, if the functions f and g are bounded, as it certainly holds for probability distributions, then the measure $\mathcal{M}$ can be defined by the ratio of these functions; such possibility will be used in the succeeding text. Notice that both if $\mathbb{F}_\alpha(f,g)\equiv\mathbb{G}_\alpha(f,g)$ and if the system is divided to two linear diffusion equations, that is, $\mathbb{F}_\alpha(f,g)\equiv 0$ and $\mathbb{G}_\alpha(f,g)\equiv 0$, this measure depends on the diffusion rates $s_f(\mathfrak{x},t)$ and $s_g(\mathfrak{x},t)$ and can be used for stability analysis of the process defined by the function f with respect to g and vice versa.

The measures $\mathcal{M}_0$ and $\mathcal{M}$ for the system (5.55) are illustrated in Figure 5.25. For comparison, this figure also shows the measures for the system with the same function $\mathbb{F}_\alpha(f,g)=g$ and the function $\mathbb{G}_\alpha(f,g)=-\omega^2 f+g^3$, which includes additional nonlinear term. Initial concentrations and parameters of the system are the same as the ones used in Figure 5.20.

As it is expected, for the system (5.55), in which $\mathbb{F}_\alpha(f,g)=g$ and the function $\mathbb{G}_\alpha(f,g)=-\omega^2 f$, the measure $\mathcal{M}_0$ does not depend on time and the stability of the patterns is characterized by the measure $\mathcal{M}$. In contrast, for the system (5.54) with $\mathbb{F}_\alpha(f,g)=g$ and $\mathbb{G}_\alpha(f,g)=-\omega^2 f+g^3$, both measures are nonlinear and the stability of the patterns is defined by the values $\mathcal{M}_0(t)$ and $\mathcal{M}(t)$.

Certainly, the earlier presented techniques for the stability analysis of the Turing patterns are one of the wide range of possible methods. For the other methods, especially for consideration of the Turing systems with different boundary conditions, see, for example, Sowers (1994), Alos & Bonaccorsi, (2002), Ishikawa, Miyajima, & Tasaka (2003).

5.4.2.2 Spatial and Temporal Patterns in Target–Searcher System

Let us return to the target–searcher system (5.65), which deals with location probabilities of the target and the searcher with respect to the results of observations. Following this system (cf. algorithms of probabilistic search presented in Section 4.4 and Bayesian search [Stone, 2001; Stone, Barlow, & Corwin, 1999], the searcher changes the target's location probabilities (as they observed

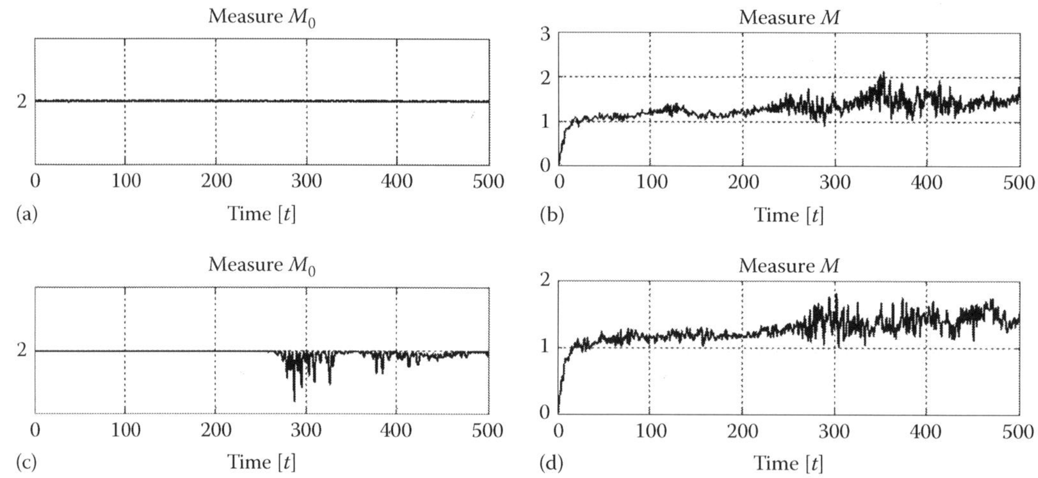

FIGURE 5.25 Measures $\mathcal{M}_0$ and $\mathcal{M}$ for the system (5.55) (figures (a) and (b)) and for the system (5.54) with the functions $\mathbb{F}_\alpha(f, g) = g$ and $\mathbb{G}_\alpha(f, g) = -\omega^2 f + g^3$ (figures (c) and (d)). Parameters of the system are $\sigma_f^2=\sigma_g^2=0.5$ and $\omega^2 = 1$, and the time is $t = 1,\ldots,500$.

the searcher) and finally, in the case of successful search, obtains the distribution such that the probability in the point, where the target is located is one and in all other points is zero. In the case of unsuccessful search, the process continues infinitely, while the target's and the searcher's location probabilities form certain spatial ordered or disordered structures over the domain, which change in time without convergence to the distribution required for successful search.

Consider the target–searcher system (5.65) as a Turing system (5.67), in which the function u represents the distribution of activator and function v the distribution of inhibitor over a domain. For simplicity, assume that the diffusion rates do not depend on the probabilities, that is, $\hat{s}(\mathfrak{x},t,u,v) \equiv \hat{s}(\mathfrak{x},t)$ and $\breve{s}(\mathfrak{x},t,u,v) \equiv \breve{s}(\mathfrak{x},t)$. In addition, assume that similar to Examples 5.8 and 5.9, the drifts are $\hat{m}(\mathfrak{x},t,u,v) = \breve{m}(\mathfrak{x},t,u,v) \equiv 0$ and the detection function ψ is such that $\psi(u,v) = u(\mathfrak{x},t)v(\mathfrak{x},t)$ for all points $\mathfrak{x} \in \mathfrak{X}$ and times $t \in [0, T]$. The target–searcher system in this case is written in the form of the Turing system:

$$\begin{cases} \dfrac{\partial}{\partial t} u(\mathfrak{x},t) = \hat{s}(\mathfrak{x},t)\Delta u(\mathfrak{x},t) + \mathbb{F}_{\mathcal{V}_0}(u,v), \\ \dfrac{\partial}{\partial t} v(\mathfrak{x},t) = \breve{s}(\mathfrak{x},t)\Delta v(\mathfrak{x},t) + \mathbb{G}_{\mathcal{U}_0}(u,v), \end{cases} \tag{5.70}$$

where

$$\mathbb{F}_\alpha(f,g) = -\mathcal{V}_0\left(u(\mathfrak{x},t)\right)^2 v(\mathfrak{x},t) \text{ and } \mathbb{G}_\alpha(f,g) = \mathcal{U}_0 \nabla u(\mathfrak{x},t) \nabla v(\mathfrak{x},t).$$

For this system, the measures $\mathcal{M}_0$ and $\mathcal{M}$ defined by Equations 5.68 and 5.69 are written as follows:

$$\mathcal{M}_0(t) = \int_{\mathfrak{X}} \left| -\mathcal{V}_0\left(u(\mathfrak{x},t)\right)^2 v(\mathfrak{x},t) - \mathcal{U}_0 \nabla u(\mathfrak{x},t) \nabla v(\mathfrak{x},t) \right| d\mathfrak{x}, \tag{5.71}$$

$$\mathcal{M}(t) = \int_{\mathfrak{X}} \left| \frac{\partial}{\partial t}\left(\nabla u(\mathfrak{x},t) - \nabla v(\mathfrak{x},t)\right) \right| d\mathfrak{x}. \tag{5.72}$$

The evolution of these measures for the system with the parameters used in Example 5.8 is shown in Figure 5.26.

It is seen that both measures are bounded by the values of the same order as the values of the functions u and v and both decrease with time to zero. Thus, the patterns provided by the system are stable and the search not necessary successful, and the targets and the searcher can infinitely move over a domain.

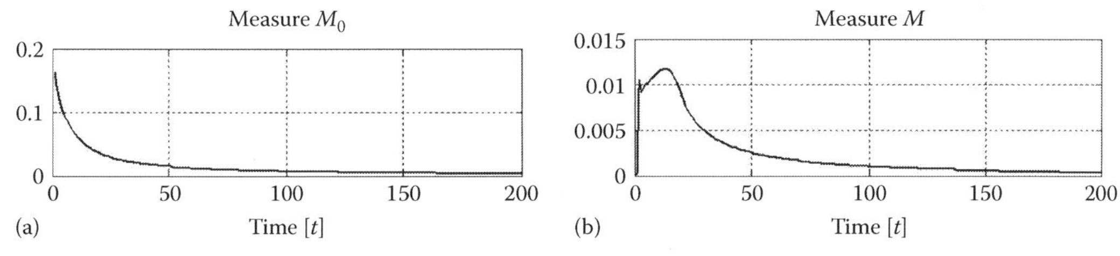

FIGURE 5.26 Measures $\mathcal{M}_0$ and $\mathcal{M}$ (figures (a) and (b), respectively) for the system (5.70) with functions $\mathbb{F}_\alpha(f, g) = g$ and $\mathbb{G}_\alpha(f, g) = -\omega^2 f + g^3$. Parameters of the system are the same as in Example 5.8, and the time is $t = 1,\ldots, 200$.

In contrast, if the diffusion rates are $\hat{s}(\mathfrak{x},t) \equiv \hat{\sigma}^2 = 1.1$ and $\breve{s}(\mathfrak{x},t) \equiv \breve{\sigma}^2 = 0.1$, then the patterns are unstable, and the searcher finds the target with necessity. The patterns provided by the system and the evolution of the measures are shown in Figure 5.27.

In the beginning of search, the values of both measures $\mathcal{M}_0$ and $\mathcal{M}$ are close to zero. Then, both measures start fluctuating, and at certain time $t^* = 173$, the measure $\mathcal{M}_0$ reaches its maximum value $\mathcal{M}_0(t^*) \sim 6 \cdot 10^4$ that corresponds to the moment of finding the target. Notice that the value of measure $\mathcal{M}$ at this time is still the same as at the beginning of fluctuations. The target's and the searcher's locations before finding the target at $t = 100$ are shown in Figure 5.27e and f, respectively. After finding the target, the locations of the target and the searcher do not change and both agents slightly fluctuate around the target's location; Figure 5.27g and h correspondingly present the locations of the target and the searcher after finding the target in the point $\mathfrak{x} = (70, 70)$.

Notice that the applied measures $\mathcal{M}_0$ and $\mathcal{M}$ are not unique, and since the functions u and v define the probability distributions of the target and the searcher over the domain, the stability of the patterns can be considered using the Kullback–Leibler distance between these distributions

$$\mathcal{KL}(t) = \int_{\mathcal{X}} u(\mathfrak{x},t) \ln \frac{u(\mathfrak{x},t)}{v(\mathfrak{x},t)} d\mathfrak{x},$$

where, by convention, it is assumed that $0 \ln 0 = 0$. Because of the entropic nature of this distance, also it is often called the Boltzmann–Kullback measure (Klimontovich, 1991).

The presented target–searcher system provides a general framework for modeling the processes of probabilistic search and foraging. Certainly, for specific tasks, the functions $\mathbb{F}_\alpha$ and $\mathbb{G}_\alpha$ as well as the functions $\hat{s}$ and $\breve{s}$, which define the diffusion rates, and the functions $\hat{m}$ and $\breve{m}$, which define the drifts, should be specified with respect to the locomotion abilities of the target and the searcher. In addition, this system together with the model of canonical active Brownian motion (see Section 5.3.2) forms a basis for the model of social foraging; the next section includes some useful remarks on such model.

5.4.3 Model of Social Foraging

In Section 5.4.1, the target–searcher system was defined as a Turing system, which governs an evolution of the location probabilities without specific stressing of the number of searchers; however, the examples addressed the case of search by a single searcher. In the succeeding text, the system is formulated for the case of search by several searchers, and the obtained equations complete the consideration of the multiagent search discussed in Section 5.3.2.

Similar to Section 5.3.2, assume that the search is conducted by the group of m searchers. The evolution of the searcher's location probabilities in the group can be considered in two ways. The first one follows the canonical approach with the Hamiltonian H representing the energy of the swarm and corresponding probability density function $v_{1\ldots m}$, which specifies the mutual distribution of m searchers (see Equation 5.37). Then, the target–searcher system for m searchers is defined as follows:

$$\begin{cases} \dfrac{\partial}{\partial t} u(\mathfrak{x},t) = \hat{s}(\mathfrak{x},t,u,v)\Delta u(\mathfrak{x},t) - \hat{m}(\mathfrak{x},t,u,v)\nabla u(\mathfrak{x},t) - \mathcal{V}_0 \psi(u, v_{1\ldots m}) u(\mathfrak{x},t), \\ \dfrac{\partial}{\partial t} v_{1\ldots m}(\mathfrak{x},t) = \breve{s}(\mathfrak{x},t,u,v)\Delta v_{1\ldots m}(\mathfrak{x},t) - \breve{m}(\mathfrak{x},t,u,v)\nabla v_{1\ldots m}(\mathfrak{x},t) + \mathcal{U}_0 \nabla u(\mathfrak{x},t)\nabla v_{1\ldots m}(\mathfrak{x},t), \end{cases} \tag{5.73}$$

with appropriate diffusion rates and drifts defined with respect to the function $v_{1\ldots m}$.

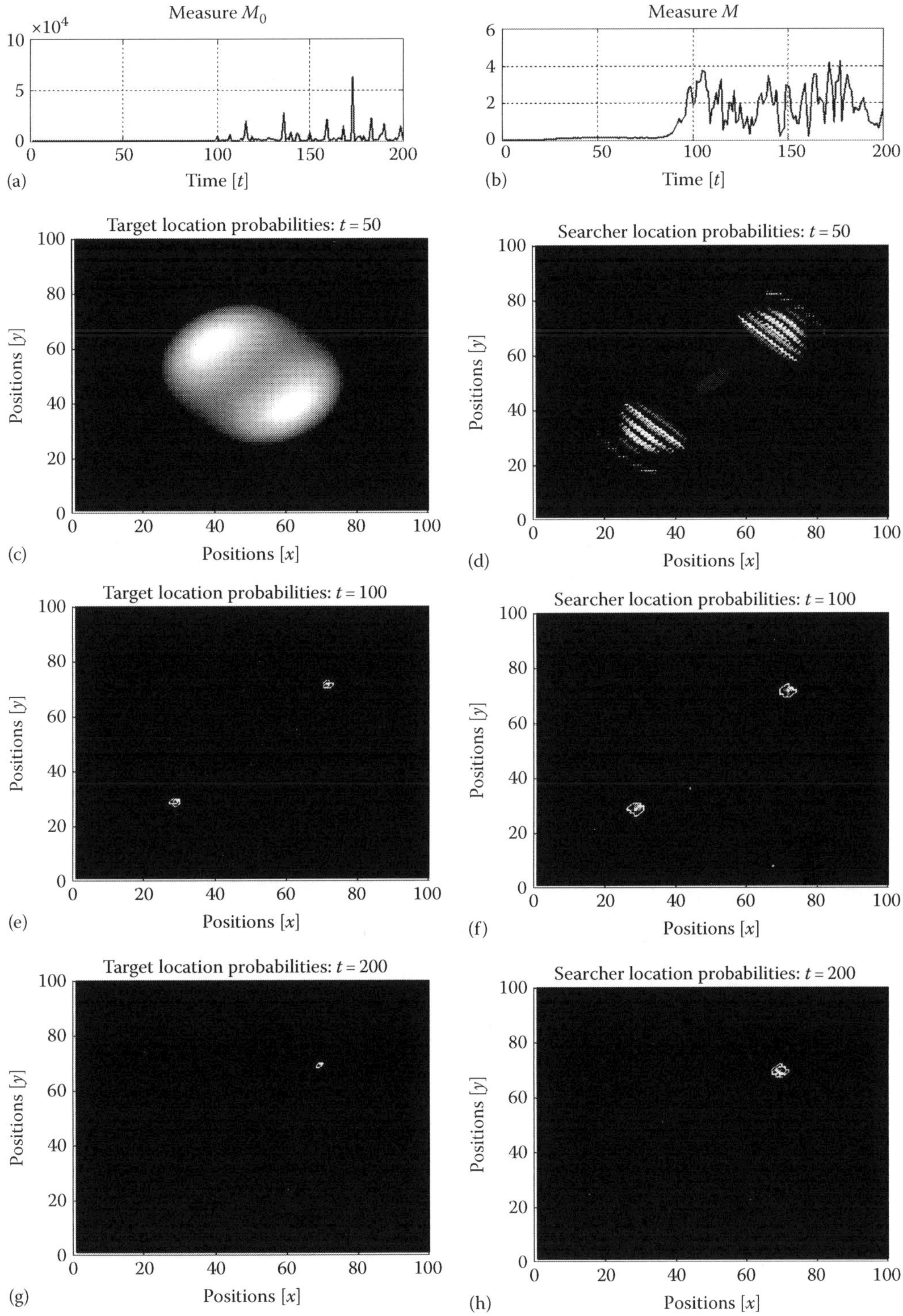

FIGURE 5.27 Measures $\mathcal{M}_0$ and $\mathcal{M}$ (figures (a) and (b), respectively) and spatial patterns for the target (figures (c), (e), and (g)) and for the searcher (figures (d), (f), and (h)) provided by the system with the functions $\mathbb{F}_\alpha(f, g) = g$ and $\mathbb{G}_\alpha(f, g) = -\omega^2 f + g^3$. The diffusion rates are $\hat{s}(\mathfrak{x},t) \equiv \hat{\sigma}^2 = 1.1$ and $\breve{s}(\mathfrak{x},t) \equiv \breve{\sigma} = 0.1$, and the other parameters of the system are the same as in Example 5.8, and the time is $t = 1,\ldots, 200$.

Following the second approach, the probability density functions v_j of each jth searcher, $j = 1, 2,\ldots, m$, are defined separately, and the detection function $\psi : \mathbb{R}^{m+1} \to \mathbb{R}^+$ is such that the value $\psi(u, v_1, v_2,.., v_m)$ is a potential of detecting the target, while the probability of target's location in the point $\mathfrak{x}$ at time t is $u(\mathfrak{x},t)$ and the probability of searchers' locations in the point $\mathfrak{x}$ at time t are $v_j(\mathfrak{x},t)$. The target–searcher system in this case includes $m + 1$ equations and is defined as follows:

$$\begin{cases} \dfrac{\partial}{\partial t}u(\mathfrak{x},t) = \hat{s}(\mathfrak{x},t,u,v)\Delta u(\mathfrak{x},t) - \hat{m}(\mathfrak{x},t,u,v)\nabla u(\mathfrak{x},t) - \mathcal{V}_0\psi(u,v_1,v_2,..,v_m)u(\mathfrak{x},t), \\ \dfrac{\partial}{\partial t}v_j(\mathfrak{x},t) = \breve{s}(\mathfrak{x},t,u,v_j)\Delta v_j(\mathfrak{x},t) - \breve{m}(\mathfrak{x},t,u,v_j)\nabla v_j(\mathfrak{x},t) + \mathcal{U}_0\nabla u(\mathfrak{x},t)\nabla v_j(\mathfrak{x},t) \end{cases} \tag{5.74}$$

with appropriate target's diffusion rates $\hat{s}(\mathfrak{x},t,u,v)$ and drifts $\hat{m}(\mathfrak{x},t,u,v)$ defined with respect to m functions v_j, $j = 1,2,\ldots,m$.

Notice that in both cases, communication and interactions between the agents are defined by the same manner as in Section 5.3.2, and are based on the attraction and repulsion potentials.

The evolution of the location probabilities defined by the systems (5.73) and (5.74) is illustrated by the following example.

Example 5.10

As in the previous examples, assume that the search is conducted in the square domain $\mathcal{X}$ of the size $n = 100 \times 100$ with the impermeable bounds and that initial location probabilities of the target at time $t = 0$ are defined by the binormal distribution with the central point $\mathfrak{x}(0) = (50,50)$ and variances over the axes $\sigma_1 = \sigma_2 = 10$.

The search is conducted by $m = 10$ searchers, and initial location probabilities of each searcher are defined by the binormal distribution with random central point $\breve{\mathfrak{x}}_j(0)$ such that its x and y coordinates are drawn uniformly. The variances over the axes for each searcher are also $\sigma_1 = \sigma_2 = 10$.

Assume that the location probabilities are governed by the system (5.74). Following the settings of Example 5.8, it is assumed that the diffusion rates of both the target and the searchers are constant $\hat{s}(\mathfrak{x},t,u,v) = \hat{\sigma}^2 = 1$ and $\breve{s}(\mathfrak{x},t,u,v) = \sigma^2 = 1$ and the drift coefficients are $\hat{m}(\mathfrak{x},t,u,v) = \breve{m}(\mathfrak{x},t,u,v) \equiv 0$. The implemented detection function ψ is a direct extension of the detection function used in Example 5.8, and its values are $\psi(u,v_1,v_2,..,v_m) = u(\mathfrak{x},t)\sum_{j=1}^{m} v_j(\mathfrak{x},t)$

The evolution of the target's location probabilities $u(\mathfrak{x},t)$ and of the probabilities $v_{1\ldots m}(\mathfrak{x},t) \sim \sum_{j=1}^{m} v_j(\mathfrak{x},t)$ and the searchers' location in the points $\mathfrak{x} \in \mathcal{X}$ of the domain $\mathcal{X}$ is illustrated in Figure 5.28. The figure shows the target's and the searchers' probabilities at times $t = 0$, $t = 100$, and $t = 500$.

Notice that in the simulation it is assumed that the searchers do not communicate and each of them acts independently with respect to the target's location probabilities. In contrast, the target's location probabilities are defined using the detection function, which considers the location probabilities of all searchers.

The trajectories of the target and the searchers are defined by the system of $m + 1$ Langevin equations, which is a direct extension of the system (5.66), namely (cf. the Langevin equation [5.36]),

$$\begin{cases} \dfrac{d}{dt}\hat{\mathfrak{x}}(t) = \hat{v}(t), & \dfrac{d}{dt}\hat{v}(t) = -\hat{\gamma}_0\hat{v}(t) + \nabla U(\mathfrak{x},t) + \sqrt{2\hat{S}}\hat{\mathcal{G}}_t, \\ \dfrac{d}{dt}\breve{\mathfrak{x}}_j(t) = \breve{v}_j(t), & \dfrac{d}{dt}\breve{v}_j(t) = -\breve{\gamma}_{0j}\breve{v}_j(t) + \nabla V(\mathfrak{x},t) + \sqrt{2\breve{S}}\breve{\mathcal{G}}_{jt}, \end{cases} \tag{5.75}$$

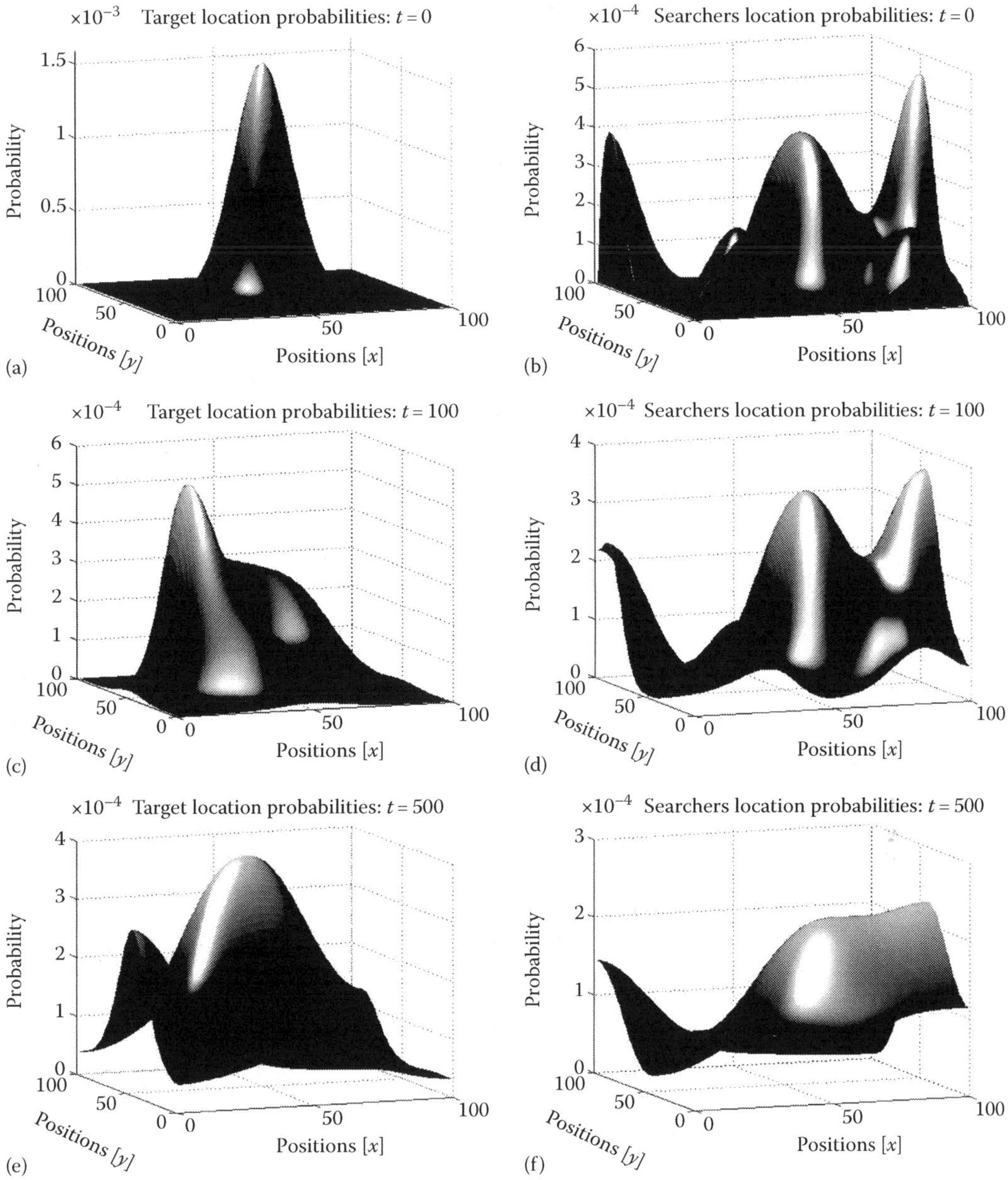

FIGURE 5.28 Evolution of the target's location probabilities and the location probabilities of $m = 10$ searchers governed by the target–searcher system (5.74). Figures (a) and (b) show the initial distributions of the target and the searchers, and figures (c) and (d) show the distributions at time $t = 100$ and figures (e) and (f) at time $t = 500$. The diffusion rates of both the target and the searchers are $\hat{s}(\mathfrak{x},t,u,v) = \breve{s}(\mathfrak{x},t,u,v) \equiv 1$, drift coefficients are $\hat{m}(\mathfrak{x},t,u,v) = \breve{m}(\mathfrak{x},t,u,v) \equiv 0$, and the normalization coefficients are $\mathcal{V}_0 = 10^5$ and $\mathcal{U}_0 = 10^4$.

where $\mathcal{U}$ and $\mathcal{V}$ are potential fields defined by the target's and the searchers' location probabilities, that is, $\mathcal{U}(\mathfrak{x},t) \sim u(\mathfrak{x},t)$ and $\mathcal{V}(\mathfrak{x},t) \sim \sum_{j=1}^{m} v_j(\mathfrak{x},t)$ for all $\mathfrak{x} \in \mathfrak{X}$ and $t \in [0,T]$; $\hat{\mathfrak{x}}(t)$ stands for the location of the target and $\hat{v}(t)$—for its velocity at time t; and $\breve{\mathfrak{x}}_j(t)$ denotes the location of jth searcher and $\breve{v}_j(t)$ its velocity at time t, $j = 1, 2,..., m$. The values $\hat{\gamma}_0$ and $\breve{\gamma}_{0j}$ denote the friction coefficients for the target and jth searchers, $\hat{\mathcal{G}}_t$ and $\hat{\mathcal{G}}_{jt}$ are independent Gaussian white noise processes for the target and jth searcher, and $\hat{S}$ and $\breve{S}$ are intensities of the noise for the target and the searcher. ■

The trajectories of the target and the searchers governed by the system (5.75) with respect to the location probabilities defined by the systems (5.73) and (5.74) are illustrated by the following example.

Example 5.11

Assume that the target and $m = 10$ searchers act following the same settings as in Example 5.10, and that the target's and the searchers' location probabilities are the same as shown in Figure 5.28. As mentioned earlier, in the system (5.75) it is also assumed that the friction coefficients are $\hat{\gamma}_0 = \breve{\gamma}_{0j} = 0.5$, $j = 1, 2,..., m$, and the noise strengths are $\hat{S} = \breve{S} = 0.5$. The values $\mathcal{U}(\mathfrak{x},t)$ and $\mathcal{V}(\mathfrak{x},t)$ of the potential fields $\mathcal{U}$ and $\mathcal{V}$ are specified by the normalized location probabilities $u(\mathfrak{x},t)$ and $\sum_{j=1}^{m} v_j(\mathfrak{x},t)$, respectively.

The target starts in the center of the domain, that is, $\hat{\mathfrak{x}}(0) = (50,50)$, and the searchers start in randomly chosen points $\breve{\mathfrak{x}}_j(0)$, which correspond to the centers of distributions used in Example 5.11. The systems (5.74) and (5.75) are evaluated in parallel for times $t = 1,...,500$, and the trajectories of the target and the searchers for times $t = 0$, $t = 100$, and $t = 500$ are shown in Figure 5.29.

It is seen that according to its Langevin equation, the target follows the maximum gradient defined by the searchers' location probabilities, and without communication and interactions, each searcher follows its potential field. Notice that for the searchers, which are able to communicate using attraction and repulsion potentials, the trajectories will differ from the ones shown in Figure 5.29; such case was considered in Example 5.7 and illustrated in Figure 5.18. ■

The presented example finalizes the consideration of the target–searcher system. As indicated earlier, this system, both for the search by the single searcher and for the search by the group of searchers, provides a basis for consideration of different search scenarios. In the case of communicating and interacting searchers, the system can include the corresponding potentials, which define interactions between the searchers (see Section 5.3.2), and for the searchers interacting with the environment, the model of active Brownian motion with energy depot can be implemented.

Finally, notice again that for the successful search, the Turing patterns appearing in the target–searcher system should be unstable, and the search time depends on the divergence of the stability measures of these patterns. The optimization of search can be conducted following the approach that was used in the algorithms of search and screening presented in Sections 2.2 and 2.3, where the distribution of the search efforts corresponds to the distribution of the searchers' locations. The algorithmic implementation of the presented models and specification of the agents' trajectories follows the same scheme as already presented algorithms of probabilistic search and foraging (see Section 5.2).

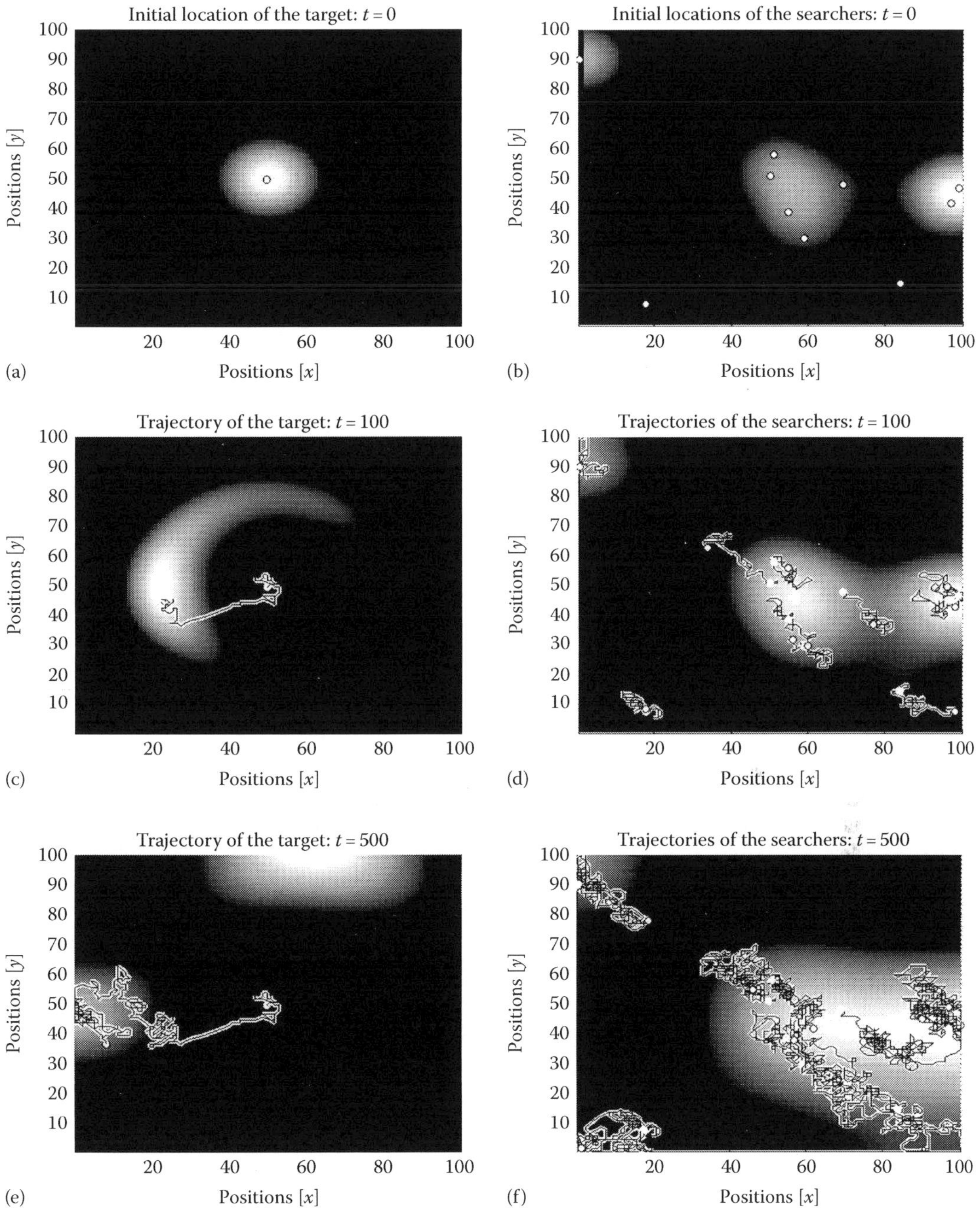

FIGURE 5.29 Trajectories of the target (a), (c) and (e) and m = 10 searchers (b), (d), (f) defined by the target–searcher system (5.74) and the system (5.75) of Langevin equations for different times: (a), (b) time $t = 0$, (c), (d) time $t = 100$, and (e), (f) time $t = 500$. The settings of the system are the same as in Example 5.10. In the system (5.75) the friction coefficients are $\hat{\gamma}_0 = \breve{\gamma}_{0j} = 0.5, j = 1, 2,\ldots, m$, and the noise strengths are $\hat{S} = \breve{S} = 0.5$. The starting point of the target is $\hat{\mathfrak{x}}(0) = (50, 50)$, and the starting points $\breve{\mathfrak{x}}_j(0) = (50, 50)$ of the searchers are random and correspond to the centers of distributions shown in Figure 5.28b.

5.5 SUMMARY

The multiagent search and foraging is based on general swarming techniques that imply the methods of flocking, information exchange, and interactions between the members of the group (Section 5.1).

The information exchange in the multiagent systems follows two main approaches (see Section 5.2.1):

1. Direct communication using certain protocol that allows sending and receiving information between the agents with further corrections of their activities. An example of such communication based on biologically inspired signaling and donating techniques is provided by Algorithm 5.1.
2. Indirect communication by stigmergy, which is based on the changes of the environment and corrections of the agents behavior with respect to these changes. Such communication that mimics the behavior of the ants is illustrated by Example 5.2.

The algorithms of search and foraging, which implement different types of communication, are considered in the following settings (Section 5.2.2):

1. The multiagent search with common target's location probabilities (Algorithm 5.2) assumes that each searcher acts independently following the available probability map, while the map is common for all searchers and is changed simultaneously with respect to their observation results.
2. The multiagent search with biologically inspired signaling and donation (Algorithm 5.3) considers the activity of the group, in which the agents act according to predefined process (Brownian walk, Lévy flight, or Lévy walk) and are allowed to inform other searchers about their environment and to collaborate with other searchers if needed.
3. The multiagent search with indirect communication (Algorithms 5.4 and 5.5) illustrates the stigmergic principle using the pheromone-based communication.

General models of swarm dynamics are based on active Brownian motion (Section 5.3). These models define the motion of the agents using the appropriate Langevin equations, and the evolution of the agents' location probabilities are specified by the corresponding Fokker–Planck equations (see Sections 5.3.1 and 5.3.2). For the movements, which do not depend on the environment, the energy dissipation remains constant during the mission (see, e.g., Equations 5.10 and 5.11), while destructive search scenario, which models the probabilistic search, is represented by the models with nonlinear friction and energy depot (see, e.g., Equations 5.16, 5.17, and 5.22). Notice that in the most cases it is assumed that the motion is conducted in external potential field.

The model of multiagent search implements the canonical active Brownian motion with attraction and repulsion (see Equation 5.24 with corresponding definitions of the dissipative forces and attraction and repulsion potentials). The illustration of such motion is presented by Example 5.7.

Following the approach provided by the active Brownian motion, the target–searcher system is defined in the form of the Turing system (see Equation 5.65), which allows the existence of spatial and temporal Turing patterns. In this system, the searcher acts in the potential field specified by the target's location probabilities and change these probabilities with respect to detection function. The target, in its turn, acts with respect to its location probabilities obtained after the searcher's observations.

The target–searcher system for multiagent search includes either the Fokker–Planck equation, which defines joint probability density function of all searchers (see Equation 5.73) or the separate Fokker–Planck equations, which define the probability density of each searcher in the group (see Equation 5.74). The interactions of the searchers in both cases are defined by the attraction and repulsion potentials.

Finally, notice that for the successful search, the Turing patterns appearing in the system should be unstable.

6 Remarks on Swarm Robotic Systems for Search and Foraging

This short chapter briefly indicates the contemporary directions of the studies of the multirobot systems. Some of the implementations are indicated in the corresponding preceding sections; the other systems require additional considerations, and their relations with the methods of search and foraging are only remarked. A complete review of contemporary studies in multirobot systems is presented in the recently published handbook (Kernbach, 2013).

Collective behavior of robots is a basis for any automated system, especially, of computer-integrated manufacturing systems, which require synchronized activity of a large number of controlled manipulators. However, when the autonomous mobile robots are considered, the main questions are concentrated on swarming itself and self-organization of the swarm with respect to the agents' abilities and the task to be solved. In this concern, Winfield indicates that (Winfield, 2009a, p. 3682; see also Winfield, 2009b, p. 185)

> Foraging is a benchmark problem for robotics, especially for multi-robot systems. It is a powerful problem for several reasons: (1) sophisticated foraging observed in social insects... provides both inspiration and system level models for artificial systems. (2) Foraging is a complex task involving the coordination of several – each also difficult – tasks... (3) Efficient multi-robot foraging requires cooperation between individuals involving either communication to signal to others where the objects may be found and/or cooperative transport of objects too large for a single individual to transport.

Because of the complexity of the problem of search and foraging by the robot swarms, the variety of methods and techniques, which are used for its consideration, are often considered as a distinct theory of social foraging (Andrews, Passino, & Waite, 2007a,b). For a very brief overview of mathematical models used in this theory and in swarm robotics in general, see the report by Muniganti and Pujol (2010) and a survey by Chung, Hollinger, and Isler (2011).

The most popular taxonomy of the multirobot systems was suggested in 2001 by Iocchi, Nardi, and Salerno (2001). Following this taxonomy, the systems are considered with respect to the cooperation level as it is shown in Figure 6.1.

Notice that if the robots are not aware about the other group members, then the actions of each robot can be considered separately and the group behavior is a result of parallel independent activities of the members. If the robots are not coordinated, they execute their tasks in parallel, but the actions of one robot can depend on the results of the actions of the other robot, for example, as in the production lines.

In the case of weak coordination, the robots do not apply the coordination protocol and act in parallel with certain corrections of the behavior with respect to the other robots, for example, for collision avoidance. Strong coordination, in opposite, implies that the robots support the coordination protocol and consider the actions with respect to their influence to the behavior of the other robots.

In the strongly centralized systems, the decision making is conducted by a single leading robot, which obtains information about the other robots and prescribes their behavior, and this leading status remains during the mission. Weak centralization also assumes that the leading robot controls the activity of the group but allows changing the leader during the mission. Finally, in the distributed systems, the robots make their decisions autonomously according to the activities of the other robots.

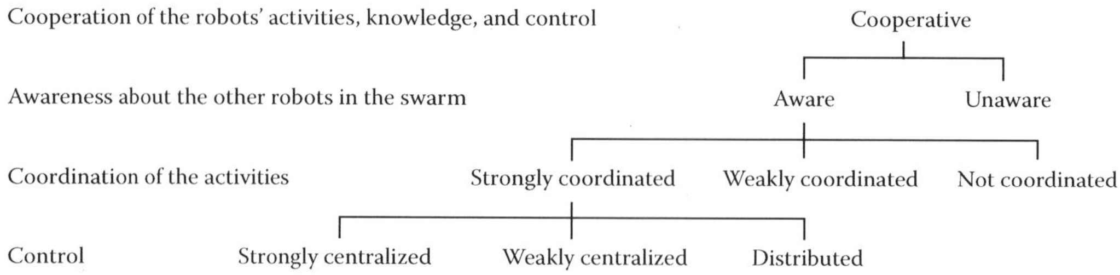

FIGURE 6.1 Taxonomy of the multirobot systems with respect to the cooperation level. (From Iocchi, L. et al., Reactivity and deliberation: A survey on multi-robot systems, in: M. Hannebauer, J. Wendler, and E. Pagello (eds.), *Balancing Reactivity and Social Diliberation in Multi-Agent Systems. From Robocup to Real-World Applications*, Springer, Berlin, Germany, 2001, pp. 9–32.)

A detailed review of coordination and control methods and corresponding task and motion planning is presented in recent paper by Yan, Jouandeau, and Cherif (2013).

For all indicated types of the systems, additional taxonomies consider the manner of information exchange between the robots that include direct and indirect communication (see Section 5.2.1). In the first case, the robots support communication protocol for direct transmitting and receiving data regarding their actions and observations. In the second case, the robots communicate by stigmergy, that is, by changing the environmental states (Iocchi, Nardi, & Salerno, 2001; Mohan & Ponnambalam, 2009, 2010). In addition, Mohan and Ponnambalam (2010) stress the biological inspirations used in the studies of multirobot systems and swarm robotics.

The other taxonomy addresses the design and analysis methods and the problems considered in the framework of the multirobot systems and swarm robotic systems (Bayindir & Sahin, 2007). These taxonomies are shown in Figure 6.2.

Sensor-based modeling considers the sensing abilities of the robots and the behavior of both the robots and the system in general that are modeled with respect to these abilities. In microscopic modeling, decision making and actions of each robot are defined separately, and the system's dynamics are resulted by individual activities of the robots. In contrast, in macroscopic modeling, the system is considered "large" and the robots' activities are specified with respect to the system's dynamics. The model-based cellular automata follow the methods of network design, and since the multirobot systems include a finite number of elements, such models often provide simple but rather efficient modeling techniques.

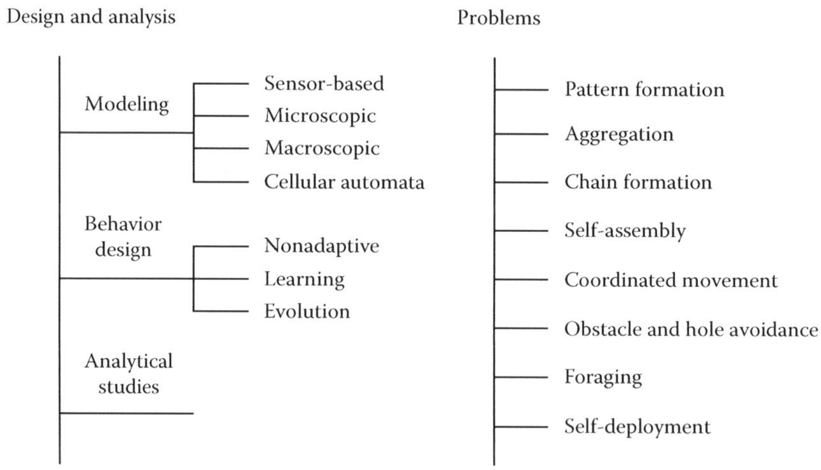

FIGURE 6.2 Taxonomies of the design and analysis methods and of the problems considered in the framework of multirobot systems. (From Bayindir, L. and Sahin, E., *Turkish J. Electric. Eng.*, 15(2), 115, 2007.)

The design of the system dynamics and individual behavior of the robots can follow a nonadaptive approach, in which the specified decision-making procedures are still unchangeable during the mission of the system. In the learning-based behavior, the robots correct their activities during the mission with respect to the obtained results and changes of the environment. Evolution of the system's dynamics can be considered as a kind of learning, but, in opposition to the previous case, this dynamics does not require certain predefined goal function, and the changes of the system's behavior follow the robots' communication and sensing abilities.

The problems presented in Figure 6.2 certainly do not exhaust all possible problems appearing in the considerations of swarm dynamics. Some of these problems are considered in the aforementioned discourse and possible solutions are included into the presented algorithms.

In particular, Chapters 2 and 5 address path planning and dynamics of multiagent systems and swarms, in which cooperative search and foraging require strong coordination between the searchers; however, in specific tasks, different types of control can be applied. In fact, the methods of search and screening considered in Chapter 2 assume that a definite central unit specifies the distribution of search efforts and activities of the agents follow this prescribed plan and are controlled by this unit following strongly centralized scheme. In contrast, the search and foraging methods considered in Chapter 5 imply that agents act independently without centralized control and coordinate their activities with respect to the interactions, changes of the environment, and observation results.

Some of the considered methods follow pure analytical techniques based on the predefined random processes, while the other methods implement biologically inspired methods based on the observed animals' behavior with and without gradient sensing (Nurzaman, Matsumoto, Nakamura, Koizumi, & Ishiguro, 2009); such techniques in individual search are considered in Chapter 4 and in cooperative search—in Section 5.3. In some cases, the searchers communicate indirectly by stigmergy and by deploying pheromone trails (Payton, Daily, Estowski, Howard, & Lee, 2001) and—in the other cases—by direct sharing information representing the target's location probabilities (see Section 5.2). Aggregation of the agents and pattern formation are considered in Sections 5.3 and 5.4, respectively. The closest studies in swarm robotic systems, which follow the evolutionary approach, are presented in the indicated books by Trianni (2008) and by Hamann (2010). Aggregation methods and social forging by robotic systems are considered by Gazi and Passino (2011) (for robotic implementation of foraging methods, see also the paper by Pavlic and Passino (2009)).

In practice, the presented search and foraging algorithms can be implemented using different robotic systems. Detailed reviews of the contemporary projects in swarm robotics are presented in recent papers by Navarro and Matia (2013) and by Tan and Zheng (2013). Especially, the robotic systems, which can be used for direct simulations of the suggested algorithms, are the Harvard system (the recent simulations with this system are presented by Rubenstein, Cornejo, and Nagpal (2014)) and the ALICE system used for modeling the movements following the gradient field (Garnier, Combe, Jost, & Theraulaz, 2013) and for simulations of the prey–predator behavior (Floreano & Keller, 2010). Useful information on the software simulation platforms can be found in the work by Shi et al. (2012).

As indicated earlier, this brief consideration does not aim to give a complete review of the robotic systems; its goal is to link the presented algorithms and techniques with the methods of swarm robotics and provide some useful references. More detailed information and additional sources are presented in introducing remarks in each chapter.

7 Conclusion

This book describes how to employ the limited resources for the most effective search of hidden objects and addresses the methods and algorithms of probabilistic search by the agents acting individually and in groups. It considers the theory of search and screening and the theory of optimal foraging attempting to provide a formal mathematical and algorithmic framework that allows considerations of the search and foraging behavior using the unified models and methods. Such approach allows application of the methods of one theory to the tasks considered in the other theory and results both in effective algorithms of search and in novel insights regarding the behavior of natural foragers.

This book starts with the models and optimization techniques used in the search and screening theory and overviews the main algorithms of search, including the algorithms of search for moving target. These methods provide the distributions of search efforts over a domain, which describe the search by a group of abstract agents with strongly centralized coordination and control. Such methods were originated during World War II and nowadays are applied in different military and security applications for navigation and positioning of automatic agents and for prediction of the enemies' and crimes' behavior.

The individual search is considered following several approaches. The first approach implements the methods used in the foraging theory, which describes the search activities of living organisms. In this direction, this book reviews the methods of optimal foraging in deterministic and probabilistic settings and presents algorithmic implementations of these techniques. The second approach follows the heuristic methods of online path planning in the probabilistic search for static and moving targets. The book presents novel algorithms of such search with errorless and erroneous detections and an algorithm of search with shadowing. In spite of different techniques, in both methods, the motion of the agent is planned with respect to the environmental states and observation results. The third method of individual search follows another approach, in which the motion of the agent is defined by a certain stochastic process. The book considers the main processes, which are used for governing the searcher's motion, as well as an intermittent search that implies the searcher changes the strategies of search and corresponding processes. In addition to military and security tasks, during the last decades, the presented methods and algorithms are intensively studied in the framework of robotic systems and in biology for deeper understanding the animals' behavior.

The collective search with distributed control is considered following three approaches. The first implements direct communication, which implies that the agents are able to send and receive messages and to share information about their actions and search results. In this direction, this book presents the algorithm of probabilistic search with common target's location probabilities and the algorithm of search with biologically inspired signaling and donating techniques. The second approach implements indirect communication using the specific changes of the environment that are used for corrections of the activities of the searchers. In this direction, the algorithm, which mimics the foraging behavior of the ants, is presented. Finally, the third approach applies the attraction and repulsion potentials that define the group formation and collision avoidance. The descriptive and predictive abilities of these methods form a basis for the studies in mathematical biology, multirobot systems, and swarm dynamics. In the last years, the methods and algorithms of collective search and foraging formed a basis for cybernetic research of swarm intelligence that led to novel ideas in computation machinery and biological computation.

The consideration of both individual and collective search and foraging deals with the trajectories of the agents and with the probability distributions of the agents' location in the domain.

The trajectories of the searchers are defined by certain decision-making techniques, by underlying stochastic processes, or by the dynamical equations corresponding to the evolution of the agents' location probabilities. The probability distributions, which define the target's and the searchers' location probabilities, are defined either by search and screening algorithms or by appropriate target–searcher systems, which govern their spatial dynamics.

The theoretical methods and ready-to-use algorithms presented in this book can be implemented in systems dealing with the search problems, in mathematical biology for modeling and analysis of the animals' foraging activities, in the navigation and control of mobile robots and multirobot systems, and in the cybernetic studies of swarm intelligence. In addition, the materials presented in the book can be used in the courses in optimization, algorithms, probabilistic models, cybernetics, and robotics and can be useful for the courses in mathematical biology. Since the book presents a contemporary overview of the methods developed in the search and screening theory and in the optimal foraging theory and includes novel algorithms and techniques, it can form a basis and provide insights for further research in both theories that hopefully will lead to better understanding of living organisms and to the development of new computational techniques.

Bibliography

Abrahams, J. (1994). Parallelized Huffman and Hu-Tucker searching. *IEEE Transactions on Information Theory, 40*(2), 508–510.

Adamatzky, A., Costello, B. D., & Asai, T. (2005). *Reaction-Diffusion Computers*. Amsterdam, the Netherlands: Elsevier.

Ahlswede, R., & Wegener, I. (1987). *Search Problems*. New York: John Wiley & Sons.

Alos, E., & Bonaccorsi, S. (2002). Stochastic partial differential equations with Dirichlet white-noise boundary conditions. *Annales de l'Institut Henri Poincare (B) Probability and Statistics, 38*, 125–154.

Alpern, S., & Gal, S. (2003). *The Theory of Search Games and Rendezvous*. New York: Kluwer.

Andrews, B. W., Passino, K. M., & Waite, T. A. (2007a). Social foraging theory for autonomous vehicle decision-making system design. *Journal of Intelligent and Robotic Systems, 49*(1), 39–65.

Andrews, B. W., Passino, K. M., & Waite, T. A. (2007b). Social foraging theory for robust multiagent design. *IEEE Transactions on Automation Science and Engineering, 4*(1), 79–86.

ANTS. (1998–2014). *Proceedings of International Conference on (Ant Colony Optimization and) Swarm Intelligence*. Berlin, Germany: Springer-Verlag.

Assaf, D., & Zamir, S. (1985). Optimal sequential search: A Bayesian approach. *Annals of Statistics, 13*(3), 1213–1221.

Balckarey, J. E. (1987). Autowaves and dissipative structures in two-stratum system of Van-der-Pol oscillators. *Journal of Theoretical Physics, 57*(2), 209–218.

Baldassarre, G., Nolfi, S., & Parisi, D. (2003). Evolving mobile robots able to display collective behaviors. *Artificial Life, 9*, 255–267.

Bartumeus, F. (2009). Behavioral intermittence, Levy patterns, and randomness in animal movement. *Oikos, 118*, 488–494.

Bartumeus, F., & Catalan, J. (2009). Optimal search behavior and classic foraging theory. *Journal of Physics A: Mathematical and Theoretical, 42*, 434002.

Bartumeus, F., Da Luz, M. G., Viswanathan, G. M., & Catalan, J. (2005). Animal search strategies: A quantitative random-walk analysis. *Ecology, 86*(11), 3078–3087.

Bassler, K. E., Gunaratne, G. H., & McCauley, J. L. (2006). Markov processes, Hurst exponents, and nonlinear diffusion equations: With application to finance. *Physica A, 369*, 343–353.

Batursin, I., Kaynak, O., & Rudas, I. (2002). Fussy modeling based on generalized conjunction operators. *IEEE Transactions on Fuzzy Systems, 10*(5), 678–683.

Bayindir, L., & Sahin, E. (2007). A review of studies in swarm robotics. *Turkish Journal of Electrical Engineering, 15*(2), 115–147.

Beckers, R., Holland, O. E., & Deneubourg, J. L. (1994). From local actions to global tasks: Stigmergy and collective robotics. In R. A. Brooks & P. Maes (Eds.), *Artificial Life IV* (pp. 181–189). Cambridge, MA: MIT Press.

Bell, W. J. (1990). *Searching Behavior: The Behavioral Ecology of Finding Resources*. London, U.K.: Chapmann & Hall.

Benichou, O., Chevalier, C., Klafter, J., Meyer, B., & Voituriez, R. (2010). Geometry-controlled kinetics. *Nature Chemistry, 2*, 472–477.

Benichou, O., Coppey, M., Moreau, M., Suet, P.-H., & Voituriez, R. (2005). Optimal search strategies for hidden targets. *Physical Reviews Letters, 94*, 198101, 1–4.

Benichou, O., Loverdo, C., Moreau, M., & Voituriez, R. (2006). Two-dimensional intermittent search processes: An alternative to Lévy flight strategies. *Physical Review E, 74*, 020102(R), 1–4.

Benichou, O., Loverdo, C., Moreau, M., & Voituriez, R. (2007). A minimal model of intermittent search in dimension two. *Journal of Physics: Condensed Matter, 19*, 065141, 1–17.

Benichou, O., Loverdo, C., Moreau, M., & Voituriez, R. (April 1, 2011). Intermittent search strategies. Retrieved February 13, 2013, from arXiv.org:1104.0639.

Ben-Jacob, E., Cohen, I., Golding, I., & Kozlovsky, Y. (2001). Modelling baranching and chiral colonial patterning of lubricating bacteris. In P. Maini, & H. Othmer (Eds.), *Mathematical Models for Biological Pattern Formation* (pp. 211–253). New-York: Springer.

Ben-Jacob, E., & Garik, P. (1990). The formation of patterns in non-equilibrium growth. *Nature, 343*, 523–530.

Ben-Jacob, E., & Levine, H. (2001). The artisty of nature. *Nature, 409*, 985–986.

Berg, H. C. (1993). *Random Walks in Biology*. Princeton, NJ: Princeton University Press.

Bergstrom, C. T., & Lachmann, M. (1997). Signaling among relatives I. Is signaling too costly? *Philosophical Transcations of London Royal Society B, 352*, 609–617.

Bergstrom, C. T., & Lachmann, M. (1998a). Signaling among relatives II. Beyond the tower of Babel. *Theoretical Population Biology, 54*, 146–160.

Bergstrom, C. T., & Lachmann, M. (1998b). Signaling among relatives III. Talk is cheap. *Proceedings of the National Academy of Sciences of the United States of America, 95*, 5100–5105.

Berkolaiko, G., & Havlin, S. (1997). Territory covered by N Levy flights on d-dimensional lattices. *Physical Review E, 55*(2), 1395–1400.

Bonabeau, E., Dorigo, M., & Theraulaz, G. (1999). *Swarm Intelligence: From Natural to Artificial Systems.* New York: Oxford University Press.

Bonabeau, E., Theraulaz, G., Deneubourg, J.-L., Aron, S., & Camazine, S. (1997). Self-organization in social insects. *Trends in Ecology and Evolution, 12*(5), 188–193.

Bourgault, F., Furukawa, T., & Durrant-Whyte, H. F. (2003). Coordinated decentralized search for a lost target in a Bayesian world. In *IEEE/RSJ International Conference on Intelligent Robots and Systems* (pp. 48–53). Las Vegas, NV.

Bovet, P., & Benhamou, S. (1988). Spatial analysis of animals' movements using a correlated random walk model. *Journal of Theoretical Biology, 131*, 419–433.

Braitenberg, V. (1984). *Vehicles: Experiments in Synthetic Psychology*. Cambridge, MA: The MIT Press.

Brillinger, D. R. (2010). Modeling spatial trajectories. In A. E. Gelfand, P. J. Diggle, M. Fuentes, & P. Guttorp (Eds.), *Handbook of Spatial Statistics* (pp. 465–477). Boca Raton, FL: CRC Press.

Brown, S. S. (1980). Optimal search for a moving target in discrete time and space. *Operations Research, 28*(6), 1275–1289.

Buchanan, M. (2008). The mathematical mirror to animal nature. *Nature, 453*, 714–716.

Bulinsky, A. V., & Shiryaev, A. N. (2003). *Theory of Stochastic Processes*. Moscow, Russia: Nauka.

Burrow, J. F., Baxter, P. D., & Pitchford, J. W. (2008). Levy processes, saltatory foraging, and superdiffusion. *Mathematical Modelling of Natural Phenomena, 3*(3), 115–130.

Calengea, C., Draya, S., & Royer-Carenzia, M. (November 19–20, 2009). The concept of animals' trajectories from a data analysis perspective. *Ecological Inforamtics, 4*, 34–41.

Cetnarowicz, K., Turek, W., Zabinska, M., & Sosnicki, T. (2013). Decentralized algorithm for tasks assignment to robots. In *The Israeli Conference on Robotics* (*ICR'13*) (p. 46). Tel-Aviv, Israel.

Charnov, E. L. (1976). Optimal foraging, the marginal value theorem. *Theoretical Population Biology, 9*(2), 129–136.

Chechkin, A., Sokolov, I. M., & Klafter, J. (2012). Natural and modified forms of distributed-order fractional diffusion equations. In J. Klafter, S. C. Lim, & R. Metzler (Eds.), *Fractional Dynamics. Recent Advances* (pp. 107–127). Hackensack, NJ: World Scientific.

Chernikhovsky, G., Kagan, E., Goren, G., & Ben-Gal, I. (2012). Path planning for sea vessel search using wideband sonar. In *Proceedings of the 27th IEEE Convention EEEI,* Eilat, Israel. DOI 10.1109/EEEI.2012.6377122.

Chew, M. C. (1967). A sequential search procedure. *Annals of Mathematical Statistics, 38*(2), 494–502.

Choset, H., Lynch, K., Hutchinson, S., Kantor, G., Burgard, W., Kavraki, L., et al. (2005). *Principles of Robot Motion: Theory, Algorithms, and Implementation.* Cambridge, MA: Bradford Books/The MIT Press.

Chung, T., Hollinger, G., & Isler, V. (2011). Search and pursuit-evasion in mobile robotics: A survey. *Autonomous Robots, 31*(4), 299–316.

Ciervo, A. P. (1976). Search for moving targets. PSR Report 619. Santa-Monica, CA: Pacific-Sierra Research Corporation.

Clark, A. (1997). *Being There: Putting Brain, Body and World Together Again.* Cambridge, MA: The MIT Press.

Clark, A. (2003). *Natural-Born Cyborgs—Minds, Technologies, and the Future of Human Intelligence.* Oxford, U.K.: Oxford University Press.

Codling, E. A., Plank, M. J., & Benhamou, S. (2008). Random walk models in biology. *Journal of the Royal Society Interface, 5*, 813–834.

Collett, M., & Collett, T. S. (2000). How do insects use path integration for their navigation? *Biological Cybernetics, 83*, 245–259.

Collett, M., & Collett, T. S. (2009). Local and global navigational coordinate systems in desert ants. *Journal of Experimental Biology, 212*, 901–905.

Condamin, S., Benichou, O., Tejedor, V., Voituriez, R., & Klafter, J. (2007). First-passage times in complex scale-invariant media. *Nature, 450*, 77–80.

Condat, C. A., & Sibona, G. J. (2002). Diffusion in a model for active Brownian motion. *Physica D, 168–169*, 235–243.

Cooper, D. C., Frost, J. R., & Robe, R. Q. (2003). *Compatibility of Land SAR Procedures with Search Theory*. Washington, DC: US Department Homeland Security.

Couzin, I. D., Krause, J., Franks, N. R., & Levin, S. A. (2005). Effective leadership and decision-making in animal groups on the move. *Nature, 433*, 513–516.

Cover, T. M., & Thomas, J. A. (1991). *Elements of Information Theory*. New York: John Wiley & Sons.

Da Luz, M., Buldyrev, S. V., Havlin, S., Raposo, E., Stanley, H. E., & Viswanathan, G. (2001). Improvements in the statistical approach to random Levy flight searches. *Physica A, 295*, 89–92.

Danaher, S. (June 4, 1998). MATLAB Huffman. Retrieved February 27, 2013, from http://webscripts.softpedia.com/script/Scientific-Engineering-Ruby/Signal-Processing/matlab-huffman-35063.html.

De Gennes, P. G. (1976). La percolation: Un concept unificateur. *La Recherche, 7*, 919–927.

DeGroot, M. H. (1970). *Optimal Statistical Decisions*. New York: McGraw-Hill.

DeLima, P., Daniel, P., & Sciortino Jr, J. C. (2007). Optimizing a search strategy for multiple mobile agents. In *Evolutionary and Bio-Inspired Computation*. 65630B, SPIE: Orlando, FL.

Distributed Autonomous Robotic Systems. (1992–2013). *Proceedings of International Symposia*. Berlin, Germany: Springer.

Dobbie, J. M. (1974). A two-cell model of search for a moving target. *Operations Research, 22*(1), 79–92.

Dorfman, R. (1943). The detection of defective members of large population. *Annals of Mathematical Statistics, 14*, 436–440.

Dorigo, M., & Stutzle, T. (2004). *Ant Colony Optimization*. Cambridge, MA: The MIT Press/A Bradford Book.

Dorigo, M., Trianni, V., Sahin, E., Gross, R., Labella, T. H., Baldassarre, G., et al. (2004). Evolving self-organizing behaviors for a swarm-bot. *Autonomous Robots, 17*, 223–245.

Du, D.-Z., & Hwang, F. (1993). *Combinatorial Group Testing and Its Applications*. Singapore, Singapore: World Scientific Publishing.

Eagle, J. N. (1984). The optimal search for a moving target when the search path is constrained. *Operations Research, 32*, 1107–1115.

Ebeling, W. (2004). Nonlinear Brownian motion—Mean square displacement. *Condensed Matter Physics, 7*(3(39)), 539–550.

Ebeling, W. (2007). Active Brownian motion of pairs and swarms of particles. *Acta Physica Polonica B, 38*(5), 1657–1671.

Edwards, A. M., Phillips, R. A., Watkins, N. W., Freeman, M. P., Murphy, E. J., Afanasyev, V., et al. (2007). Revisiting Levy flight search patterns of wandering albatrosses, bumblebees and deer. *Nature, 449*, 1044–1049.

Emlen, J. M. (1966). The role of time and energy in food preference. *American Naturalist, 100*, 611–617.

Erdmann, U., Ebeling, W., Schimansky-Geier, L., & Schweitzer, F. (2000). Brownian particles far from equilibrium. *European Physical Journal B, 15*, 105–113.

Fahimi, F. (2009). *Autonomous Robots: Modeling, Path Planning, and Control*. New York: Springer.

Feinerman, O., Korman, A., Lotker, Z., & Sereni, J.-S. (2012). Collaborative search on the plane without communication. In *Proceedings of the ACM Conference PODC* (pp. 77–86).

Feller, W. (1970–1971). *An Introduction to Probability Theory and Its Applications* (Vols. I, II). New York: John Wiley & Sons.

Ferreira, A., Raposo, E., Viswanathan, G., & da Luz, M. (2012). The influence of the environment on Lévy random search efficiency: Fractality and memory effects. *Physica A, 391*, 3234–3246.

Fewell, J. H. (2003). Social insect networks. *Science, 301*, 1867–1870.

Fiorini, P., & Shiller, Z. (1998). Motion planning in dynamic environments using velocity obstacles. *International Journal of Robotics Research, 17*(7), 760–772.

Floreano, D., & Keller, L. (2010). Evolution of adaptive behavior in robots by means of Darwinian selection. *PLoS Biology, 8*(1), e1000292.

Fodor, J., Rudas, I., & Bede, B. (2004). Uninorms and absorbing norms with applications to image processing. In *Proceedings of the SISY*, Subotica, Serbia (pp. 59–73).

Fodor, J., Yager, R., & Rybalov, A. (1997). Structure of uninorms. *International Journal of Uncertainty, Fuzziness and Knowledge-Based Systems, 5*, 411–427.

Frank, T. D. (2005). *Non-Linear Fokker-Plank Equations. Fundamentals and Applications*. Berlin, Germany: Springer-Verlag.

From Animals to Animats. (1991–2012). *Proceedings of International Conference on Simulation and Adaptive Behavior*. Cambridge, MA: Bradford Books/MIT Press.

Frost, J. R., & Stone, L. D. (2001). *Review of Search Theory: Advances and Applications to Search and Rescue Decision Support*. Groton, CT: US Coast Guard Research and Development Center.

Gal, S. (1980). *Search Games*. New York: Academic Press.

Garnaev, A. Y. (2000). *Search Games and Other Applications of Game Theory*. Berlin, Germany: Springer-Verlag.

Garnier, S., Combe, M., Jost, C., & Theraulaz, G. (2013). Do ants need to estimate the geometrical properties of trail bifurcations to find an efficient route? A swarm robotics test bed. *PLoS Computational Biology*, *9*(3), e1002903.

Gasparri, A., Priolo, A., Oriolo, G., & Ulivi, G. (2012). A swarm aggregation algorithm based on local interactions for multi-robot systems with actuator saturation. In *Proceedings of IEEE/RSJ Conference on Intelligent Robots and Systems* (pp. 539–544). Vilarmoura, Portugal.

Gaubert, L., Redou, P., Harrouet, F., & Tisseau, J. (2007). A first mathematical model of brood sorting by ants: Functional self-organization without swarm-intelligence. *Ecological Complexity, 4*, 234–241.

Gazi, V., & Passino, K. M. (2011). *Swarm Stability and Optimization*. Berlin, Germany: Springer.

Glück, A., Hüffel, H., & Ilijić, S. (2009). Canonical active Brownian motion. *Physical Review E 79*, 021120, 1–7.

Golding, I., Kozlovsky, Y., Cohen, I., & Ben-Jacob, E. (1998). Studies of bacterial branching growth using reaction-diffusion models for colonial development. *Physica A, 260*(3–4), 510–554.

Gordon, D. M. (2010). *Ant Encounters: Interaction Networks and Colony Behavior*. Princeton, NJ: Princeton University Press.

Goren, G., Kagan, E., & Ben-Gal, I. (November 10–11, 2010). Probabilistic double-distance search by autonomous mobile agent in discrete time and space. In *The Israeli Conference on Robotics* (*ICR'10*), (p. 61). Herzlia, Isreal.

Gorenflo, R., & Mainardi, F. (1998). Random walk models for space-fractional diffusion processes. *Fractional Calculus & Applied Analysis, 1*, 167–191.

Gorenflo, R., Mainardi, F., Moretti, D., Pagnini, G., & Paradisi, P. (2002). Discrete random walk models for space-time fractional diffusion. (R. Hilfer, R. Metzler, A. Blumen, & J. Klafter, Eds.) *Chemical Physics* (*Special issue "Stange kinetics"*), *284*(1/2), 521–541.

Grasse, P.-P. (1959). La reconstruction du nid et less coordinations interindividuelles chez bellicositermes natalensis et cubitermes sp. la theorie de la stigmergie: Essai d'interpretation du comportement des termites constructeurs. *Insectes Sociaux*, 6(1), 41–83.

Green, R. F. (1980). Bayesian birds: A simple example of Oaten's stochastic model of optimal foraging. *Theoretical Population Biology*, *18*, 244–256.

Green, R. F. (2006). A simpler, more general method of finding the optimal foraging strategy for Bayesian birds. *OIKOS, 112*, 274–284.

Haefner, J. W., & Crist, T. O. (1994). Spatial model of movement and foraging in harvester ants (Pogonomyrex) (I): The role of memory and communication. *Journal of Theoretical Biology*, *166*, 299–313.

Hahn, M. G., Kobayashi, K., & Umarov, S. (2010). Fokker-Planck-Kolmogorov equations associated with time-changed fractional Brownian motion. *Proceedings of the American Mathematical Society*, *139*(2), 691–705.

Hajek, O. (1975). *Pursuit Games. An Introduction to the Theory and Applications of Differential Games of Pursuit and Evasion*. New York: Academic Press.

Haken, G. (2012). *Synergetics: Introduction and Advanced Topics*. Berlin, Germany: Springer.

Haken, H. (1977). *Synergetics: An Introduction. Nonequlibrium Phase Transitions and Self-Organization in Physics, Chemistry and Biology*. Berlin, Germany: Springer-Verlag.

Hamann, H. (2010). *Space-Time Continuous Models of Swarm Robotic Systems: Supporting Global-to-Local Programming*. Berlin, Germany: Springer.

Havlin, S., Bunde, A., Eisenberg, E., Lee, J., Roman, H. E., Schwarzer, S., et al. (1993). Multifractional fluctuations in the dynamics of disordered systems. *Physica A, 194*, 288–297.

Hayashi, Y., Yuki, M., Sugawara, K., Kikuchi, T., & Tsuji, K. (2008). Analysis and modeling of ant's behavior from single to multi-body. *Artificial Life and Robotics, 13*, 120–123.

Hellman, O. (1972). On the optimal search for a randomly moving target. *SIAM Journal of Applied Mathematics, 22*(4), 545–552.

Hellman, O. (1985). *Introduction of the Theory of Optimal Search*. Moscow, Russia: Nauka.

Hoffmann, G. (1983). Optimization of Brownian search strategies. *Biological Cybernetics, 49*, 21–31.

Holldobler, B., & Wilson, E. O. (1990). *The Ants*. Cambridge, MA: Belknap Press/Harvard University Press.

Holldobler, B., & Wilson, E. O. (2009). *The Superorganism: The Beauty, Elegance, and Strangeness of Insect Societies*. New York: W. W. Norton & Company.

Holmes, E., Lewis, M., Banks, J., & Veit, R. (1994). Partial differential equations in ecology: Spatial interactions and population dynamics. *Ecology, 75*(1), 17–29.

Huffman, D. A. (1952). A method for the construction of minimal-redundancy codes. *Proceedings of the I.R.E., 40*, 1098–1101.

Hughes, B. D. (1995–1996). *Random Walks and Random Environments* (Vols. I, II). Oxford, U.K.: Oxford University Press.

Humphries, N. E., Queiroz, N., Dyer, J. R., Pade, N. G., Musyl, M. K., Schaefer, K. M., et al. (2010). Environmental context explains Levy and Brownian movement patterns of marine predators. *Nature, 465*, 1066–1069.

Humphries, N. E., Weimerskirch, H., Queiroz, N., Southall, E. J., & Sims, D. W. (2012). Foraging success of biological Levy flights recorded in situ. *PNAS, 109*(19), 7169–7174.

Hutteger, S. M., & Zollman, K. J. (2009). Dynamic stability and basins of attraction in the Sir Philip Sydney game. *Proceedings of the Royal Society B, 277*(1689), 1915–1922.

Ibe, O. C. (2013). *Elements of Random Walk and Diffusion Processes*. Hoboken, NJ: John Wiley & Sons.

Iida, K. (1992). *Studies on the Optimal Search Plan*. Berlin, Germany: Springer-Verlag.

Indelman, V., Carlone, L., & Dellaert, F. (November 19–20, 2013a). Towards planning in generalized belief space with applications to mobile robotics. In *The Israeli Conference on Robotics (ICR'13)* (p. 94). Tel-Aviv, Israel.

Indelman, V., Carlone, L., & Dellaert, F. (December 2013, 2013b). Towards planning in the generalized belief space. In *International Symposium on Robotics Research (ISRR'13)* (pp. 1–16). Singapore.

Iocchi, L., Nardi, D., & Salerno, M. (2001). Reactivity and deliberation: A survey on multi-robot systems. In M. Hannebauer, J. Wendler, & E. Pagello (Eds.), *Balancing Reactivity and Social Diliberation in Multi-Agent Systems. From Robocup to Real-World Applications* (pp. 9–32). Berlin, Germany: Springer.

Isaacs, R. (1954). *A Game of Aiming and Evasion: General Discussion and the Marksman's Strategies*. Research memorandum, 1385. Santa Monica, CA: RAND Corporation.

Isaacs, R. (1965). *Differential Games: A Mathematical Theory with Applications to Warfare and Pursuit, Control and Optimization*. New York: John Wiley & Sons.

Isaacs, R., & Karlin, S. (1954). *A Game of Aiming and Evasion*. Research memorandum, 1316. Santa Monica, CA: RAND Corporation.

Ishida, T., & Korf, R. E. (1995). Moving target search: A real-time search for changing goals. *IEEE Transactions of Pattern Analysis and Machine Intelligence, 17*(6), 609–619.

Ishikawa, M., Miyajima, K., & Tasaka, T. (2003). Analysis of Turing patterns by stochastic reaction-diffusion systems. In *International Conference of SIGE'02* (pp. 2897–2901). Osaka, Japan.

Israel, M., Khmelnitsky, E., & Kagan, E. (2012). Search for a mobile target by ground vehicle on a topographic terrain. In *Proceedings of the 27th IEEE Convention of EEEI*, Eilat, Israel. DOI 10.1109/EEEI.2012.6377123.

James, A., Plank, M. J., & Edwards, A. M. (2011). Assessing Levy walks as models of animal foraging. *Journal of the Royal Society Interface, 8*, 1233–1247.

Jespersen, S., Metzler, R., & Fogedby, H. C. (1999). Levy flights in external force fields: Langevin and fractional Fokker-Plank equations and their solutions. *Physical Review E, 59*(3), 2736–2745.

Johnstone, R. A., & Grafen, A. (1992). The continuous Sir Philips Sydney game: A simple model of biological signaling. *Journal of Theoretical Biology, 156*, 215–234.

Kadane, J. B. (1968). Discrete search and the Neyman-Pearson lemma. *Journal of Mathematical Analysis and Applications, 22*(1), 156–171.

Kagan, E. (1995). The model of concentration structures on the semiconductor surface. *Izvestiya TRTU, 1*, 43–45.

Kagan, E. (1998). On one principle of controlled patterns formation on the semiconductor surface. In *Proceedings of the Third International Arctic Seminar on Physics* (pp. 48–51). Murmansk, Russia.

Kagan, E. (2008). Stability of Turing patterns in a reaction-diffusion system with boundary control. In M. B. Belonenko (Ed.), *Nanomaterials: Methods and Approaches* (pp. 134–147). Volgograd, Russia: PrinTerra.

Kagan, E. (2010). Turing systems, entropy, and kinetic models for self-healing surfaces. *Entropy, 12*(3), 554–569.

Kagan, E., & Ben-Gal, I. (2006). An informational search for a moving target. In *Proceedings of the 24th IEEE Convention of EEEI*, Eilat, Israel (pp. 153–155).

Kagan, E., & Ben-Gal, I. (2013a). Moving target search algorithm with informational distance measures. *The Open Applied Informatics Journal, 6*, 1–10.

Kagan, E., & Ben-Gal, I. (2013b). *Probabilistic Search for Tracking Targets*. Chichester, U.K.: John Wiley & Sons.

Kagan, E., & Ben-Gal, I. (2014). A group-testing algorithm with online informational learning. *Transactions of IIE, 46*, 164–181.

Kagan, E., Goren, G., & Ben-Gal, I. (2010). Probabilistic double-distance algorithm of search after static or moving target by autonomous mobile agent. In *Proceedings of the 26th IEEE Convention of EEEI*, Eilat, Israel (pp. 160–164).

Kagan, E., Goren, G., & Ben-Gal, I. (2012). Algorithm of search for static or moving target by autonomous mobile agent with erroneous sensor. In *Proceedings of the 27th IEEE Convention of EEEI*, Eilat, Israel. DOI 10.1109/EEEI.2012.6377124.

Kagan, E., Rybalov, A., Sela, A., Siegelmann, H., & Steshenko, J. (2014). Probabilistic control and swarm dynamics in mobile robots and ants. In S. A. Burki, G. Dobbie, & Y. S. Koh (Eds.), *Biologically-Inspired Techniques for Knowledge Discovery and Data Mining*. Hershey, PA: IGI Global.

Kagan, E., Rybalov, A., Siegelmann, H., & Yager, R. (2013). Probability-generated aggregators. *International Journal of Intelligent Systems, 28*(7), 709–727.

Kennedy, J., Eberhart, R. C., & Shi, Y. (2001). *Swarm Intelligence*. San Francisco, CA: Morgan Kaufmann.

Kernbach, S. (Ed.). (2013). *Handbook of Collective Robotics: Fundamentals and Challenges*. Boca Raton, FL: CRC Press/Taylor & Francis.

Kerner, B., & Osipov, V. (1994). *Autosolitons: A New Approach to Problems of Self-Organizations and Turbulence*. Berlin, Germany: Springer-Verlag.

Kilbas, A. A., Srivastava, H. M., & Trujillo, J. J. (2006). *Theory and Applications of Fractional Differential Equations*. Amsterdam, theNetherlands: Elsevier.

Klafter, J., Lim, S. C., & Metzler, R. (Eds.). (2012). *Fractional Dynamics. Recent Advances*. Hackensack, NJ: World Scientific.

Klafter, J., & Sokolov, I. M. (2005). Anomalous diffusion spreads its wings. *Physics World, 18*, 29–32.

Klimontovich, Y. L. (1987). Entropy evolution in self-organization processes. H-theorem and S-theorem. *Physica A, 142*, 390–404.

Klimontovich, Y. L. (1989). Problems in the statistical theory of open systems: Criteria for the relative degree of order in self-organization processes. *Soviet Physics Uspekhi, 32*(5), 416–433.

Klimontovich, Y. L. (1991). *Turbulent Motion and the Structure of Chaos*. Dordrecht, the Netherlands: Kluwer.

Klimontovich, Y. L. (1994). Nonlinear Brownian motion. *Physics-Uspekhi, 37*(8), 737–770.

Knuth, D. (1998). *Sorting and Searching. The Art of Computer Programming* (2 ed., Vol. 3). Reading, MA: Addison-Wesley.

Kohonen, T. (1978). *Associative Memory: A System-Theoretical Approach*. Berlin, Germany: Springer-Verlag.

Kolmogorov, A. H. (1940). Wienersche Spiralen und einige andere interessante Kurven in Hilbertschen Raum. *Dokladi Academy of Science of the USSR, 26*, 115–118.

Koopman, B. O. (1946). Search and screening. Operation Evaluation Research Group Report, 56. Arlington, VA: Center for Naval Analysis.

Koopman, B. O. (1956–1957). The theory of search. *Operations Research, 4, 5*, 324–346, 503–531, 613–626.

Koopman, B. O. (1980). *Search and Screening: General Principles with Historical Applications*. New York: Pergamon Press.

Korf, R. E. (1990). Real-time heuristic search. *Artificial Intelligence, 42*(2–3), 189–211.

Landa, P. (1996). *Nonlinear Oscillations and Waves in Dynamical Systems*. Dordrecht, the Netherlands: Kluwer Academic Publishers.

Langevin, P. (1908). Sur la theorie du mouvement brownien. *Comptes Rendus L'Academie des Sciences, 146*, 530–533.

Larralde, H., Trunfio, P., Havlin, S., Stanley, H. E., & Weiss, G. H. (1992a). Number of distinct sites visited by N random walkers. *Physical Review A, 45*(10), 7128–7138.

Larralde, H., Trunfio, P., Havlin, S., Stanley, H. E., & Weiss, G. H. (1992b). Territory covered by N diffusing particles. *Nature, 355*, 423–426.

Lawler, G. F. (2010). *Random Walk and Heat Equation*. Providence, RI: American Mathematical Society.

Lawler, G. F., & Limic, V. (2010). *Random Walk: A Modern Introduction*. Cambridge, U.K.: Cambridge University Press.

Leccese, A., Gasparri, A., Priolo, A., Oriolo, G., & Ulivi, G. (2013). A swarm aggregation algorithm based on local interaction with actuator saturations and integrated obstacle avoidance. In *IEEE International Conference on Robotics and Automation* (pp. 1865–1870). Karlsruhe, Germany.

Leipata, T. (1976). *On Discretized Stationary Search*. Turku, Finland: University of Turku, Institute for Pure and Applied Mathematics.

Leppanen, T. (2004). Computational studies of pattern formation in turing systems (PhD thesis). Espoo, Finland: Helsinki University of Technology.

Leppanen, T. (2005). The theory of Turing pattern formation. In K. Kaski, & R. Bario (Eds.), *Current Topics in Physics* (pp. 190–227). London, U.K.: Imperial College Press.

Lerman, K., & Galstyan, A. (2002). Mathematical model of foraging in a group of robots: Effect of interference. *Autonomous Robots, 13*(2), 127–141.

Lerman, K., Martinoli, A., & Galstyan, A. (2005). A review of probabilistic macroscopic models for swarm robotic systems. In E. Sahin, & W. M. Spears (Eds.), *Swarm Robotics, LNCS 2004, 3342* (pp. 143–152). Heidelberg, Germany: Springer.

Liu, W., Winfield, A. F., & Sa, J. (2007). Modeling swarm robotic systems: A case study in collective foraging. In *Towards Autonomous Robotic Systems* (pp. 25–32). Aberystwyth, U.K.

Lukka, M. (1974). *On the Optimal Searching Tracks for a Stationary Target.* Turku, Finland: University of Turku, Institute for Pure and Applied Mathematics.

Lukka, M. (1977). On the optimal searching tracks for a moving target. *SIAM Journal of Applied Mathematics, 32*(1), 126–132.

MacArthur, R. H., & Pianka, E. R. (1966). On optimal use of patchy environment. *American Naturalist, 100*(916), 603–609.

Mandelbrot, B. B. (1983). *The Fractal Geometry of Nature.* New York: W. H. Freeman and Company.

Mandelbrot, B. B., & van Ness, J. W. (1968). Fractional Brownian motions, fractional noises and applications. *SIAM Review, 10*(4), 422–437.

Mangel, M. (1981). Search for a randomly moving object. *SIAM Journal of Applied Mathematics, 40*(2), 327–338.

Mangel, M. (1982). Probability of success in the search for a moving target. *Operations Research, 30*(1), 216–222.

Mangel, M., & Clark, C. W. (1986). Towards a unified foraging theory. *Ecology, 67*(5), 1127–1138.

Marthaler, D., Bertozzi, A. L., & Schwartz, I. B. (2004). Levy searches based on a priori information: The biased Levy walk. University of California, Los Angeles, Computational Applied Mathematics Publications Report 04-50.

McCauley, J. L., Gunaratne, G. H., & Bassler, K. E. (2007). Hurst exponents, Markov processes, and fractional Brownian motion. *Physica A, 379*, 1–9.

McElreath, R., & Boyd, R. (2007). *Mathematical Models of Social Evolution: A Guide for the Perplexed.* Chicago, IL: The University of Chicago Press.

McFarland, D., & Bösser, T. (1993). *Intelligent Behavior in Animals and Robots.* Cambridge, MA: The MIT Press.

McNamara, J. (1982). Optimal patch use in a stochastic environment. *Theoretrical Population Biology, 21*, 269–288.

Metzler, R., & Klafter, J. (2000). The random walk's guide to anomalous diffusion: A fractional dynamics approach. *Physics Reports, 339*, 1–77.

Metzler, R., & Klafter, J. (2004). The restaurant at the end of the random walk: Recent developments in the description of anomalous transport by fractional dynamics. *Journal of Physics A: Mathematics and General, 37*, R161–R208.

Mohan, Y., & Ponnambalam, S. G. (2009). An extensive review of research in swarm robotics. In *IEEE World Congress on Nature and Biologically Inspired Computing* (pp. 140–145). Coimbatore, India.

Mohan, Y., & Ponnambalam, S. G. (2010). Swarm robotics: An extensive research review. In I. Fuerstner (Ed.) Advanced Knowledge Application in Practice. Rijeka, Croatia: InTech, 259–278.

Morelande, M. R., Kreucher, C. M., & Kastella, K. (2007). A Bayesian approach to multiple target detection and tracking. *IEEE Transactions on Signal Processing, 55*(5), 1589–1604.

Morters, P., & Peres, Y. (2010). *Brownian Motion.* Cambridge, U.K.: Cambridge University Press.

Muniganti, P., & Pujol, A. O. (2010). A survey on mathematical models of swarm robotics. In *XI Workshop of Physical Agents* (p. 10). Valencia, Spain.

Nahin, P. J. (2007). *Chases and Escapes: The Mathematics of Pursuit and Evasion.* Princeton, NJ: Princeton University Press.

Navarro, I., & Matia, F. (2013). An introduction to swarm robotics. *ISRN Robotics, 2013*, 608164, 10.

Nosonovsky, M., & Mortazavi, V. (2014). *Friction-Induced Vibrations and Self-Organization.* Boca Raton, FL: CRC/Taylor & Francis.

Nouvellet, P., Bacon, J. P., & Waxman, D. (2009). Fundamental insights into the random movement of animals from a single distance-related statistic. *American Naturalist, 174*(4), 506–514.

Nouyan, S. (2004). Path formation and goal search in swarm robotics. Universite Libre de Bruxelles, Faculte de Sciences Appliquees. Brussels, Belgium: IRIDIA.

Nurzaman, S. G., Matsumoto, Y., Nakamura, Y., Koizumi, S., & Ishiguro, H. (2009). Biologically inspired adaptive mobile robot search with and without gradient sensing. In *IEEE/RSJ International Conference on Intelligent Robots and Systems* (pp. 142–147). St. Louis, MO.

Nurzaman, S. G., Matsumoto, Y., Nakamura, Y., Shirai, K., Koizumi, S., & Ishiguro, H. (2011). From Levy to Brownian: A computational model based on biological fluctuation. *PloS One, 6*(2), e16168.

Oaten, A. (1977). Optimal foraging in patches: A case for stochasticity. *Theoretical Population Biology, 12*(3), 263–285.

Ohsumi, A. (1991). Optimal search for a Markovian target. *Naval Research Logistics, 38*, 531–554.

Okubo, A. (1986). Dynamical aspects of animal grouping: Swarms, schools, flocks, and herbs. *Advances in Biophysics, 22*, 1–94.

Opoytzev, V. I. (1977). *Equilibrium and Stability in the Models of Collective Behavior*. Moscow, Russia: Nauka (In Russian).

Ornstein, D. S. (1971). *Ergodic Theory, Randomness and Dynamical Systems*. New Haven, CT: Yale University Press.

Othmer, H. G., Dunbar, S. R., & Alt, W. (1988). Models of dispersal in biological systems. *Journal of Mathematical Biology, 26*, 263–298.

Panigrahi, B. K., Shi, Y., & Lim, M.-H. (Eds.). (2011). *Handbook of Swarm Intelligence: Concepts, Principles and Applications*. Berlin, Germany: Springer-Verlag.

Papoulis, A., & Pillai, S. U. (2002). *Probability, Random Variables and Stochastic Processes*. New York: McGraw-Hill.

Passino, K. M. (2005). *Biomimicry for Optimization, Control, and Automation*. London, U.K.: Springer.

Patlak, C. S. (1953). Random walk with persistence and external bias. *Bulletin of Mathematical Biophysics, 15*, 311–338.

Pavlic, T. P., & Passino, K. M. (2009). Foraging theory for autonomous vehicle speed choice. *Engineering Applications of Artificial Intelligence, 22*, 482–489.

Pavlyukevich, I. (2007). Levy flights, non-local search and simulated annealing. *Journal of Computational Physics, 226*, 1830–1844.

Payton, G. H., Daily, M., Estowski, R., Howard, M., & Lee, C. (2001). Pheromone robotics. *Autonomous Robots, 11*, 319–324.

Peltier, R. F., & Levy Vehel, J. (1995). Multifractional Brownian motion: Definition and preliminary results. INRIA Research Report No. 2645. Le Chesnay, France: INRIA.

Petrosjan, L. A. (1993). *Differential Games of Pursuit*. Singapore, Singapore: World Scientific.

Pirolli, P. (2007). *Information Foraging Theory. Adaptive Interaction with Information*. New York: Oxford University Press.

Plank, M. J., & James, A. (2008). Optimal foraging: Levy pattern or process? *Journal of the Royal Society Interface, 5*, 1077–1086.

Pollock, S. M. (1970). A simple model of search for a moving target. *Operations Research, 18*, 883–903.

Pontryagin, L. S., Boltiansky, V. G., Gamkrelidze, R. V., & Mishchenko, E. F. (1961). *Mathematical Theory of Optimal Processes*. Moscow, Russia: Fizmatgiz.

Prigogine, I. (1980). *From Being to Becoming: Time and Complexity in the Physical Sciences*. San Francisco, CA: W. H. Freeman and Company.

Prigogine, I., & Nicolis, G. (1977). *Self-Organization in Non-Equilibrium Systems: From Dissipative Structures to Order Through Fluctuations*. New York: John Wiley & Sons.

Pyke, G. H. (1984). Optimal foraging theory: A critical review. *Annual Review of Ecology and Systematics, 15*, 523–575.

Rabinovich, M., Ezersky, A., & Weidman, P. (2000). *The Dynamics of Patterns*. Singapore, Singapore: World Scientific.

Reynolds, A. M. (2009). Adaptive Lévy walks can outperform composite Brownian walks in non-destructive random searching scenarios. *Physica A, 388*, 561–564.

Reynolds, C. W. (1987). Flocks, herds, and schools: A distributed behavioral model. *Computer Graphics (ACM SIGGRAPH'87 Conference Proceedings), 21*(4), 25–35.

Richardson, T. O., Christensen, K., Franks, N. R., Jensen, H. J., & Sendova-Franks, A. B. (2011a). Ants in a labyrinth: A statistical mechanics approach to the division labor. *PLoS One, 6*(4), e18416.

Richardson, T. O., Christensen, K., Franks, N. R., Jensen, H. J., & Sendova-Franks, A. B. (2011b). Group dynamics and record signals in the ant Temnothorax albipennis. *Journal of the Royal Society Interface, 8*, 518–528.

Risken, H. (1989). *The Fokker-Plank equation. Methods of solution and applications*. Berlin, Germany: Springer-Verlag.

Robinson, E. J., Feinerman, O., & Franks, N. R. (2009). Flexible task allocation and the organization of work in ants. *Proceedings of the Royal Society B, 276*, 4373–4380.

Romanczuk, P., Bar, M., Ebeling, W., Lindner, B., & Schimansky-Geier, L. (2012). Active Brownian particles. *European Physical Journal-Special Topics, 202*, 1–162.

Romanovsky, Y., Stepanova, N., & Chernavsky, D. (1984). *Mathematical Biophysics*. Moscow, Russia: Nauka.

Rubenstein, M., Cornejo, A., & Nagpal, R. (2014). Programmable self-assembly in a thousand-robot swarm. *Science, 345*(6198), 795–799.

Russell, S., & Norvig, P. (2010). *Artificial Intelligence. A Modern Approach* (3 ed.). Upper Saddle River, NJ: Pearson Education.

Rybalov, A., Kagan, E., & Yager, R. (2012). Parameterized uninorm and absorbing norm and their application for logic design. In *Proceedings of the 27th IEEE Convention of EEEI*, Eilat, Israel. DOI 10.1109/EEEI.2012.6377125.

Ryvkina, J. (2013). Fractional Brownian motion with variable Hurst parameter: Definition and properties. Retrieved October 7, 2013, from *Journal of Theoretical Probability*: link.springer.com/article/10.1007%2Fs10959-013-0502-3.

Samko, S. G., Kilbas, A. A., & Marichev, O. I. (2002). *Fractional Integrals and Derivatives: Theory and Applications*. London, U.K.: Taylor & Francis.

Schimansky-Geier, L., Ebeling, W., & Erdmann, U. (2005). Stationary distribution densities of active Brownian particles. *Acta Physica Polonica B, 36*(5), 1757–1769.

Schonhage, A. (1980). Storage modification machines. *SIAM Journal on Computing, 9*(3), 490–508.

Schweitzer, F. (2000). Active Brownian particles with internal energy deport. In D. Helbing, H. Herrmann, M. Schreckenberg, & D. Wolf (Eds.), *Traffic and Granular Flow '99: Social, Traffic and Granular Dynamics* (pp. 161–172). Berlin, Germany: Springer.

Schweitzer, F. (2003). *Brownian Agents and Active Particles. Collective Dynamics in the Natural and Social Sciences*. Berlin, Germany: Springer.

Schweitzer, F., Ebeling, W., & Tilch, B. (1998). Complex motion of Brownian particles with energy depots. *Physical Reviews Letters, 80*(23), 5044–5047.

Schweitzer, F., Ebeling, W., & Tilch, B. (2001). Statistical mechanics of canonical-dissipative systems and applications to swarm dynamics. *Physical Review E, 64*, 021110, 1–12.

Schweitzer, P. J. (1971). Threshold probabilities when searching for a moving target. *Operations Research, 19*(3), 707–709.

Shi, Z., Tu, J., Qiao, Z., Liu, L., & Wei, J. (2012). A survey of swarm robotics system. In Y. Tan, Y. Shi, & Z. Ji (Eds.), *Advances in Swarm Intelligence*. Lecture Notes in Computer Science (Vol. 7331, pp. 564–572). Berlin, Germany: Springer.

Shiller, Z., & Gwo, Y.-R. (1991). Dynamic motion planning of autonomous vechicles. *IEEE Transactions of Robotics and Automation, 7*(2), 241–249.

Shiller, Z., Sharma, S., Stern, I., & Stern, A. (2013). Online obstacte avoidance at high speeds. *International Journal of Robotics Research, 32*(9–10), 1030–1047.

Shiryaev, A. N. (1995). *Probability*. New York: Springer.

Shoji, H., Yamada, K., Ueyama, D., & Ohta, T. (2007). Turing patterns in three dimensions. *Physical Review E, 75*, 046212.

Siegelmann, H. (1998). *Neural Networks and Analog Computation: Beyond the Turing Limit*. Boston, MA: Birkhuser.

Siegelmann, H., Kagan, E., & Ben-Gal, I. (2014). Honest signaling in the cooperative search. In *Proceedings of the 28th IEEE Convention of EEEI*. Eilat, Israel. DOI 10.1109/EEEI.2014.7005774.

Singh, S., & Krishnamurthy, V. (2003). The optimal search for a moving target when the search path is constrained: The infinite-horizon case. *IEEE Transactions of Automatic Control, 48*(3), 493–497.

Sirazetdinov, T. K. (1987). *Stability of Distributed Systems*. Novosibirsk, Russia: Nauka.

Skellam, J. G. (1951). Random dispersal in theoretical populations. *Biometrica, 38*(1/2), 196–218.

Smith, J. M. (1991). Honest signaling: The Philip Sydney game. *Animal Behavior, 42*, 1034–1035.

Sobel, M., & Groll, P. (1958). Group-testing to eliminate efficiently all defectives in a binomial sample. *Bell System Technical Journal, 38*, 1179–1252.

Sosnicki, T., Turek, W., Cetnarowicz, K., & Zabinska, M. (2013). Dynamic assignment of tasks to mobile robots in presence of obstacles. In *18th International Conference on Methods and Models in Automation and Robotics* (*MMAR'13*) (pp. 538–543). Miedzyzdroje, Poland: IEEE.

Sowers, R. (1994). Multidimensional rection-diffusion equations with white noise boundary perturbation. *Annals of Probability, 22*, 2071–2121.

Srokowski, T. (2009). Fractional Fokker-Plank equation for Levy flights in nonhomogeneous environments. *Physical Review E, 79*, 040104(R), 1–4.

Stanley, H. E., & Meakin, P. (1988). Multifractional phenomena in physics and chemistry. *Nature, 335*, 405–409.

Stauffer, D., & Aharony, A. (2003). *Introduction to Percolation Theory*. London, U.K.: Taylor & Francis.

Stephens, D., & Charnov, E. L. (1982). Optimal foraging: Some simple stochastic models. *Behavioral Ecology and Sociobiology, 10*, 251–263.

Stephens, D. W., Brown, J. S., & Ydenberd, R. C. (Eds.). (2007). *Foraging: Behavior and Ecology*. Chicago, IL: The University of Chicago Press.

Stephens, D. W., & Krebs, J. R. (1986). *Foraging Theory*. Princeton, NJ: Princeton University Press.

Sterrett, A. (1957). On the detection of defective members of large populations. *The Annals of Mathematical Statistics, 28*, 1033–1036.

Steshenko, J., Kagan, E., & Ben-Gal, I. (2011). A simple protocol for a society of NXT robots communicating via Bluetooth. In *Proceedings of the IEEE 53rd Symposium ELMAR*, Zadar, Croatia (pp. 381–384).

Stewart, T. (1979). Search for a moving target when searcher motion is restricted. *Computers and Operations Research, 6*, 129–140.

Stone, L. D. (1975). *Theory of Optimal Search*. New York: Academic Press.

Stone, L. D. (2001). A Bayesian approach to multiple-target tracking. In D. L. Hall, & J. Llinas (Eds.), *Handbook of Multisensor Data Fusion* (pp. 265–298). Boca Raton, FL: CRC Press.

Stone, L. D., Barlow, C. A., & Corwin, T. L. (1999). *Bayesian Multiple Target Tracking*. Boston, MA: Artech House Inc.

Sumpter, D. J. (2010). *Collective Animal Behaviour*. Princeton, NJ: Princeton University Press.

Tan, Y., & Zheng, Z.-y. (2013). Research advance in swarm robotics. *Defence Technology, 9*, 18–39.

Tarasov, V. E. (2010). *Fractional Dynamics. Application of Fractional Calculus to Dynamics of Particles, Fields and Media*. Heidelberg, Germany: Springer.

Theraulaz, G., Bonabeau, E., Nicolis, S. C., Sole, R. V., Fourcassie, V., Blanco, S., et al. (2002). Spatial patterns in ant colonies. *Proceedings of the National Academy of Science of the United States of America, 99*(15), 9645–9649.

Theraulaz, G., Gautrais, J., Camazine, S., & Deneubourg, J. L. (2003). The formation of spatial patterns in social insects: From simple behaviours to complex structures. *Philosophical Transactions of the Royal Society of London Series A, 361*, 1263–1282.

Toner, J., & Tu, Y. (1998). Flocks, herds, and schools: A quantitative theory of flocking. *Physical Review E, 58*(4), 4828–4858.

Trianni, V. (2008). *Evolutionary Swarm Robotics: Evolving Self-Organising Behaviors in Groups of Autonomous Robots*. Berlin, Germany: Springer-Verlag.

Tsetlin, M. L. (1973). *Automaton Theory and Modeling of Biological Systems*. New York: Academic Press.

Tsimring, L., Levine, H., Aranson, I., Ben-Jacob, E., Cohen, I., Shochet, O., et al. (1995). Aggregation patterns in stressed bacteria. *Physical Review Letters, 75*(9), 1859–1862.

Turchin, P. (1998). *Quantitative Analysis of Movement: Measuring and Modeling Population Redistribution in Animals and Plants*. Sunderland, MA: Sinauer Associates.

Turchin, P. (2003). *Complex Population Dynamics: A Theoretical/Empirical Synthesis*. Princeton, NJ: Princeton University Press.

Turing, A. M. (1952). The chemical basis of morphogenesis. *Philosophical Transactions of the Royal Society of London Series B, 237*, 37–72.

Vasiliev, V., Romanovsky, Y., Chernavskii, D., & Yakhno, V. (1987). *Auto-Wave Processes in Kinetic Physics: Spatial and Temporal Self-Organization in Physics, Chemistry, Biology, and Medicine*. Dordrecht, the Netherlands: D. Reidel.

Viswanathan, G., Afanasyev, V., Buldyrev, S. V., Havlin, S., da Luze, M., Raposo, E., et al. (2000). Levy flights in random searches. *Physica A,* 282, 1–12.

Viswanathan, G., Raposo, E., & da Luz, M. (2008). Lévy flights and superdiffusion in the context of biological encounters and random searches. *Physics of Life Reviews, 5*, 133–150.

Viswanathan, G. M., Afanasyev, V., Buldyrev, S. V., Murohy, E. J., Prince, P. A., & Stanley, H. E. (1996). Levy flight search patterns of wandering albatrosses. *Nature, 381*, 413–415.

Viswanathan, G. M., Buldyrev, S. V., Havlin, S., da Luz, M. G., Raposo, E. P., & Stanley, H. E. (1999). Optimizing the success of random searches. *Nature, 401*, 911–914.

Viswanathan, G. M., da Luz, M. G., Raposo, E. P., & Stanley, H. E. (2011). *The Physics of Foraging*. Cambridge, MA: Cambridge University Press.

von Neumann, J. (1966). *Theory of Self-Reproducing Automata*. (A. W. Burks, Ed.) Champaing, IL: University of Illinois Press.

Wang, L., Shi, H., Chu, T., Zhang, W., & Lin, Z. (2005). Aggregation of foraging swarms. In *Advances in Artificial Intelligence, AI 2004* (pp. 766–777). Berlin, Germany: Springer.

Washburn, A. & Kress, M. (2009). *Combat Modelling*. Dordrecht, the Netherlands: Springer.

Washburn, A. R. (1980). On a search for a moving target. *Naval Research Logistics Quarterly, 27*, 315–322.

Washburn, A. R. (1983). Search for a moving target: The FAB algorithm. *Operations Research, 31*(4), 739–751.

Washburn, A. R. (1989). *Search and Detection*. Arlington, VA: ORSA Books.

Washburn, A. R. (1998). Branch and bound methods for a search problem. *Naval Research Logistics, 45*, 243–257.

Weiss, G. (Ed.). (1999). *Multiagent Systems. A Modern Approach to Distributed Artificial Intelligence*. Cambridge, MA: The MIT Press.

Wiener, N. (1948). *Cybernetics, or Control and Communication in the Animal and the Machine*. Cambridge, MA: MIT Press.

Wilkinson, D. H. (1952). The random element in bird navigation. *Journal of Experimental Biology, 29*, 532–560.

Wilson, E. (1971). *The Insect Societies*. Cambridge, MA: Belknap Press/Harvard University Press.

Winfield, A. F. (2009a). Foraging robots. In R. A. Meyers (Ed.), *Encyclopedia of Complexity and Systems Science* (pp. 3682–3700). New York: Springer-Verlag.

Winfield, A. F. (2009b). Towards and engineering science of robot foraging. In H. Asama, H. Kurokawa, J. Ota, & K. Sekiyama (Eds.), *Distributed Autonomous Robotic Systems 8* (pp. 185–192). Heidelberg, Germany: Springer.

Wittlinger, M., Wehner, R., & Wolf, H. (2006). The ant odometer: Stepping on stilts and stumps. *Science, 312*, 1965–1967.

Wittlinger, M., Wehner, R., & Wolf, H. (2007). The desert ant odometer: A stride integrator that accounts for stride length and walking speed. *Journal of Experimental Biology, 210*, 198–207.

Wohlgemuth, S., Ronacher, B., & Wehner, R. (2001). Ant odometry in the third dimension. *Nature, 411*, 795–798.

Wong, E.-M., Bourgault, F., & Furukawa, T. (2005). Multi-vehicle Bayesian search for multiple lost targets. In *IEEE International Conference on Robotics and Automation* (pp. 3180–3185). Barcelona, Spain.

Yadin, M., & Zacks, S. (1988). Visibility probabilities on the line segments in three-dimensional spaces subject to random Poisson fields of obscuring spheres. *Naval Research Logistics, 35*, 555–569.

Yager, R. R., & Rybalov, A. (1996). Uninorm aggregation operators. *Fuzzy Sets and Systems, 80*, 111–120.

Yan, Z., Jouandeau, N., & Cherif, A. A. (2013). A survey and analysis of multi-robot coordination. *International Journal of Advanced Robotic Systems, 10*, 399.

Zabinska, M., Sosnicki, T., Turek, W., & Cetnarowicz, K. (2013). Robot task allocation using signal propagation model. *Procedia Computer Science, 18*, 1505–1514.

Zacks, S., & Yadin, M. (1994). The survival probability function of a target moving along a straight line in a random field of obscuring elements. *Naval Research Logistics, 41*, 689–706.

Zimmerman, S. (1959). An optimal search procedure. *American Mathematics Monthly, 66*, 690–693.

Index

T

V

W

Z

Made in the USA
Las Vegas, NV
24 October 2026

4867c589-f34a-48b8-832b-723ee5a43f53R01